Physicists at Fault

Why you don’t understand quantum mechanics, yet

Nathan Armstrong, Ph.D.

Nathan Armstrong, *Physicists at Fault: Why you don't understand quantum mechanics, yet*, Canada: Otter River Publishing, 2025.

ISBN: 978-1-0693099-0-7 (softcover)
978-1-0693099-1-4 (e-book)
978-1-0693099-3-8 (hardcover)

Cover by Nathan Armstrong.

Nathan's website is drnatephysics.com

Contents

Acknowledgments

Foremost I would like to thank Geoffrey Fitzgerald for reading every part of this book with enthusiasm and dedication. His passion for learning quantum mechanics has served *me* well. He caught many errors and awkward explanations that would have confused the readers. Next, I would like to thank Prof. (ret'd) John de Bruyn for reading the first several chapters. His suggestions allowed me to tighten up some arguments, and he too caught some errors. Finally, Sae-in Park deserves much thanks for her unwavering belief in and continuous support for my book. Countless times, I referred to her for aesthetic advice, and even when I thought I found them all, she still found many typos.

Preface

> *"Science is the belief in the ignorance of experts."*
>
> — Richard Feynman

Confusing and contradictory

I wrote this book with two groups of physics enthusiasts in mind. The first group is comprised of the lay students of quantum mechanics who have had no choice but to be constrained in their learning by the vagaries of pop-science authors. You have read books by physicists and watched endless videos by popularizers of quantum mechanics. You have tried to get closer to an understanding, but there are always confusing terms, contradictions, and vague language. With you in mind, I quote and paraphrase pop-sci edutainers, who are often physicists, and deconstruct their explanations. At other times, I elevate your understanding of quantum mechanics.

With that description, you can see that this book is not for a rank novice. This book is meant to give you the next step beyond what you have read before. The reader should have some knowledge of science and math and have been exposed to many topics related to quantum mechanics from videos and books. I more or less wrote as if the following would readily come to your mind's eye when you read them: absolute value symbol, center of mass, the Cartesian coordinate system, constructive and destructive interference of waves, electric charges, electric fields, electrons, even and odd function addition rules, kinetic energy, light as an electromagnetic wave, light polarization, magnetic fields, momentum, neutrinos, neutrons, photons, potential energy, protons, sine/cosine functions, square of a number, speed of light in vacuum vs. materials, traveling waves, vectors, wave packets, and weighted averages.

For those who know calculus and linear algebra, the math in this book will be easy for you. In fact, you might get a little frustrated at the times when I talk about math rather than show the equations in all their terrible glory.

So that you have more context to what I discuss and to find this book interesting, it's better, but not necessary, to have also been introduced to most of the following things: complex numbers, the collapse of the wave function, the Copenhagen interpretation, the double-slit experiment, Feynman's path-integral formulation, hydrogen energy levels, names of the famous founders of quantum mechanics, quantum probabilities, quantum spin, quantum tunneling, Schrödinger's cat, superpositions, transitions (quantum jumps), the uncertainty principle, and wave-particle duality. However, I do explain all those things to the level of detail needed.

The second group of readers I have in mind is composed of those who will take, are taking, or have already taken a more formal study of quantum mechanics. You may be a student about to start your first serious course in quantum mechanics. Or, you may already have some experience, having already completed your study of quantum mechanics but did not become an expert in the subject. With the overview of the physics from this book, your math-dense quantum mechanics textbook will make more sense. More importantly, you will learn the dirty little secret about quantum mechanics: a not insignificant part of what you learn in an introductory textbook is not how the universe works.

These introductory textbooks convey the message that the math of quantum mechanics is settled, we have an axiomatic system from which other truths can be infallibly deduced, and all that is left is to tie up some loose ends. And if you don't understand the quantum mechanics, it's not the textbook author's fault.

In reality, quantum mechanics is not a faultless axiomatic system. In fact, physicists who calculate quantum things ignore parts of it.

The primary problem with the axiomatic system is that it grafts measurements onto the physics at a fundamental level. However, measurements are still the *least* understood part of quantum mechanics. Researchers in the field of quantum foundations have identified inconsistencies with some of the postulates of the axiomatic system. Yet, the majority of quantum textbooks do not adjust to the times. Throughout the various chapters, I will detail ten postulates that are the foundation of the not-quite-faultless axiomatic system.

Problems with textbooks are not restricted to attempts at an axiomatic system. Many other problems result from pushing a corpuscular particle picture of quantum objects and using non-rigorous mathematics to accomplish that picture.

Ultimately, even if you forget most of the details I have laid out in this book, you will be left with a core lesson. In your previous endeavors to learn more, your confusion did not signal that you were inadequate or that quantum mechanics is magic. It signaled that you were being sold something quite far from the plain truth. Amplifying the problem is that the plain truth, physics, is not easy.

The scope and limits of this book

In the late 1980's, the celebrated science-fiction and non-fiction author Isaac Asimov received a letter from a young English Lit major who, while quoting Socrates, tried to convince Asimov that our current scientific knowledge is wrong because all of our previous scientific theories had turned out wrong century after century. In response, Asimov wrote the essay *The Relativity of Wrong* in which he says:

> When people thought the Earth was flat, they were wrong. When people thought the Earth was spherical, they were wrong. But if you think that thinking the Earth is spherical is just as wrong as thinking the Earth is flat, then your view is wronger than both of them put together.

It's wrong to say the Earth is spherical because, with the equator farther from the center of the Earth than the poles are, it's more accurate to say the Earth resembles an oblate spheroid. That, too, is wrong still because one pole is closer to the equator than the other. Again, that simple picture is wrong too. We could be more correct in our description because, with its mountains, hills, valleys, and ocean trenches, the Earth is bumpy.

In quantum mechanics, each explanation of some phenomenon or theorem is somewhere on the spectrum of being relatively wrong. In this book, I counter those physicists who teach students and laypeople the quantum-mechanical version of a flat Earth. However, I make no claim that you will know the heights of mountains or the depths of the oceans by the end of this book. Some of the things I say *will still be incomplete* in some fashion. They will just be less incomplete than what you were told before. I will illustrate ideas of quantum mechanics that are, figuratively speaking, as accurate as a spherical model of the Earth or better. With this knowledge, you will be better able to set sail in the right direction in your future quests to learn the heights and depths of quantum mechanics.

There were many topics I wanted to write about, but, in the end, they weren't all a good fit for this book. Most of the topics I cover are found in non-relativistic quantum mechanics courses. In the chapter summaries below, you can see what those topics are. Two of the biggest absences are photons and entanglement. They do make brief appearances, so they are not completely absent, but they are not the focus of any chapter. The absence of photons means that this book is primarily focused on quantum objects with mass.

With any physics topic, there are a great many details, caveats, and exceptions. For some topics in this book, I might teach you the spherical Earth version first and sometime later reveal the oblate spheroid version. When I do this, I do so to avoid cognitive overload for the novice reader or as a rhetorical technique that allows me to contrast the two levels of understanding. I take a layered approach with the math, too. If you find you don't quite understand some math, do not worry. For the most part, the type of math I show you is quite repetitive and will appear anew with another explanation.

Unlike in books by other authors, I give you an extensive technical vocabulary. A minority of the terms are not elaborated on and do not appear in the glossary. They are still beneficial to you because with the correct word for a concept you may look it up online.

Chapter summaries

Below are summaries of the chapters, appendices, and the glossary. For the most part, the chapters are meant to be read in order. This is also true for experts because I often refer back to things that have been said previously. The first appendix should be read as soon as possible after the first chapter and definitely before the *Gallery* chapter. The second appendix is a helpful simplification of the postulates. The third and fourth appendices could have been made into chapters but are instead optional because they are more math-focused.

Ch. 1: On interpretations The biggest impediment to understanding quantum mechanics is having nothing to visualize because the predominant philosophy in quantum mechanics is anti-reality and tells us not to ask questions about what actually exists. With this chapter, I lay bare part of the shaky foundation of the anti-reality position because I want to dispel the idea that you are not allowed to think of what happens in the quantum realm. I'm not fond of rehashing the history you've read a few times in other pop-sci accounts, but in this chapter, we must go back in time and see the arguments of a most influential physicist who thought of himself as an able philosopher.

Ch. 2: A first look at quantum angular momentum You are undoubtedly aware of spin angular momentum (spin) but may be unaware of its counterpart, orbital angular momentum (OAM). Together, they are the two types of quantum angular momentum (QAM). QAM will be one of the most difficult topics to comprehend in this book, so I lay out how you should mentally picture these concepts in this chapter.

Ch. 3: Misconceptions about the electron in particular Some illogical and naive ideas would have us picture quantum objects as tiny, corpuscular particles or as geometric points. They usually come up in discussions that include electrons. These naive ideas stop many from appreciating that the wave function is the center of it all. I show why the pictures are illogical and end with a rudimentary visualization of wave functions.

Ch. 4: Waves and superposition Mathematically, superposition is a simple idea. It is the addition of fields of numbers. This short chapter gives you some insight into how to understand superpositions mathematically and physically.

Ch. 5: The cat, the keg, and the cut This is the second and final chapter with a significant call back to history. Pop-sci authors and click-bait websites are fond of mischaracterizing what Schrödinger said about his cat experiment. I correct the record on this and also give Einstein's version of the cat experiment. At this point in history, the physicist-philosopher of the first chapter wanted to defend his Copenhagen interpretation from criticisms by Einstein and Schrödinger, so I have a go at another aspect of his anti-reality claims. I end with an exploration of the idea of macroscopic wave functions.

Ch. 6: Mathematics vs measurements This is where a gentle introduction to using Dirac's mathematical notation of kets, operators, and numerical quantities begins. Additionally, several postulates of axiomatic quantum mechanics are introduced here. You will see that one of the postulates, which could only have been invented for its mathematical convenience, often does not match the physical situation that occurs in actual experiments.

Ch. 7: Wave-particle duality and the double-slit experiment The term wave-particle duality is so abused that I must give a critical analysis of it. Most explanations of the double-slit experiment make it seem like quantum mechanics is magical. I walk through a description of the double-slit experiment by a British professor and show

how it misleads you into thinking that quantum objects behave magically.

Ch. 8 and 9: The mysterious hieroglyphics of Schrödinger's equation On a few occasions, you have probably been presented with the mass of symbols that is the Schrödinger equation. Yet, it still seems undecipherable. I take two chapters to show you what the math means, first **As an energy balance** and then **As a wave equation**.

Ch. 10: Gallery of atomic orbitals The Schrödinger equation gives us the wave functions for hydrogen's atomic orbitals. Via dozens of colorful images, I examine the matter density of hydrogen and the complex-valued nature of its wave functions.

Ch. 11: Spins, superpositions, and abstract spaces Spin and superpositions are two of the topics that novices find particularly counterintuitive. I provide a comprehensive explanation of spin and its superpositions. I also use abstract mathematical spaces to illustrate why spins and superpositions are so counterintuitive. To give you a more complete understanding, I do not restrict myself to spin-$\frac{1}{2}$ states. I end the chapter with examples of spin-0, spin-1, and spin-$\frac{3}{2}$ states.

Ch. 12: Spring cleaning Some misconceptions that I have to clear up are delayed until this chapter because you need the foundational concepts from previous chapters. The final three postulates of axiomatic quantum mechanics are introduced here.

Ch. 13: The uncertainty principle is not what you think it is Far too often, the uncertainty principle is confused with a measurement process or is used to justify fantastical stories of a frantic quantum realm. In this chapter, I show you what the uncertainty principle is fundamentally about, and it's certainly not measurements.

Ch. 14: The postulates of quantum mechanics Students are often told that an axiomatic system of quantum mechanics exists. Having such a system should mean unerring truths follow from a foundational set of postulates. Throughout the book, I have pointed out where some of these postulates don't work as universally as they are claimed to. In this chapter, while keeping the math to a minimum, I teach you what it means to have an axiomatic system and give an extensive commentary on the postulates and what goes unsaid.

Afterword Some parting words to give you some direction.

Appendix A: Atomic orbitals, their labels and transitions This appendix teaches you the numbering and lettering system for orbitals. It also explains the types of quantum states and gives a cursory explanation of atomic transitions. Before starting the *Gallery of atomic orbitals* chapter, you should understand this labeling system.

Appendix B: The postulates of quantum mechanics, a non-mathematical overview I pepper introductions to ten postulates across various chapters. You should often refer to the alternative wording in this appendix to keep the ideas straight in your mind.

Appendix C: Matrix mechanics and spin At various points in the book, I presented some spin superpositions and expectation values and promised to show you how these and some other math objects came about. I keep my promise with this appendix. It is for those who want a peek into a small subset of the math of matrix mechanics hidden behind Dirac's notation.

Appendix D: The energy-time uncertainty principle doesn't exist It is important to know that the so-called energy-time uncertainty principle cannot be rigorously derived from quantum mechanics. I illustrate that to you with this appendix.

Glossary of technical terms A comprehensive glossary with detailed entries. Sometimes I explain a concept differently here, so the reader should refer to the glossary often. After finishing a chapter, I suggest reviewing the chapter's important terms here.

1 On interpretations

> *Judging by its social impact, quantum weirdness will never go away as long as highly reputed scientists are willing to play the role of quantum magician. But only the presentation makes quantum mechanics appear weird. It is fully rational to the mind sufficiently trained in mathematics and theoretical physics.*
> *Does quantum mechanics have to be weird? It sells much better to the general public if it is presented that way, and there is a long history of proceeding that way. But, in fact, it is an obstacle for everyone who wants to understand quantum mechanics, and to physics students who have to unlearn what they were told as lay persons.*
>
> — Arnold Neumaier

There is a philosophical choice

The orthodoxy

I have a lot of misconceptions to correct throughout this book, and it will help you immensely if the **Copenhagen interpretation** of quantum mechanics does not weigh you down. You see, some physicists make quantum mechanics more confusing to you than necessary because they must engage in verbal acrobatics to avoid landing on a forbidden wording. The result is tortuous sentences and confused readers.

Depending on which tenets of the Copenhagen interpretation a physicist subscribes to, they may be forbidden from saying reasonable things like "Between measurements, the electron exists", "The spin vector points in a particular direction", "Energy is conserved in every process", "Momentum is conserved in every process", and "A cat cannot be both alive and dead." This way of working comes from a radical version of scientific empiricism. It seems that these physicists don't accept the possibility of making inferences from other data. Instead, they must maintain odd views of the universe and say things like "We can only say what our instruments tell us, and that information is only good at the time of measurement", "Energy and momentum are conserved only on average over a large number of measurements", and "A cat can be literally both alive and dead."

I have seen physicists write in books and on internet forums as if the Copenhagen interpretation is the single allowable interpretation of quantum mechanics. They have been misinformed, and they misinform you. As long as an interpretation does not contradict known facts and is internally consistent, then you are allowed to subscribe to it. Facts are the foundation; interpretation is the scaffolding over this that builds up our view. Of course, the interpretations may be differentiated by new facts in the future, but until then, you have freedom of choice.

The three views covered

That being said, I don't want to dwell on a long list of ways you may view quantum physics. In this book, I only refer to three views consistently: 1) textbook **axiomatic quantum mechanics**, 2) the Copenhagen interpretation, and 3) **the reality interpretation**. There are, in fact, many reality interpretations, but in this book, I will refer to the one I prefer as *the* reality interpretation.

Axiomatic quantum mechanics is commonly assumed to be a sterile description of quantum mechanical facts. It is based upon a short list of postulates (aka axioms) that are supposed to serve as foundational statements of a consistent mathematical system that describes our experimental observations of quantum-mechanical phenomena. I eventually lay out all of the postulates for you in the final chapter but introduce all of them on our way there. Additionally, I have included a short appendix introducing them in a non-mathematical way which you should read well before you get to the *Postulates* chapter. Note that there are many moderately different versions of the postulates. In this book, I present an amalgamated list that characterizes what an average undergraduate student might encounter.

In passing, so that you aren't left wondering, I want to mention how physicists who do quantum-mechanical calculations use it to move science forward. Some may say that those physicists use axiomatic quantum mechanics. However, since it leads to inconsistencies and absurdities – which you will understand well by the end of the book – it cannot always give correct results. Roughly speaking, physicists (generally, not the ones who write lists of postulates) use axiomatic quantum mechanics minus the parts that don't work plus other mathematical tricks or workarounds that make things work. In other words, it's not genuinely axiomatic quantum mechanics. Textbooks rarely acknowledge the actual state of the field.

According to the Copenhagen interpretation, the **wave function** Ψ is nothing more than a mathematical tool. (The Ψ is pronounced "sigh" and spelled "psi".) We are supposed to just "shut up and calculate" results, not picture the wave function as part of reality. In much of the philosophical literature, they call this the **anti-reality** view. It is a rather dull way to be a physicist.

Apparently, the Copenhagen interpretation is the most common interpretation. It is much looser than axiomatic quantum mechanics. It roughly follows the modern postulates and has a few other ideas added on. In the *The cat, the keg, and the cut* chapter, I bring up a point about the Copenhagen interpretation because it fits the historical context (and gives my chapter title a nice, third alliterating word). Here in this first chapter, the only rule from the Copenhagen interpretation I am concerned with is the anti-reality claim, which says that we cannot talk about reality between measurements. One reason given for this rule is that we don't have direct knowledge of the happenings between measurements. The other reason given is quite surprising: they say that reality *does not* exist between measurements. These two claims are the main focus of this chapter, and I will return to them soon enough.

The position taken in this book

I certainly did not become a physicist simply to confirm by experiment quantum-mechanical calculations with no thought given to what lies deeper than the numbers on a machine. I did it all to comprehend the nature of reality and the wonders of the universe.

Most days of the week, I am a realist and believe that the wave function of a quantum object represents an element of objective reality. The term *objective reality* refers to things that truly exist and are independent of a mind or anyone's opinion. In contrast, the perception of how warm a room feels is part of subjective reality. To an obese husky, it's too warm. To a naked mole-rat, it is too cold. To Goldilocks, it's just right. Philosophers would say I most often follow an ontological interpretation (etymologically, *ontology* is the study of objective reality).

The position I take in resisting the anti-reality view is not an uncommon one. There was a small poll taken of scholars who specifically research the foundations of quantum mechanics. Sean Carrol says that one of the bar graphs in this study is "The most embarrassing graph in modern physics" because we still don't agree on a single interpretation. In the graph that Carrol mentions, the largest group, with 42% of respondents, prefers the Copenhagen interpretation. The many-worlds interpretation came in at 18% (I do not discuss this interpretation). And 12% of those polled had no preference. In another question, 24% of the researchers said they believe that the wave function is real, and a further 33% believe that it is a mixture of a real and informational thing.

If you are relatively new to quantum mechanics or are confused by the "Whole Sort of General Mish Mash" of it all, I suggest you read this book thinking that the wave function written on the page reflects a quantum object that is an element of reality. Realism has helped me from being totally confused by quantum mechanics. Ultimately, I chose to write this book as if wave functions are part of reality.

I try to avoid discussing the measurement process whenever I can (although I still talk about it plenty). My reason for this is the **measurement problem**, which arises due to our lack of understanding of the process of going from a quantum wave to a measurement result. Different interpretations deal with the problem differently.

The most famous partial hypothesis that attempts to solve the measurement problem is that, somehow, a spread-out wave becomes localized when it interacts with matter or a photon. This localization then allows for a click in a detector or a registration in a pixel. Going from a spreadout wave function to a localized wave function is often described as **the collapse of the wave function**. We know neither the details nor the mechanism of how this would work. That being said, I will assume throughout this book that the collapse of the wave function occurs at some point during the measurement process.

By only bringing up how we measure the quantum realm when it's useful or when I am pointing out a contradiction, I will be walking around this source of confusion, which other popularizers of quantum mechanics capitalize on to bring you a sense of awe and wonder. Instead, I concentrate on the cause-and-effect quantum realm that obeys the Schrödinger equation. In the reality interpretation, the quantum realm has the mass and electric charge of a single electron distributed throughout space according to the squared magnitude of the wave function (i.e., $|\Psi|^2$. See Fig. 1.1). Everything will be clearer for you via this route. Though, it is not essential for you to believe this in your heart. Once you have finished the book, you can go back to your preferred interpretation. It's your choice of philosophy.

For decades, scholars have been discussing the different positions regarding reality in quantum mechanics. I'm sure you could find numerous books, dissertations, and articles with thorough, detailed, and long analyses on this. However, I want you to be confident, here and now, that you can let go of the no-questions-allowed burden of the

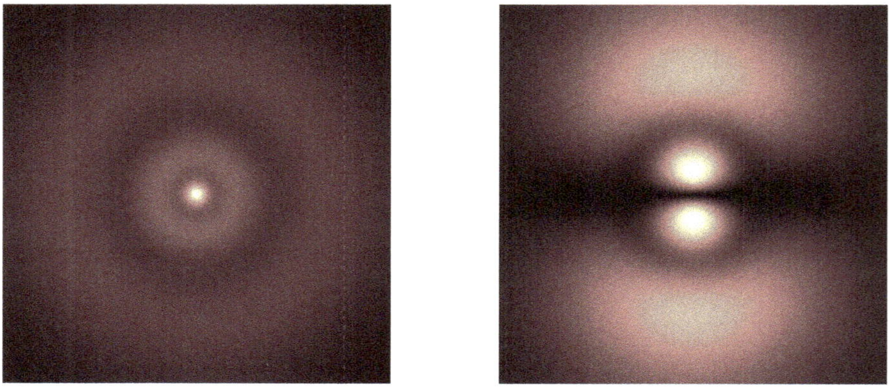

Figure 1.1: How to picture the mass and charge distributed in space in the reality interpretation. The images are of an electron's matter density in hydrogen (which is proportional to the probability density $|\Psi|^2$) when it is in a 3*s* state (left) and in a 3*p* state (right). Appendix A explains this number and letter labeling system. The squares are 1.6 nm on a side (1 nanometer = 1 millionth of a millimeter). You can interpret the images as if they were semi-transparent clouds. I created them by adding up slices through the matter density. To get the charge or mass density from $|\Psi|^2$, multiply it by the charge q or the mass m. Those densities are $q \cdot |\Psi|^2$ and $m \cdot |\Psi|^2$. Because it is brighter, the right image deceptively appears to have more matter in it. Both images represent one electron. The brightness scales are set by the maximum pixel value in each image for optimal visibility. Since the right image has less variation in values, it has large bright regions. I chose these states because I will mention sodium's 3*s* and 3*p* states in the next chapter. We can only find exact atomic wave functions for hydrogen-like atoms/ions, and the researchers who do the numerical calculations don't show the wave functions or electron densities for other atoms. In sodium, the matter density likely would retain the same prominent features. I estimate the square's edges would be double that of hydrogen's (I used the ratio of Na and H's van der Waal radii).

Copenhagen interpretation. So, I give you a cursory case against anti-reality and another for reality below.

Note that, after all is said and done, the Copenhagen crew could retort back with "well, the reality interpretation can't explain x, y, and z!" Sure. That's fine. For example, I don't know if it's better to say "Ψ is an object" or say "$|\Psi|^2$ is the object and Ψ is a property it possesses". This shows there is an asymmetry between the anti-reality and reality positions. Schrödinger, Einstein, and other realists don't claim that our understanding of quantum mechanics is complete. To us, the evidence suggests that there is more to it.

In contrast, some of those who invented the Copenhagen interpretation did claim that their version of quantum mechanics is complete, and there is no need to burrow down further to find deeper truths. That is a much stronger claim and, as such, should have no cracks in its foundation. In the cursory case that follows, I will supply you with some history surrounding and some arguments for the Copenhagen interpretation. In doing this, I do not definitively refute the Copenhagen interpretation. But, I do hope you will see that what has been historically said of this interpretation isn't quite the truth and that the reasoning for the interpretation is not as solid as some say.

Our antagonist

It seems that we physicists collectively have a false and imagined history of a well-established Copenhagen interpretation and have been quoting sacred scripture that doesn't exist. The story usually includes the detail that Heisenberg and Bohr established their united framework while at the University of Copenhagen. However, historians and philosophers who study this in detail tell us that Heisenberg and Bohr never fully agreed on the interpretation and that there is no single universally accepted Copenhagen interpretation. This is partly why I will only focus on one person and one aspect of this interpretation.

You can thank Werner Heisenberg, the anti-realist, for the widespread belief that we cannot know anything about a quantum object if it isn't measured. His influence on what is taught to all new physicists can hardly be understated. He took a conglomeration of mystical and philosophical ideas and solidified it into his version of the Copenhagen interpretation by 1955 when he introduced the term to the public in *Niels Bohr and the Development of Physics.* Of course, there were other giants of physics that contributed to the various versions of the Copenhagen interpretation. Nonetheless, Heisenberg was the most energetic proselytizer.

Adam Becker in *What is Real?: The Unfinished Quest for the Meaning of Quantum Physics* reveals why Heisenberg wrote so many books after World War II that make him appear as a pillar of a community of wise sages. Becker characterizes Heisenberg as a liar and a revisionist of his personal history as well as that of quantum mechanics. His books were part of a "public-relations campaign" to distract from his detestable history of trying to produce an atomic bomb for the Nazis: "Heisenberg's invention of a single unified Copenhagen interpretation may have been an attempt to rework the history of quantum physics to his benefit."

To declare Heisenberg a through-and-through anti-realist might be a bit much. He did, when being careful with his words, talk of objective reality being inaccessible between measurements rather than reality not existing. It is then more accurate to say his stated views varied over the years, and we can't know if he was adjusting his message for his audience. For completeness, I will have to deconstruct both of his stated views.

Against Heisenberg's anti-reality

The narrower stance: reality is not fully knowable

His first reason

In his 1971 book *Physics and Beyond: Encounters and Conversations*, Heisenberg tells the story of how, in the spring of 1919, during a hike in the mountains before his high school final exam, over conversations with friends about physics, philosophy, and religion, he concluded that the objective reality of atomic phenomena would be forever inaccessible to science because "our experience of atoms can only be indirect."

Across all his writings, Heisenberg is fond of showing his familiarity with philosophy and dropping the names of numerous philosophers. In the story of his youth, he had been reading Plato's *Timaeus* in the original Greek, and his acquaintance gave him his epiphany by mentioning the French theologian-philosopher Malebranche's ideas. Specifically, it was the realization that "as they impinge on our senses, objects produce concepts directly in our minds" which planted the seed of anti-reality in his mind.

When Heisenberg rejects indirect knowledge of atoms, he rejects standard scientific inference and embraces a narrow, radical version of scientific empiricism. It is often said that Heisenberg follows some brand of philosophical positivism, which is consistent with the stance I discuss here.

Heisenberg's anecdote is quite revealing because, if this story is accurate, Heisenberg's philosophical position of a partially inaccessible objective reality is a worldview he adopted after a single conversation in his youth. In the text, it is clear that as a high school student, he wasn't even yet aware of Bohr's model of the atom with its orbits of electrons. The story illustrates that his general view of atomic phenomena is based upon a general philosophical opinion and is not one based on quantum mechanics.

His other reasons

In contrast to the story of his youth, in many other places, he gave a more elaborate set of reasons for us not having access to objective reality. One of the earliest accounts is in a 1930 book meant for a technical audience, which is based on lectures he gave in Chicago in 1929, titled *The Physical Principles of the Quantum Theory.*

Part of his reason for saying it is verboten to discuss what happens between measurements is because of the wave and particle pictures. Neither picture completely describes the quantum realm. He claims each picture is needed in turn to describe a photon as it progresses through an experiment. He says the two pictures contradict each other in explaining the same object. Therefore, according to Heisenberg, things in quantum mechanics are not consistently visualizable, and as such, we cannot claim to visualize what truly happens between measurements.

During these explanations he also invokes his famous **measurement-disturbance relation** ($\Delta x \cdot \Delta p_x \approx h$) to deny that reality is knowable. (This relation is often erroneously called the **uncertainty principle** because the math symbols of the two relations appear somewhat similar. I will sort these two out in a later chapter.)

Heisenberg's measurement-disturbance relation tells us we cannot simultaneously determine a quantum object's momentum and position. This is because, to measure either quantity, we must do some violence to the object's state by throwing a second quantum object at it. He says there is a destruction of knowledge and we can never obtain perfect knowledge of the present state of a quantum object. He takes it as a principle that we may only speak of things existing to the degree we have knowledge of them. Since we can't exactly predict the outcomes of measurement, we can't say atoms exist before the measurements.

He protects his peculiar principles with the declaration that our description of reality will be forever limited by the natural languages we use to describe the macroscopic world. He's quite explicit that we aren't allowed to adopt a more precise language because we use this language to describe the classical physics of our measurement devices.

Should we accept any or all of Heisenberg's reasons for the view that we can't discuss the existence of things between measurements? One would have to accept that there are only two acceptable ways to picture quantum objects: with a pure particle or a pure wave picture. No blending of the two is allowed. The possibility of a future, alternative picture that could make sense of the experimental data would have to be declared to be zero either by a sense of hopelessness or by brazen self-assuredness that one knows there can be no other.

Heisenberg's requirement of perfect knowledge comes from a philosophy of radical scientific empiricism (originating from philosophical postivism). I ask if this is a viable position. Do we ever interpret our experimental data with no models in our minds? Do we not make countless assumptions in every experiment we do? If neither models nor assumptions were used to interpret what we do, scientists would simply need to describe their experiments and present their data in journal articles. The introductory, theory, discussion, and conclusion sections would be superfluous because bare data would be sufficient for us to know what is knowable. If one cannot accept such radical empiricism, I believe one should not accept the Copenhagen interpretation.

As for his stance on language, it too suffers from either a sense of hopelessness or brazen self-assuredness that humans will never find words to adequately describe quantum mechanics. Doesn't natural language evolve with our needs? Suppose Heisenberg's personality was put into an ancient Greek philosopher and shown a modern automobile. It seems he would say that we would never be able to describe the inner workings because ancient Greeks did not have the words to adequately describe an automobile's physical principles. The possibility that greater ideas may yet come are absent in Heisenberg's thinking.

The anti-realist's stronger stance: measurements produce reality

From some modern physicists, it is common to hear something like, "the wave function is just mathematics and is not an element of reality." They are making a stronger statement, and one couldn't be considered wrong for saying that it sounds like they are rejecting the existence of the quantum realm. How can our instruments interact with anything if no real thing is "down there" before the measurement?

These decrees by modern physicists can be traced to Heisenberg and other Copenhagenists. In a few places, Heisenberg asserts that "the possible becomes actual" due to a measurement. In *Physics and Beyond*, he is more descriptive:

> ... on the atomic scale, this objective world of time and space [does] not even exist and...the mathematical symbols of theoretical physics [refer] to possibilities rather than facts.

This more descriptive statement comes from a story about the famous Einstein-Bohr debates at the Solvay conference of 1927 in Brussels. Heisenberg's justification for this view relies on imperfect knowledge, duality, and probability. The new ingredient here is probability. In these cases, he is no longer talking about our knowledge of reality. He is now saying that reality only exists during measurements. He speaks very similar words during his Nobel lecture:

> The very fact that the formalism of quantum mechanics cannot be interpreted as [a] visual description of a phenomenon occurring in space and time shows that quantum mechanics is in no way concerned with the objective determination of space-time phenomena.

This time, his justification for denying reality is the supposed lack of visualizability of quantum phenomena.

Naturally, one can ask the obvious question, "what about when we aren't making measurements?" By using measurement to determine reality, Heisenberg has inserted people into the determination of reality.

Locations in space far from us, or even in the time before the existence of people, cause tremendous difficulty. Objects far from humans in time and space would not exist because they can't be measured. Do they then not produce gravitational, electric, and magnetic fields? Einstein recognized this obvious conclusion. He tried to illustrate to Bohr the folly of the Copenhagen interpretation's anti-reality stance when he asked Bohr if he thought the moon existed when no one was looking. We can also ask whether the moon's gravitational field exists when the moon is unobserved or unmeasured.

The answers have obvious consequences, which are at odds with Heisenberg's strong conclusion. If the gravitational field appears and disappears in lockstep with measurements on the moon, the stroboscopic changes in the field would presumably be measurable as pulses.

Certainly, the pulses would be earth shaking, but these are never observed, so this answer is false. If, alternatively, the gravitational field is always there, then it must have a source, the moon, that always exists.

You can also ponder whether you accept that the universe was not real before there were observers of it and how the first measurement occurred if it needed a real person to conduct it. Heisenberg's outlandish conclusion is thus easily discredited if you believe the macroscopic world is built up from the microscopic world.

I anticipate some readers might resist my characterization of Heisenberg and suspect that he didn't *really* mean measurements. Oh, dear reader, I assure you he did. His talk of experiments and measurements is in all his works on the foundations of quantum mechanics. It is essentially his trademark.

He needs measurements because his philosophy is based upon radical empiricism: you must only speak of what is measured, and no scientific inference is allowed. Measurements require a measurement apparatus. The particularly odd thing is that this apparatus must be classical according to Heisenberg. This then requires a freestanding classical world as a foundational assumption. Heisenberg says a dividing line, a cut, between the macroscopic and microscopic worlds is necessary to make his interpretation complete. In the fifth chapter, I analyze Heisenberg's justification for this cut. To accept the Copenhagen interpretation, one must then subscribe to radical empiricism and must also deny that macroscopic objects are made up of quantum objects.

A short case for reality

No nuance needed

For a moment, let us suppose that wave functions are elements of reality and that the quantum realm exists. Consequently, space-time exists at all times. Not too radical of assumptions for you, I hope. Further, let's say mass, charge, and spin are smeared throughout space in proportion to the values of $|\Psi|^2$. Let us also allow the quantum randomness that occurs when the wave function collapses into a new state so that UV radiation can break bonds, nuclear radiation can produce us energy, etc. We can then ask what our world would look like.

Atoms would absorb and emit light. Molecules would form. Metals would be solids with very predictable properties. You would feel what you feel. We would have exactly the universe we see around us. Water would be wet and honey would be viscous because of quantum mechanics. Any of the myriad chemical reactions you have seen in your life would occur because of quantum mechanics. It all works together in a great unity.

I think this makes the reality interpretation a useful choice because if it walks like a duck and quacks like a duck, then we can infer that it must be a duck even when we aren't looking. The only major issue with this interpretation is that the measurement problem is not explained. But that is a problem common to all interpretations. No one has convinced the scientific community that the issue has been resolved yet.

In fact, were you to read only a selection of atomic physics, solid-state physics, and chemistry textbooks, you may never have become aware of the anti-reality view. These books discuss the wave function of one molecule attracting other molecules because each has a distribution of charge due to the wave function. They might also discuss how clouds of charges from some electrons in an atom screen the charge of the nucleus. This then requires other wave functions in the atom to be more spread out because they feel less net charge. The discussions are all exactly causal and based on the existence of charge and mass distributed through space.

The language in these textbooks treats the electron wave functions as if they are responsible for the charged clouds, and it all makes sense. They lack footnotes warning the reader, "But we all know reality doesn't exist in the quantum realm. We're just using the easiest language we can as a convenient tool."

In short, that's it. There's no need for nuance. Entire textbooks read as if the quantum realm exists. There's no constant reminder of anti-realism. The authors are not trying to sell books with the hook of quantum mystery as some do with books for lay people.

Visualizable wave mechanics

To hammer home how useful the picture is, I'll use an example that surprised me while writing this book. Before getting to it, I must provide some historical and physical background of the early decades of quantum mechanics.

Much of the impetus to develop quantum mechanics came from the desire to explain stellar and laboratory atomic spectra. In emission-spectrum experiments, a hot gas of atoms emits a colorful glow. Neon lights are a perfect example of this glow. When this light is collimated into a beam and dispersed through a glass prism, it will create a series of bright lines on a wall in a dark room. The lines indicate which individual colors went into making the colorful glow.

In Fig. 1.2, I show you an example of hydrogen's emission spectrum in the visible and near-ultraviolet bands of wavelengths. This particular collection of spectral lines is called the Balmer series. It is named after the mathematician who discovered the equation that describes their wavelengths.

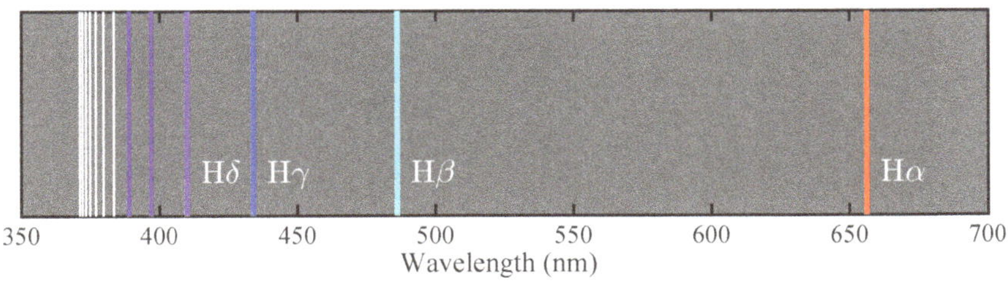

Figure 1.2: Visible and near-UV spectrum of atomic hydrogen. The gas discharge tube that would create this spectrum glows pink. The α, β, γ, and δ are the first four lowercase letters of the Greek alphabet and are pronounced "alpha", "beta", "gamma", and "delta". They represent the four colors most easily seen by early researchers. The white lines are in the ultraviolet.

Before Schrödinger found his equation, the best physical modeling of hydrogen spectra was first done by Bohr with his shell model and then by Heisenberg with his **matrix mechanics**. While Bohr's model had many unsatisfactory characteristics, matrix mechanics is correct and powerful. However, it is highly unintuitive. The math involves moving numbers around in square blocks called *matrices* (singular: *matrix*). For example, Heisenberg could mathematically find the values for the simple Balmer series spectra, but matrix mechanics did not give a visual picture of the quantum realm.

Schrödinger discovers **wave mechanics** and is hailed a hero, quite literally. He is said to have saved quantum mechanics because he gave physicists a picture of the quantum world (i.e., atomic hydrogen wave functions in 3D space) and took agreement with experiments to another level.

In class and lay books, it is typical to hear that he came up with the Schrödinger equation and that he was able to predict the energy levels of the hydrogen atom at the level of accuracy of the Bohr model. In a physics class, reading the original research literature is not typical. The old terminology and notation often make it difficult for a student or an expert. I had the idea that he was one and done. I was quite surprised to learn that, in 1926 and 1927, he published, by my count, nine papers developing wave mechanics.

What surprised me the most was that the fourth paper revealed the usefulness of the charged-cloud picture. His picture allowed him to see a method to solve the problem of the *Stark effect.*

When a hot gas of hydrogen atoms is placed in an electric field, the spectral lines subtly split. This splitting is the Stark effect. Wolfgang Pauli was able to use matrix mechanics to calculate how many sublines each line would split into, but no one could calculate how bright the lines were until Schrödinger did it.

Schrödinger surmised the electron's charge is distributed in space according to $|\Psi|^2$. He figured that, during the Stark experiments, the electric field caused the electron clouds to distort a tiny bit. He also knew that oscillating charges produce electromagnetic radiation (i.e.,

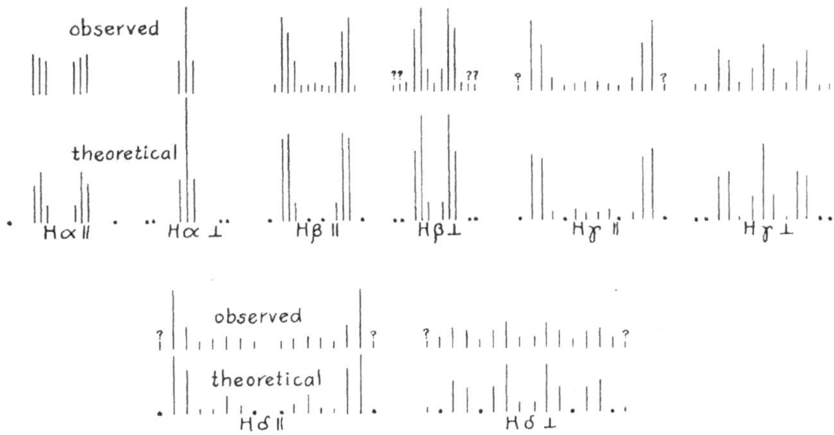

Figure 1.3: The sole figure in a paper where Schrödinger summarizes his work to date for the English-speaking world. The figure shows his match to the Stark effect. The upper set of lines labeled "observed" are Stark's experimental data. The lower set labeled "theoretical" is from Schrödinger's calculations. The height of a line corresponds to its brightness (technically, I should have been saying *luminosity* because brightness is subjective). The main thing I want you to take away from this picture is that Schrödinger was the first to explain the observed data precisely because he could visualize a quantum object that exists in our world of space and time. The rest of this caption walks through the features of the image for the curious. Notice the label below each group of line shows which member of the Balmer series it belongs to. After each Greek letter is another symbol indicating whether the light was collected through a polarizing filter oriented parallel (∥) or perpendicular (⊥) to the applied electric field. Dots in the theoretical data tell us that the line length is too short to represent on this scale. The question marks are due to Stark, who was unsure of the veracity of those lines in his paper. (The figure is used with permission by the American Physical Society [E. Schrödinger, Physical Review, 28(6), p. 1049].)

light). If we say the electron's charge cloud is a duck, he saw it walking and heard it quacking. He was going to show the world that this duck was real and that it flies.

When he calculated the frequency of light under Stark effect conditions, he assumed three-dimensional clouds of charge were oscillating between two orbital forms. He found great agreement with the experimental results. You can see how his calculations match experiments in Fig. 1.3. With his visualizable math and agreement with a complex experimental situation, it's easy to see why the community quickly accepted Schrödinger's equation and considered him a hero.

The rest of the book

When I set out to write this book, I did not do it with the purpose of promoting the reality interpretation. However, explaining quantum mechanics without an interpretation as a framework is pointless. You need something to visualize, and the reality interpretation is the most concrete, so the subtext throughout this book will be that of realism. Roughly speaking, I would say the majority of physicists will not have serious issues with 90% of what I write. After all, my knowledge comes from standard quantum mechanics texts. The other 10% could be said to be my personal analysis of the facts. Worry not; I do not wander into crackpot territory. Compared to some other physicists' views (e.g., reality is created by measurements, infinite densities in black holes, consciousness causes collapse, and 11 dimensions in string theory), my views are rather mild. Whenever a topic deviates so far into the reality interpretation that many experts might cluck their tongues at what I say, to the degree that I deviate from the norm, I will endeavor to highlight with the language I use that I am doing this.

2 A first look at quantum angular momentum

> "...*two-valuedness not describable classically.*"
>
> — Wolfgang Pauli describing what would later be known as spin angular momentum (1924).

Introduction

Quantum angular momentum (QAM), or **quantized angular momentum**, is a concept that takes up significant portions of quantum textbooks. It is as important as the other conserved quantities of energy and linear momentum. Despite that, in typical accounts for the public, the specifics of quantum angular momentum are not so common. I will not follow the typical route because the details are important for your understanding.

I anticipate that some readers will be surprised by the following fact. There are two types of QAM: **spin angular momentum** and **orbital angular momentum (OAM)**. Many pop-sci authors don't bother to mention orbital angular momentum. For simplicity, I'll often call the spin angular momentum by the name everyone knows, which is **spin**.

The physics of the two types of angular momenta are similar in almost every way. Their origins are what particularly differentiates them. The word *intrinsic* refers to a property of an object that is innate and cannot be changed. We often say spin angular momentum is **intrinsic** to the quantum object. However, it is most accurate to say the value of the spin quantum number s, which is directly related to angular momentum, is the intrinsic quantity. In electrons, $s = \frac{1}{2}$, and we say the electron is a spin-$\frac{1}{2}$ quantum object. On the other hand, the presence of orbital angular momentum is a consequence of how the wave function is distributed throughout space at the present moment. Since the amount of OAM is dependent on some changeable conditions, it is an **extrinsic** property.

This chapter will give you some understanding and mental pictures of QAM that will be useful at other points in the book. In the *Gallery* chapter, I will show you numerous images of particular types of wave functions (i.e., atomic orbitals), each of which has a particular value of OAM. The content in the first appendix will help you understand how we organize the atomic orbitals into groups based on three quantum numbers, two of which directly relate to OAM. For completeness, that appendix includes some minor commentary on spin quantum numbers. Spin will be represented by equations in this book. Most of these equations are concentrated in the *Spins, superpositions, and abstract spaces* chapter.

Spinning or not?

Whether quantum objects are spinning in the quantum realm is a popular question among physics enthusiasts. In this section, I will tell you exactly which quantum objects are spinning (rotating) and which are not.

First, let's take a step back and have a cursory look at the classical angular momentum of the macroscopic world. I want you to imagine a figure skater spinning in place. This skater has a set amount of classical angular momentum L. If the skater brings their arms and their one free leg closer to the axis of rotation, there is an increase in their rotation rate ω (Greek lowercase 'omega'). If they do the opposite and spread out their arms and free leg, the rotation rate decreases. This effect can be explained by the conservation of angular momentum.

There is a quantity called the *moment of inertia* (symbol is an I) that can be roughly said to describe how far a set of masses is from the rotation axis. When the skater pulls their arms and leg further inward, the moment of inertia decreases. When the arms and leg go outwards, the moment of inertia increases.

Now, why does the rotation rate of the skater change with the changing moment of inertia? It's because the amount of the skater's angular momentum is conserved and cannot change. (We imagine an ideal scenario where there is no friction between the skate and the ice, so no angular momentum is lost to the earth.) The relationship between the three quantities is $L = I \cdot \omega$. If I increases or decreases, ω must change in the opposite sense a proportional amount to keep L from changing.

Angular momentum also has a mathematical direction, a vector. Now, imagine looking down on the skater from above. The skater can choose to spin clockwise or counterclockwise. We associate a vector pointing up or down along the rotation axis with these two choices. We have what's called a *right-hand rule* for this purpose. Pretend the rotation axis is a pole that you grab with your right hand while your thumb is along the axis. The rule says you must orient your hand so that your fingers curl and point in the direction of rotation. If the skater is rotating counterclockwise, your right hand will give a thumbs up, and we say the angular-momentum vector points up. If the skater is rotating clockwise, you need to twist your right hand into a thumbs down so that the curl and pointing direction of your fingers match this rotation direction. We say the angular-momentum vector points down.

That completes our cursory account of the macroscopic realm's classical angular momentum. We now switch to the issue of whether things are spinning in the quantum realm. With regard to spin angular momentum, unfortunately, physicists did not have the foresight of what we know now, and they chose the word *spin* for part of its name, so until the end of time, we must constantly tell the public that nothing is spinning in space with regard to spin angular momentum.

Something similar to what is said of the non-spinning spin can often — but not always — be said of orbital angular momentum. To make sense of that statement, I must tell you that there are states of quantum objects known as **stationary states**. They are characterized by the fact that their matter densities (i.e., $|\Psi|^2$) do not change as time goes on. In the first image of the first chapter, I showed you the matter densities of two atomic states. Those states are stationary states. When quantum objects are in stationary states, the OAM of those states is not associated with the spinning of any mass or charge because, by definition, stationary states don't have matter densities that move. Be that as it may, in the *Gallery* chapter, you will see a rotation of phase occurs when a wave function (i.e., Ψ) is in a stationary state and has non-zero OAM.

Non-stationary states with non-zero OAM are another story. You see, every non-stationary state has a matter density that changes as time progresses. This opens up the possibility for matter to be rotating. In the reality interpretation, the microscopic realm smoothly scales up into the macroscopic world, both of which are quantum. Therefore, in the reality interpretation, some aspect of quantum angular momentum must smoothly transition into classical angular momentum. Because OAM comes from the distribution of the wave function in space and the wave function determines the moving matter density, when we have angular momentum due to rotating matter, it is orbital angular momentum. More specifically, in a micro-to-macro picture, what we see as classical angular momentum comes from non-stationary states with enormous amounts of OAM relative to what we consider normal in microscopic situations. In a macro-to-micro picture, when you send a macroscopic object rotating and thus give it "classical" angular momentum, you actually put it into a new, complicated, non-stationary quantum state with OAM.

Magnetism due to QAM

The close association between magnetism and quantum angular momentum often confuses those new to quantum mechanics. Please read the following statement closely: when an immobile quantum object has both electric charge and non-zero QAM, it also has a magnetic field resembling a bar magnet's field, and this field is aligned along the same axis as the QAM. This magnetic field is shown in Fig. 2.1.

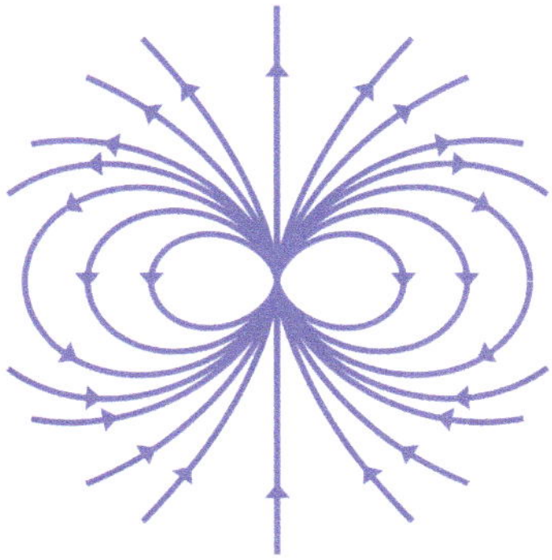

Figure 2.1: Magnetic field due to an infinitesimal bar magnet. The north end has magnetic field lines emanating from it. The south end has field lines arriving. Bar-magnet fields are described by the direction of the field along their axis. In this case, that direction is upwards.

That statement does not say that on every occasion there is spin or orbital angular momentum, there will also be magnetism. There is a second necessary condition in that statement. An electric charge must also be present. For example, neutrinos are spin-$\frac{1}{2}$ quantum objects but have no electric charge and thus have no magnetic field.

There is a pretty big hole in that rule, though. It seems to miss out on neutral quantum objects with magnetic fields. Neutrons are also neutral spin-$\frac{1}{2}$ quantum objects. Yet, they do have fields like bar magnets. Atoms are neutral quantum objects, and when they have spin or orbital angular momentum, they too have the bar-magnet fields.

Although, it is more accurate to say that while neutrons and atoms each have a net of zero electric charge, they also have equal amounts of positive and negative charge distributed unequally in space. Atoms are composed of a compact, positively charged nucleus with a negatively charged electron cloud around said nucleus. By mathematically transforming some data, researchers have found that a neutron in a nucleus is composed of a core of negative electric charge, a positive charge distribution in a middle region, and another negative charge distribution on the outside that tapers off.

With this information, it seems like the rule will now work. Neutrons can now be seen to have spin and an electric charge distribution. All atoms have an electric charge distribution, and some atoms have QAM. But in physics, details often matter. If I inform you that a hydrogen atom has a magnetic field, one could reasonably wonder why the proton's magnetic field doesn't cancel with the electron's. Couldn't one just point in the other direction from the other? I don't want to go down a rabbit hole of nuclear physics minutiae, so I will give you the experimentalist's answer: the proton and other nuclei all have magnetic fields that are several hundred times smaller than magnetic fields from electrons, so the magnetic fields from nuclei are not strong enough to cancel the fields from electrons. With that said, this is an apt time to mention that, because the electrons' bar-magnet fields dominate, I will not consider or make mention of the bar-magnet fields from nuclei in the remaining chapters.

As for neutrons, a real answer as to why it has a magnetic field is also a deep rabbit hole that I don't want to go down. I will simply say that the magnetic fields from the three quarks that blend together to make a neutron do not cancel, so there is a net magnetic field.

Looking again at the statement in the first paragraph, you will see the word *axis*. This word is not a synonym for *direction*. The distinction matters because the direction of the magnetic field along the axis is determined by 1) the direction of the QAM and 2) the sign of the electric charge that is the source of the magnetic field. A positively charged quantum object, such as the proton, has its angular momenta parallel to the axial magnetic field direction. That means if the field in Fig. 2.1 is from a proton, then the proton's QAM is pointed upward. The situation is reversed for negative charges. If the field in the figure is from an electron, then its QAM is pointed downward. As for neutrons, I must pull the answer out of the nuclear rabbit hole: the angular momentum and the axial field point in opposite directions.

The rule we have just considered only covers quantum objects at rest. It is natural to ask why we don't have magnetic fields with rotating matter, such as when we spin a basketball. The answer to this question comes from classical electromagnetism. Electric charges that move do produce magnetic fields, but the matter you see around you is made of positive and negative charges. And when this matter rotates, the magnetic fields cancel to essentially nothing.

How to visualize QAM

When you are considering a quantum object that has its matter density in a compact region compared to the scale of your scenario, a visualization of angular momentum is relatively easy. I suggest imagining QAM as a single vector arrow sitting in that location and pointing in some direction. The magnetic field can be visualized too. The central field lines from Fig. 2.1 will be centered on and aligned along the axis of the vector arrow if the quantum object has a field. You probably imagined something like that during my explanations in the previous section.

When we zoom in on the picture, more details emerge. In the caption of the first image of the first chapter, I mentioned that you should visualize mass and charge distributed over space in proportion to the matter density. By this, I mean that if the image is dim, then you imagine less mass and less electric charge density there. When visualizing QAM, it's similar. When the matter density is low, you visualize a small angular-momentum arrow there. When density is high, there is a big arrow there. As for the magnetic field visualization, you imagine superposing many bar-magnet fields throughout the matter and in proportion to the density.

This is not difficult to conceptualize in your mind. Suppose you have a piece of lodestone (aka magnetite), a mineral that produces a magnetic field. You would expect each portion of the rock to contribute to the field in proportion to the density of the rock. Some portions might have microscopic voids and thus have less density in that region. Consequently, those portions contribute less to the magnetic field. You then imagine some overall field built up from the portions of the lodestone.

At the end of the next chapter, I will give you the briefest insight into why this way of visualizing is true for spin via an analogy to the mathematics of tensor operators. And, while we are talking about the next chapter, I will say that physicists don't often say "bar-magnet field". Most often, either *magnetic dipole moment*, a quantity that characterizes the field, or *dipolar field* are used. You will see these terms in the next chapter.

QAM in spectroscopy

To end this chapter, we will see one way quantum angular momentum manifests itself to physicists. Specifically, we will see how spin and orbital angular momentum influence the spectra of atoms via their magnetic effects. In fact, both spin and orbital angular momentum were deduced to exist (at different times) because they were needed to make sense of the spectra of atoms.

In the previous chapter, I introduced the Balmer series of spectral lines, which are from atomic hydrogen. The spectrum I showed you is low resolution, though. There are finer details that can be resolved by a better experiment and be explained by further physics. When the electrons in the atoms have both spin and orbital angular momentum, the spin's magnetic field either aligns or counter-aligns with the magnetic field from OAM. In the classical world, from playing with two magnets, you know that going along with a magnetic field or fighting the magnetic field requires you to provide different levels of effort to keep them in place. Physicists say that the different alignments have different amounts of potential energy.

When we include the math for the magnetic effects in atoms, there are contributions to the potential energy. We find the spectra have very fine splittings in the energy of the spectral lines. The wavelength separations due to the split levels are fractions of a nanometer. The fine splittings in the spectra are named **fine structure**.

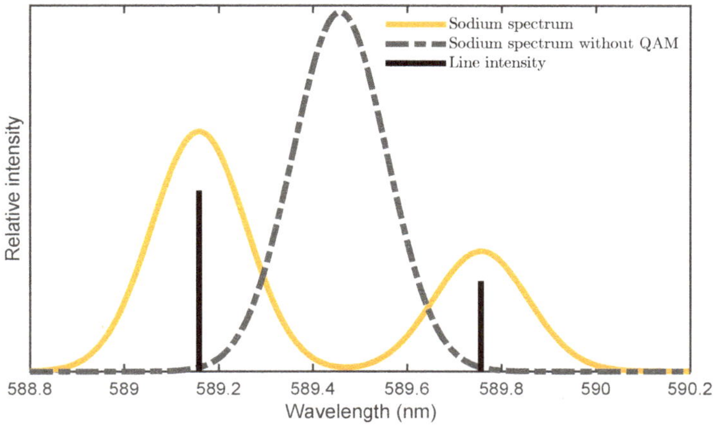

Figure 2.2: Simulated and greatly simplified intensity of the orange light from a hot sodium gas. The two vertical lines are the same types of lines you saw at the end of the first chapter in a figure created by Schrödinger. The large bump is what we would measure if the electron had no QAM. The double-bumped curve is the famous doublet that is observed. Each line of the doublet is due to a transition from a higher-energy 3*p* to a lower-energy 3*s* state. The 3*p* state is the one that gets split in energy. You may wonder why the 3*s* state does not split and give us four lines. It's because there is no OAM in the 3*s* state. Therefore, there is no magnetic field for the spin's magnetic field to align with or against in that state. For those curious, one line intensity is double the other because there are twice as many ways for those 3*p* states to be occupied in a hot gas when the orbital and spin angular momentum are aligned compared to when they are counter-aligned.

To be completely forthcoming, I should admit that my classical explanation above regarding magnetic fields is misleading. In my defense, in this book, I wouldn't be able to explain how the electric field of the nucleus couples together spin and OAM via their magnetic fields. Additionally, a second effect due to special relativity also contributes to the fine structure of the spectra we see. To keep things simple, I will not explain the effect due to relativity either.

I don't show the fine structure of any of the lines of the Balmer series because they are too complicated. Instead, I show you the most famous and simple example of fine structure in Fig. 2.2. The sodium

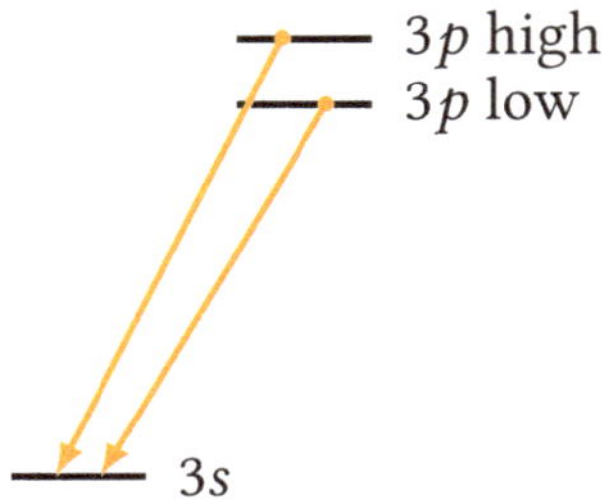

Figure 2.3: The atomic transitions of the sodium D doublet. The vertical direction is energy, and the horizontal direction has no relevance. The short horizontal lines are energy levels with multiple states represented by a single line. However, it is not to scale: the difference between the $3p$-high and the $3p$-low energy levels is 1/1000th that of the energy between them and the $3s$ level.

D doublet is famous because it has a large splitting of 0.6 nm and is in the visible spectrum, both of which helped it to be discovered by researchers who wore puffy shirts. The doublet is important in stellar astronomy, which is the field that gave it the D designation. In addition, it is also famous because your eyes are more sensitive to yellow light, which makes it an energy-efficient street light (well, it did before LED street lighting reshaped the standard of efficiency).

In the sodium spectrum, the yellow curve would be built up from numerous collected photons from many atoms. A sodium atom starts in one of two $3p$ energy levels and transitions down to the 3s energy level, emitting a photon in the process. The transitions are shown schematically in Fig. 2.3. The split in the $3p$ energy levels is due to the interaction between the spin and orbital angular momentum via magnetic fields. (Note that s in $3s$ is not the same as the spin quantum number s.)

Keeping all other things equal, without QAM, there would be no fine structure splitting of the $3p$ energy levels, and the sodium spectrum would be that of the dashed line in Fig. 2.2. It is a small difference between the two scenarios. Still, the fine structure of spectra was vital for the deduction of quantum mechanical facts in the first decades of the twentieth century.

3 Misconceptions about the electron in particular

> "*Let no one ignorant of geometry come under my roof.*"
>
> — inscription claimed to be at the entrance to Plato's academy

A simple question and wrong answers

This chapter is needed because physicists sometimes describe quantum objects in ridiculous ways. Before we handle more substantive topics, I need to erase from your mind three pictures of quantum objects and let you know why they are wrong. These pictures cling to the old ways of thinking about our universe. If we want to understand quantum mechanics, we must move past our natural intuition and learned assumptions. While I deconstruct these three pictures, it will be easier if I couch the following conversation around the electron since it is so common in explanations of quantum mechanics.

If you search the internet for "what is the shape of an electron?", you will discover a diversity of answers. For instance, some sources suggest that the electron is round like a marble, while others claim it is a point particle or a featureless point. Others say a theory shows it is possibly shaped like an egg. They can all be shown to be wrong based on simple physical or mathematical reasoning.

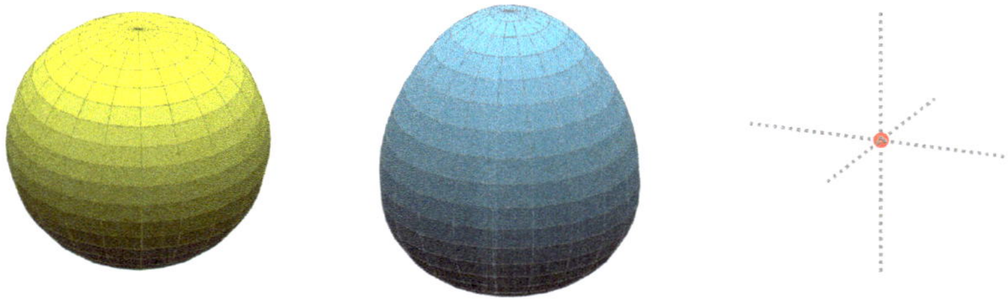

Figure 3.1: This chapter focuses on three things that don't exist: spherical electrons, egg-shaped electrons, and electrons as geometric points.

The electron is not a tiny spherical ball

I'm sure some of you would doubt that serious physicists claim that the electron or another quantum object is a sphere, so I'll have to name names. According to Dr. Don Lincoln on the Fermilab YouTube channel, "a particle is like a BB or a marble or something." Sadly, even David J. Griffiths, a renowned textbook author, tells the readers of his quantum mechanics textbook that it is okay to picture "the electron as a tiny spinning sphere". He does recover by the end of his sentence, saying, "as long as you don't take it literally." I disagree with this allowance because it certainly will and does lead to confusion among experts and non-experts.

Sometimes, the claim is not so direct, and people have implicitly learned through videos and images that the electron is a sphere. They are shown in edutainment shows and videos as if they were like tiny tennis balls going through double slits, around the nucleus, in collisions, and traveling down wires. In the case of Brian Greene's PBS edutainment show *The Elegant Universe*, he picks up a glowing CGI electron orb from a bowling ball rack and misinforms the audience: "You can picture them like this." – No, Brian. No, you cannot.

I can forgive the animators of, say, Jimmy Neutron and the Big Bang Theory for showing the ever-so-common planetary model of the atom during transitions between scenes because they are not claiming to be presenting facts. They are shows of fiction, after all. We, on the other hand, are interested in the decidedly nonfiction of the universe when we watch a PBS or BBC documentary.

To cement in their minds that the electron is not a sphere, many physics students must go through a rite of passage. Students are required to calculate the surface speed of an electron as a fictitious sphere. We deem this necessary because we bat around a factor in some of our equations called the *classical electron radius*, which would be the radius of a spherical electron by some classical physics reasoning. This radius is a rather tiny three-millionths of a nanometer.

In an imaginary world where the electron is a sphere and not quantum mechanical, the electron would need to spin on its axis to generate the magnetic field we measure. In reality, as I mentioned in the previous chapter, the electron has spin, a permanent and built-in quantum angular momentum that has nothing to do with anything spinning.

Once the students find the rate at which the sphere would have to rotate to match the spin angular momentum, they find that the surface speed at the equator exceeds the speed of light. Everyone and their dog knows that this cannot be correct. Nothing can exceed the speed of light, so the model we set up for the calculation is not correct.

You can reasonably suggest that this does not cover all possible cases. Perhaps the electron could have an intrinsic spin, not be genuinely spinning, and have some radius. Special relativity wouldn't be violated, and the electron would get its non-spinning spin. We should then have a more general reason why spherical electrons can't exist, and it's quite easy to understand.

Representations of the electron as a sphere make it out to be perfectly rigid. This is problematic. A perfectly rigid sphere of any size would transmit energy faster than the speed of light. This is because as soon as one side of a sphere is moved, the other side would move simultaneously. One could send messages or transmit energy with this movement. However, Einstein's special relativity famously and quite clearly tells us that absolute simultaneity is an illusion and quantities traveling faster than light are forbidden. Thus, there can be no perfectly rigid spheres in our universe.

Beware Wikipedia

Please be aware that you can't fully trust Wikipedia. It has numerous problems across many physics articles. Regarding the radius of the electron, Wikipedia's page on the electron says "Observation of a single electron in a Penning trap suggests the upper limit of the particle's radius to be 10^{-22} meters." You might think this sounds like we know the electron has a radius and it is some small number. A better understanding would be: *if* an electron has a radius, it has been excluded above 10^{-22} meters, and *if* there is a radius, it must be below this value. Therefore, with the better reading, we don't know whether there is or is not a radius.

But all of this misses some essential and interesting facts. The quote from the anonymous Wikipedia contributor(s) gives the impression that the community of physicists believes that a limit has been placed on the radius of the electron *as we know it.* The Wikipedia page cites a single researcher, Hans Dehmelt. He won one-third of the Nobel Prize in physics in 1989 for his work with Penning traps that determined to extreme accuracy what is called the *electron g-factor.* His work is in remarkable agreement with the theory of quantum electrodynamics.

No direct measurement of any radius has been undertaken. It was Dehmelt's personal project to determine if the electron is a **composite particle**, a quantum object composed of other quantum objects. Using a simple model that *assumes the electron is a composite particle of three other particles*, Dehmet put the upper limit on the radius of the electron using the value of the electron g-factor. This is not an illegitimate line of research. Still, because the standard model of particle physics says it isn't composite, few physicists expect (some string theorists may hope) the electron to be a composite electron.

As for the word *radius*, when applied to particle physics, it doesn't refer to the radius of a little ball. Let's take the neutron as an example.

It is a composite particle because it's made of three quarks. When we say the neutron has a radius, we mean that there is some simplifying model where we pretend something has a radius for some other end goal. As you know from the previous chapter, the neutron has a charge density that fades away. The neutron does not have a radius below which all the charge is inside and no charge is outside it like you would expect for a sphere. When a researcher sets a number to the "radius" of the neutron, it's a convenient fiction used to make comparisons.

Therefore, even if the electron were a composite particle, its radius wouldn't be the radius of a tiny little ball because the model Dehmet uses doesn't model a tiny little ball.

The electron is not a zero-dimensional point

Getting to the point

A rather large number of physicists describe the electron as a zero-dimensional point to each other and to their students. They say it is a **point particle.** The choice to describe it this way produces problems that involve geometry, physical units, and infinities.

Let me spell out what it would mean to have a zero-dimensional point particle come into realization as an actual physical thing. Imagine any three-dimensional object. If you have a book at hand, it would do nicely as an example. The book takes up some volume in space. Volume has units of meters cubed (i.e., m^3). Now imagine taking away one of its three dimensions, leaving you with a book of two dimensions. You will have a non-physical, two-dimensional object that belongs to the realm of cartoon physics. This conceptual object will now have an area instead of a volume, and the units of area are meters squared (i.e., m^2).

Let's repeat that procedure and take it down to one dimension. You will have a line segment, another non-physical object, with units of meters (i.e., m). Repeat one last time by removing its final dimension.

It is now zero-dimensional and has no units of length. It has no physical dimension, no physical extent.

Before I proceed further, I must make a clarification. We do use zero-dimensional mathematical points in our calculations, but these are not objects. The center of mass is such a mathematical point. We use the center of mass of objects to simplify our calculations, but it is not a thing that we could hold in our hand or capture in a detector.

Now, onto supposed zero-dimensional objects. A zero-dimensional point has no volume, as I showed above, and a zero-dimensional electron would have all the charge, mass, and spin of the electron in zero volume, which sounds awfully like we are trying to fit something into nothing. It must have some extraordinary consequences.

Let me remind you that density is some quantity divided by a volume. Consequently, because it has zero volume, the point would have infinite mass density, infinite electric charge density, and infinite spin density. More accurately, mathematically, those quantities are undefined because division by zero is undefined. Therefore, matter and photons cannot have infinite quantities.

The problem is that infinity is not a number. It is a concept. Trying to move **actual infinity** (aka **completed infinity**) around in equations in place of the numbers leads to mathematical and physical nonsense. If a physics student were to submit their solutions to a homework problem with actual infinity as a measurable value of some physical quantity, they would get a grade far from an A. I give the physicists who claim that an electron is a point particle the same grade as the students.

You see, actual infinity is a concept that some physicists confuse with **potential infinity** because they are never taught the difference. Actual infinity is the infinity of set theory. It is the imagined instance of more members in a set than any finite set. This infinity leads to absurdities when used in place of a number. Potential infinity is a *process* with absolute infinity as the destination that is worked towards as one counts upward without end. Absolute infinity is like the horizon, and potential infinity is the process of walking to the horizon. You certainly can trek towards the horizon, but you can never stop, look around, and say you've completed your journey.

Why do they think absolute infinities are allowed? My thoughts on this involve an explanation of calculus and the multiple differing definitions of something called the *Dirac delta*. I don't want to interrupt the flow with such a technical explanation, so I will append it to the end of this chapter and continue on.

A shell game with terminology

Some say, "By 'point particle', we mean a particle that doesn't have internal structure." Unhelpfully, they have added another confusing term to explain their definition.

In more precise language, they mean that point particles are not composite particles like, say, protons (which are composed of three quarks) or atoms (which are composed of nuclei and electrons). We already have a term for non-composite particles, like electrons and quarks, and it's better than *point particles*. We call them **elementary particles**. Nevertheless, those who use the unhelpful term *point particle* without meaning a zero-dimensional particle confuse others into thinking we are dealing with zero-dimensional particles.

If you are now frustrated by the lack of precision by physicists, I empathize. I too am frustrated. Unfortunately, we have not followed the chemists' lead and have yet to develop our physics version of the IUPAC Gold Book of chemistry terminology. The time investment needed to decipher physics terminology is very large. Physics terms are often sources of confusion for both novices and experts, as you will see throughout this book.

Though, the situation is often worse than a simple game of playing with synonyms. Some physicists teach students of quantum field theory that elementary particles are geometric points. Some textbook authors try to correct this point by showing that electrons have an extent throughout space:

- "It is often said that the electron is a point particle without structure in contrast to the proton, for example. We will see in this section that this is not true." — Scharf, G. *Finite Quantum Electrodynamics: The Casual Approach*, p. 236.

- "QED, or relativistic quantum field theory in general, is not based on the notion of 'point particles', as one sees stated so often and yet so erroneously." — O. Steinmann, *Perturbative quantum electrodynamics and axiomatic field theory*, p. 2.

Physicists who learn from the wrong textbooks are left unaware of the truth and pass on the point-particle misconception to the public. In my view, the public is misinformed in the book *The Quantum Universe (And Why Anything That Can Happen, Does)* by Brian Cox and Jeff Forshaw:

- "Dealing with particles 'of no size' sounds problematic, and perhaps impossible."
- "The notion of truly point-like objects need not be wrong, even if it flies in the face of common sense ..."
- "... there is no place for 'size' in the fundamental equations of particle physics."
- "To be emphatic: point-like particles are really of no size ..."

With the emphatic declaration of point-like particles existing, a fitting time to quote a textbook by Arnold Neumaier is now:

> When presented in the right way, quantum mechanics is not at all weird, but very close to classical mechanics. Much of the weirdness comes from forcing quantum mechanics into the straightjacket of a particle picture. The particle picture breaks down completely in the subatomic domain, as witnessed by the many weird things such a view leads to. On the other hand, the field picture remains valid at all length and time scales.

On infinitesimals and zero-dimensional points

For those with some calculus exposure, you may be aware of infinitesimals and may incorrectly equate a zero-dimensional point with an **infinitesimal volume**, or **volume element**. An infinitesimal volume is still a volume and has units of meters cubed. In contrast, as I illustrated with the example of the book and the number of its dimensions, a geometric point has no dimensions or units because it has no physical extent.

We say an infinitesimal volume approaches zero volume, but it never gets there. We can tell when it is too large based on the situation, but we don't say exactly how small it is. It is considered plenty small enough for the problem at hand. An infinitesimal volume is so small that in our mind's eye, or on paper, it appears to be a point. That is to say, it is arbitrarily small but still finite.

Sometimes, a situation permits us to approximate a physical object as having an infinitesimal volume, and we might say the object is an infinitesimal particle. We use infinitesimal particles in some of our calculations to ignore the additional layers of mathematical complexity that come with having a large, extended object. They allow us to simplify many problems and concepts in physics so that we can make progress. Alternatively, when we can't use the infinitesimal particle approximation, extended objects are treated as being composed of a multitude of infinitesimal volumes.

Let me end this section with a warning. What I mentioned is not a true infinitesimal. Let's think of it as the **physicist's infinitesimal**. The true infinitesimal, which causes logical absurdities, can be thought of as the **mathematician's infinitesimal**. It is a number greater than zero and smaller than the smallest real number. The problem is that there is no smallest real number. Supposing to the contrary that you had the smallest real number, you could divide by two and have an even smaller real number, which, of course, is inconsistent with our starting point. Mathematicians aren't fond of contradictions, so you can see why an actual infinitesimal gets them agitated.

The nonsense of egg-shaped electrons

Over the past dozen years, a claim appeared in the media that says scientists (actually string theorists) think it's possible that the electron is "egg-shaped" and physicists have been doing atomic physics experiments to measure whether this is true.

I don't think I need to convince you that an electron cannot be egg-shaped. But the physicists who contributed the idea and those who passed it on as a possible truth have irritated me with this one. In my opinion, the idea is cartoonish and will mislead the public.

Had the physicists and journalists spoken and written so that articles were written along the themes of "Physicists put a limit on the hypothetical, intrinsic electric dipole of an electron" or "The electron still appears to have an innate electric monopole within limits of experiment," then this section would be unneeded. (I decode those *-pole* terms with pictures of monopole and dipole fields below.)

According to nearly every news article about these experiments, physicists supposedly think that the electron should be egg-shaped. To me, it's evident that the media are all copying from the same press release and it must have had input from someone who understands mathematical symmetries. I could not track down who told the journalist an electron might be egg-shaped, but I can say the experimental research group is not completely innocent in their description of the electron. Have a look at what was said by the first author of a paper called *Improved measurement of the shape of the electron*:

> Conventionally, people think that the electron is round like a little ball. But some advanced theories of physics speculate that it's not round, and so what we've done is designed an experiment to check [that] with a very, very high degree of precision.

The quote, the title, and the reference to egg-shaped electrons are good illustrations of how cavalier physicists are with terminology, even when communicating with each other. Interestingly, in the published scientific paper, the only place they mention shape is in the title. I guess the unscientific title was meant to develop interest in the article. Throughout the rest of the paper, they instead refer to a theoretical, intrinsic **electric dipole moment (EDM)** of the electron. The EDM is supposed to be an incredibly weak addition to the already present *electric monopole moment*, which is a way to characterize the well-known radial electric fields.

An EDM describes a possible pattern of the electric field lines due to electric charges. You are already familiar with this pattern, as it is the same as a bar magnet's field (see Fig. 3.2). Let me quickly break down the term in the context of a magnet.

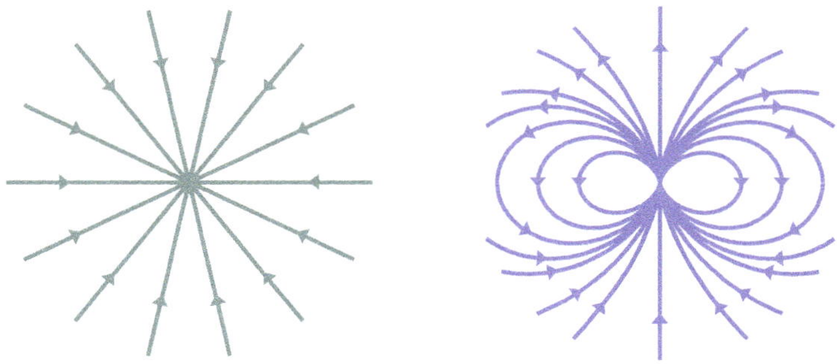

Figure 3.2: The field shapes that emanate from an electron with a wave function that is very compact relative to the dimensions of the image (i.e., in an infinitesimal volume). The left image is a radial field due to a monopole. The electron's electric and gravitational fields have this radial field distribution. The right image is a dipolar field due to a dipole. At present, the only *intrinsic* dipolar field produced by the electron that we know about is a magnetic one, and it occurs because the electron has spin angular momentum and electric charge.

The strength and orientation of a bar magnet's field is called the **magnetic dipole moment**. The word *dipole* refers to the field having two poles, north and south, which means the magnetic field arrow goes in one end and comes out the other. The word *moment* refers to the strength and orientation of the field. You will see that physicists sometimes shorten the whole term to either i) *magnetic moment*, ii) *dipole moment*, iii) *magnetic dipole*, iv) *dipole*, or v) *moment* when the context is clear (to the physicist) what the missing words are.

If you can picture a bar magnet's field or an idealized version of Earth's magnetic field, you can picture the field due to an electric dipole moment. It's the same pattern of lines in three-dimensional space, just a different flavor, so to speak.

It's important to understand that the experiment was trying to find an electric field due to a weak intrinsic EDM on top of the well-known intrinsic monopolar electric field. They were not searching for an egg-shaped charge density distribution. The claims are odd because we are already well aware of the distribution of electric charges in various fields of physics. Some readers will already have encountered charge distributions of electrons and their *extrinsic* electric dipole moments during high school chemistry.

I don't see why the public wasn't spoken to honestly and told that physicists are searching for a new configuration of an electric field. After all, the average person has some awareness of magnetic and electric fields, and science enthusiasts, who would be interested in the news, have even more knowledge at the ready.

Charge distributions and extrinsic EDMs are seen in high school chemistry

It's standard for chemistry students to encounter polar molecules and their dipoles (see Fig. 3.3). These are electric dipole moments from electron and proton charge distributions. These charge distributions can be said to be shapes in the sense of having a form. They are like clouds with varying densities and no actual surface. From this description, you may recognize that I am referring to the matter density $|\Psi|^2$.

The EDMs in molecules differ from one type to the next because the electron distributions are not the same. Any dipole that depends on the form, or shape, of the charge distribution will be an extrinsic property of that object because it is not a property of the object that exists independent of other conditions.

When someone told the media that the experimenters were measuring the electron's shape, it was quite misleading because the search

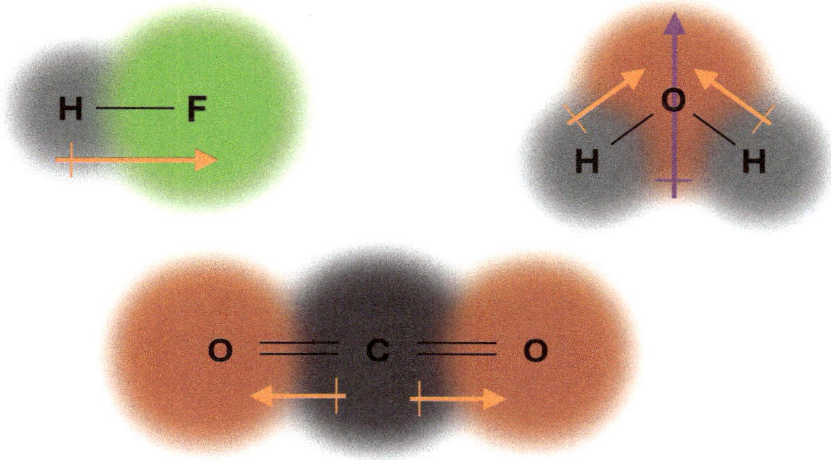

Figure 3.3: Standard high-school level molecular polarity. The point of this figure is to show that the distribution of electron clouds in space determines the extrinsic electric dipole. Molecules are needed as examples because atoms can be rotated 180 degrees and have the same electron charge distribution; thus, they do not have electric dipoles. Clockwise from top left: hydrogen fluoride, water, and carbon dioxide molecules. The orange arrows point in the direction of the dipole due to a bond between two atoms. It's too messy to include the electric field lines, but you could imagine overlaying and orienting a dipole field overtop and aligned with the arrows. The length and direction of the arrow is determined by the electronegativities of the two atoms. (Due to historical convention, the dipole arrows in high school chemistry point in the opposite direction from the physicist's arrows and the chemist's IUPAC Gold Book definition.) We know F is much more electronegative than H, so we say it steals a bit of the electron cloud from H. This makes the F end more negatively charged and the H end more positively charged in the HF molecule. C and O have similar electronegativities, so the arrows are smaller. The two bonds point in opposite directions and cancel, so there is no overall dipole moment. Water molecules have bonds that cancel in the horizontal direction but reinforce in the vertical direction, so they have a strong overall dipole, which is represented by the long, purple arrow.

was for a modified electric field. Furthermore, the hypothesis came from string theory, yet the various EDMs associated with electron charge distributions ("electron shapes") belong to the realms of chemistry, atomic physics, and condensed-matter physics.

On the opening question of the chapter

I want to end this chapter with something for you to consider. Why do the three erroneous and easily dismissed pictures of the electron exist? I think it is because a sizable number of physicists are starting with a philosophical position that gives them a blind spot. They subscribe to the Copenhagen interpretation's philosophy of not asking questions about reality when measurements are not involved. This interpretation confuses their mental pictures and halts inquiry into whether their view is consistent with everything they know.

I opened this chapter by mentioning an internet search for "What shape is the electron?" This question might be fine if one does not reflect on it. Certainly, all classical objects we have encountered have shapes, do they not? It's obvious. Chairs are chair-shaped, bottles are bottle-shaped, and wheels are wheel-shaped. I could ask someone to draw those shapes, and they could likely draw something everyone would recognize as a chair, a bottle, or a wheel.

We should pause that line of thought and reassess, though. We should consider the possibility that the opening question makes a faulty assumption. If we continue with another real-world example and say water is water-shaped, we run into difficulty. Water will take on the shape of the container it is in. It will be bucket-shaped in a bucket, and it will be bottle-shaped in our bottle. Processes also affect the shape or form of water. It could be free of any containment and be falling as a rain droplet, where its size and shape are determined by how it started in the cloud and the characteristics of the atmosphere as it falls. Alternatively, water in the ocean could be driven by wind to become an ever-changing wave.

If someone answers the question "What shape is water?" without specifying the container it is in or the process it is going through, you know that person does not understand what water is and how it behaves. They are certainly not an expert that understands water.

I say the same is true of physicists and electrons. If they don't say, "Well, it depends on the situation," when asked about the shape of an electron, then they have not thought that much about this aspect of quantum mechanics. I repeat my suggestion from the first chapter that, during the time you are reading this book, you should think of the wave function as an element of reality. In this case, the best answer regarding the electron's shape is to say the electron is its wave function or matter density and their smoothly changing distributions through space are the closest analogs to having a shape.

Some mental pictures

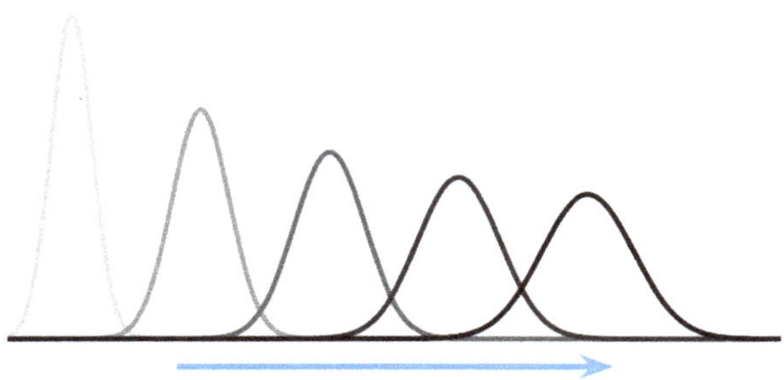

Figure 3.4: Evolution of a traveling electron. Physicist's minds are limited like any other. We think in simplified models whenever it is appropriate. This image represents the matter density (aka probability density) of an electron. The arrow indicates that the electron is moving to the right. Snapshots of five different times are shown. We call this shape a *Gaussian* (after Carl Friedrich Gauss). In general, wave functions and their matter density disperse (spread out) as they travel. An important detail is that the areas under the curves of matter density are all equal.

For now, if you don't have much exposure to the concept of a wave function, you may picture a freely moving electron like a solitary ocean wave traveling along that disperses (spreads) as time goes on (see Fig. 3.4). This is how some physicists might sketch a freely moving electron. More specifically, because the values never go below the horizontal line, this image would be a representation of the matter density $|\Psi|^2$ (aka probability density) of the electron propagating through space. If we instead want to show a wave function, we should add oscillations throughout the wave that go positive and negative, which change incredibly rapidly with time (see Fig. 3.5).

These two figures have limitations, though. They display one dimension of space along the horizontal axis, and time is displayed through a series of snapshots. The value of the wave function at a point is a real number and is shown as the sketch's height along the vertical axis. Actual wave functions are complex valued at every point in three-dimensional space.

If electric or magnetic forces confine or alter the electron, you should expect the three-dimensional wave to adopt distributions with interesting symmetries. In the first chapter, you saw matter densities of hydrogen, which are confined by the electric field of the nucleus. In atoms, electrons are bound to nuclei and are often in special types of bound states where their matter densities don't change with time. These special states are the stationary states I first mentioned in the previous chapter.

Eventually, I will change the real-valued picture of a wave function to a complex-valued one. It will be complex valued but one-dimensional in the second chapter on Schrödinger's equation and finally three-dimensional in the *Gallery* chapter. The images I have shown you so far are usually all that you need as a rough guide for much of the conceptual understanding of the wave nature of quantum objects.

Physicists aren't good with consistency

You should be aware that we physicists use the term *wave function* to mean two closely related things, and I am guilty of using it both ways. Let's say we are considering a single electron. In one sense, *wave function* refers to only the **spatial part** Ψ_{spatial}. The spatial wave

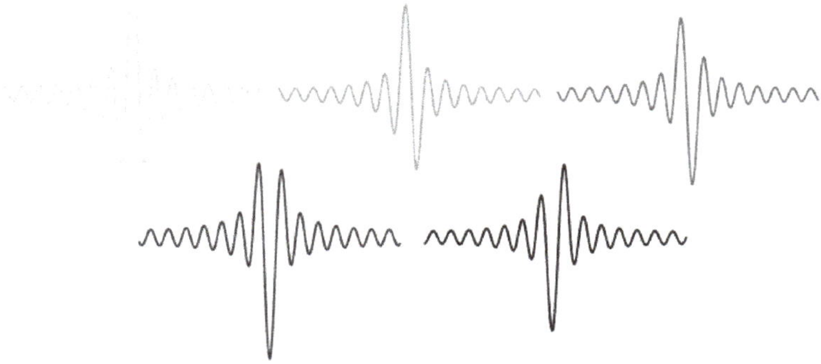

Figure 3.5: Wave functions oscillate in space and time. We add another layer of complexity and picture oscillating wave functions. Here is a collage of five snapshots in time of a one-dimensional wave. The time scale is now quite short, so you can't see the dispersion (spreading out). Wave functions oscillate rapidly compared to human time scales. The darker wave packets are later in time. This picture, too, lacks the full complexity of a true wave function.

function is what I show you in Fig. 3.5. It is not a complete description of the state of an electron. It is only half of the description.

The second way we use the word *wave function* refers to the complete description of the state of an electron. We say it is the **total wave function** of the electron Ψ_{total}. The other half contributing to the total wave function is the electron's **spin state** $|\chi\rangle$. Greek lowercase "chi", pronounced like "sky" without the "s", is commonly used as a generic symbol for a spin state. We will see the physics of this part in detail in the chapter on spin and spin superpositions.

Beware that $|\chi\rangle$ is one mathematical object, not three things. I'm using Dirac's notation, and this symbol is called a **ket**. It lets you know we are talking about a state and not another thing, like a number or an operation. So, going forward, I should put wave functions in kets, too, because they are state functions. From this point on, you should know when I discuss quantum states that $|\Psi\rangle = \Psi$. I treat the symbols as if they are identical. In actuality, they aren't exactly equal; in a quantum textbook, there are nuanced differences between the two. Despite using kets with states, I will continue to use $|\Psi|^2$ to refer to the matter density of a quantum object because it's a simpler way to write it.

The total wave function, or the state, of the electron is then a product of the spatial wave function and the spin state:

$$|\Psi_{\text{total}}\rangle = |\Psi_{\text{space}}\rangle \otimes |\chi\rangle \tag{3.1}$$

The $\otimes$ is called a **tensor product** symbol (more specifically, it represents the **Kronecker product**). It means the two things on either side of it are intimately connected, more so than even superpositions created by a + sign. Mathematically, the tensor product distributes the spin state over each position of the spatial wave function $|\Psi_{\text{space}}\rangle$. It is similar to how multiplication distributes numbers and variables For example, consider this example: $C \cdot (x + y + \cdots) = C \cdot x + C \cdot y + \cdots$. The multiplication dot is analogous to the tensor product symbol. The C is analogous to the spin ket, which leaves x and y to be analogous to positions of the spatial part of the wave function. This is why, in the caption of the first figure of the book (Fig. 1.1), I said the spin is distributed over the density you see in the images of the orbitals.

Despite the intimacy between the two, in most cases, when teaching quantum mechanics or when solving research problems, physicists work on the two parts on the right side of that equation separately to make the problems simpler. In fact, the mathematics of the parts are quite different: calculus for $|\Psi_{\text{space}}\rangle$ vs. matrix algebra for $|\chi\rangle$. Throughout most of this book, I will keep the two separate too.

Circling back to the purpose of this subsection, when I or another physicist uses the word *wave function*, it could refer to $|\Psi_{\text{total}}\rangle$ or $|\Psi_{\text{space}}\rangle$. If the distinction is important, what is meant will either be stated or is discernible from context.

Chapter appendix: Physicists and infinities

Dirac deltas

In the chapter proper, I mentioned that some physicists are okay with unphysical infinities because they are unaware of the difference between absolute and potential infinity. I repeat again that absolute infinity is a concept and is neither a number nor a representative of a number, like a variable is. It is thus forbidden to have it appear in place of a number in physics. Thus, point particles with their infinite densities are forbidden.

I surmise that the reason one can become comfortable with such an illogical thought as infinite density is that they have slowly been accustomed to the sloppy math surrounding the Dirac delta, are unaware of when math can be mapped to physical reality, and have a misunderstanding of calculus limits.

There are multiple definitions or explanations of the **Dirac delta** $\delta(x)$. The correct definition says the it picks out a single number from a function. In quantum mechanics, that means it picks out the complex number at some point in the spatial wave function. Because it is not a traditional function, it has several unintuitive properties. When physicists write the following integral, it follows the correct definition:

$$\int \delta(x-a) \cdot \Psi(x) \cdot dx = \Psi(a) \tag{3.2}$$

In this book, understanding this integral is not needed at all. I didn't even include the necessary upper and lower bounds. I include it so those who will go on to study quantum mechanics at university will recognize the correct description when they see it.

I remind you that the correct use of the Dirac delta will pick out a number from a function. It's an instruction to do something. With that definition, the Dirac delta doesn't have a defined value. That being said, so that they may picture the math that they are doing, I think it is okay to tell students that, some of the time, they can heuristically think of the Dirac delta as a spike with a height of one and no width at some point and everywhere else on the number line is zero. I have just described the *unit impulse*. Sometimes, this unit impulse is given as the definition of the Dirac delta, but it is quite wrong. However, the heuristic picture will certainly help you differentiate the correct definition from the wrong one that follows.

Let's imagine some bump, like one of the shapes in Fig. 3.4. Assume the area between the horizontal axis and the drawn bump cannot change. If we were to squeeze the bump so that it was thinner horizontally, it would have to increase its height. We can think of squeezing it thinner and thinner. Its height will continue to increase towards absolute infinity. It is said that if you could get to the end, or the limit, of this process, you would have an infinitely high spike of no thickness. This infinitely high spike is both a common and incorrect definition of a Dirac delta. You can see the spike illustrated in the bottom image of Fig. 6.1 on page 85.

Having a distribution of numbers that is zero everywhere except at a point where there is a *supposed infinite spike* is equivalent to having a distribution of numbers that is zero everywhere except at a point where *there is nothing defined.* This is so because if one wants a distribution of numbers, one must put numbers in those places, and infinity is not a number. The consequence of using the incorrect definition will lead to the integral above being undefined exactly where one wants the value of the wave function to be picked out.

Let me be clear. My intention is not to point out that students are being miseducated about the Dirac delta nor to say that the calculations physicists do are not rigorous. (The first is true. The second is also true but of less consequence.) My intention is to show the possible source of what leads to the acceptance of an infinitely dense point particle. I think physicists who don't know the difference between the Dirac deltas internalize the idea that it is okay to bandy about infinities in our math. It is a short leap from that to saying that infinities can exist in the physical world.

Now, how do I know physicists make this confusing convolution of three definitions? Because it is well-known to experts that contradictory definitions or explanations of the Dirac delta are taught in many textbooks. For examples of multiple contradictory explanations of Dirac deltas in one book, see §2.5.2 in Griffiths and Schroeter, §4.3.2 in Burkhardt and Leventhal, and §1.10 in Shankar. All three also have what I would describe as a fourth definition that I won't bother to describe.

The meta-reason for the confusion

Maybe three or four times in this book, I tell you that you don't have to follow the discussion of the math that follows and you will still understand the main point. This is one of those times.

Now, onto the question "why can't the physicists see the infinite spike makes no sense?" As I said, I surmise they must not understand the difference between absolute and potential infinity. I believe it comes from their calculus education.

Calculus legitimately uses potential infinity. It is the process of a variable increasing towards absolute infinity. The variable increasing towards its destination of absolute infinity is represented by: $x \to \infty$. A *limit* is a finite value that a math expression works toward (the math expression does not have the same journey as the variable) as its variable x works towards either absolute infinity or a finite destination. The limit is the answer that the professor wants us to find. Given to us are the variable x, its destination, and a function $f(x)$ that depends on the variable.

Calculus professors try to have students understand that the x variable doesn't need to arrive at its destination for us to figure out the value of a limit. Many times the variable can reach its destination, and many times it cannot. After all, when x journeys toward ∞, it can't reach the destination. Too often, students leave behind their calculus class thinking that, because we can find limits, a variable has gone all the way to infinity and we've achieved an instance of absolute infinity. They did not understand what is called the *epsilon-delta definition of a limit.*

Students are also confused by another issue. The common and misleading notation that follows leads students to believe that a function is allowed to truly equal absolute infinity:

$$\lim_{x \to b} f(x) = \infty \tag{3.3}$$

Math professors don't emphasize enough to their physics-professors-to-be that this expression is not supposed to be read as you would typically understand one with an equal sign. The equal sign is a lie and poor notation. The expression isn't really an equation, and it should be understood as a function diverging *towards* infinity. The *towards* is needed because we are working with potential infinity in calculus. (Why a second arrow is not standard, I do not know.) No limit was ever reached in the process above; the epsilon-delta definition of a limit tells us that a limit of a function is a particular number (and numbers are finite). Without a good understanding, students and future physics professors may leave thinking that absolute infinity is okay to use sometimes "if we're careful" with what we are doing.

If my words about calculus limits were put in front of the textbook authors, I think they would say, "Yeah, of course, I know that. Everyone knows that." I don't think they have truly internalized it, though. Their calculus explanations of Dirac deltas do not make sense.

When they think of the process to get an infinite-spike definition of the Dirac delta and an infinite-density point particle, I think it goes as follows in their head. They start with a function in their mind that resembles a bump. A continuous function like a bump feels very physical to a physicist. They could even think of it as a density distribution in space. They imagine the squeezing process I mentioned, and it can go on forever. Smoothly changing a function, like with the squeezing, also feels very physical to a physicist. I have to pause their line of thought to mention that the process has a variable that goes towards zero thickness, which isn't a potential infinity, but this process directly determines the height, so there is an implicit potential infinity involved. From their experience of getting answers from calculus limits, they unconsciously feel like they can smoothly transition from the potential-infinity process of the height increasing without limit to absolute infinity. They then feel like there can be an infinitely dense particle. It feels natural. It feels physical. Those feelings are wrong.

Physics deals with finite processes, so the last feeling they have of physicality is misplaced. They are incorrectly equating an infinitesimal volume with a geometric point. There is an immeasurable difference between picking a point along an unending process and getting to the end of the unending process.

4 Waves and superposition

> *Science is a way of trying not to fool yourself. The first principle is that you must not fool yourself—and you are the easiest person to fool.*
>
> — Richard Feynman

The essentials

By the end of this book, you should feel that much of the disagreements over the interpretations of quantum mechanics involve when, how, and whether the math on the page **corresponds** to real objects. Due to the way superpositions are presented by pop-sci physicists, many people have internalized the superposition principle as some super-powered principle that allows for ghostly behavior of objects, whether they are cats or particles imagined as tiny spherical balls. However, at the most basic level, superpositions are about math (and quantum objects are never tiny spheres). Without understanding the mathematics, you can't develop your opinion of the physics. So, with this chapter, I give you this mathematical insight. Because the superpositions are in the context of visualizable waves, they will mentally prepare you for the counterintuitive and hardly visualizable superpositions of spins in a later chapter.

So what does **superposition** mean in the context of waves? Superposition is simple addition. It's the addition of multiple functions at each position in space. Hence, the *position* part of the word. The word *super* is Latin for *in addition to* or *on top of*. So, to reiterate, a superposition is composed of functions that add one on top of another in space.

If superposition is addition, then **interference** is the result of the addition of two or more particular functions we have in mind. If you hear of interference, then a superposition is implied. There is little difference between the words. Interference is usually mentioned while examining a particular region of space. We usually talk about degrees of interference. One function can interfere with another a little or a lot in a region. We also talk of constructive interference, where two or more waves add to make a larger wave. Destructive interference is where the opposite happens: waves come together to cancel out and make a smaller or zero-amplitude wave.

Notice when I explained the meaning of the words *superposition* and *interference* that I did not use the word *quantum*. Superpositions are not necessarily quantum mechanical. We can have, for example, superpositions of acoustic, water, or electromagnetic waves. To sidestep some confusion, in the following section, let's imagine all the waves are water waves. The essential difference between this scenario and a quantum wave scenario is that the amplitudes of water waves are described by a continuum of real-valued numbers, whereas, in the quantum case, the amplitudes are described by a continuum of complex-valued numbers.

I also chose water waves to avoid some quantum mechanical technicalities. Therefore, for a book on quantum mechanics, know that this chapter's explanation of superpositions is not complete. The focus here is on the mathematical side, while we imagine water waves. The rest of the book discusses what I want to teach you about the quantum mechanical side of superpositions.

Superpositions involve some arbitrariness

In everything that follows in this section, we will say we are looking down upon a ripple tank at a snapshot in time. It is a tank of water illuminated from above so that we can infer the locations of the crests and troughs of water waves by viewing the light patterns on a flat surface below the waves. In our specific scenario, we will imagine a wide slit in a barrier. Some source outside the picture creates linear waves that are incident on the barrier. Consequently, waves emerge from the slit. (See Figs. 4.1 and 4.2.)

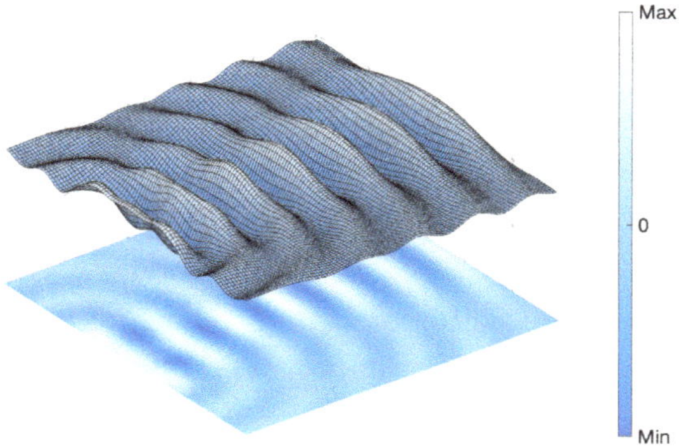

Figure 4.1: The basic idea of an illuminated ripple tank. Bright areas correspond to crests, and dark areas correspond to troughs. The zero level is the level of water in an undisturbed tank.

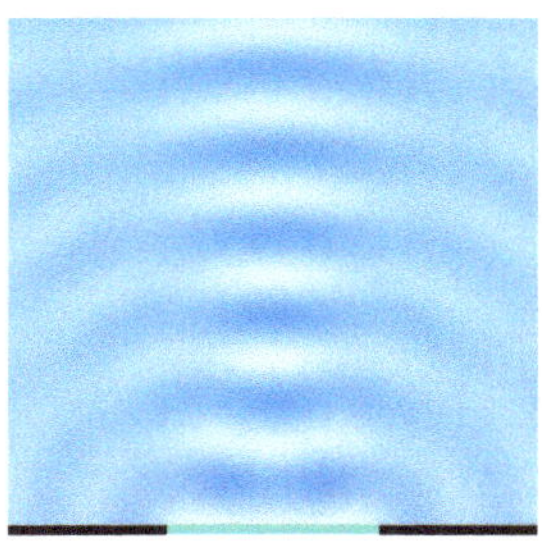

Figure 4.2: The only waves that exist in the present scenario. All subsequent images in this section are in our mathematical imaginations. When you see the symbol Ψ_{total} in this section, this is the image that it represents. The waves travel from the bottom to the top. The colored strip at the bottom indicates the location of the opening in the barrier.

We are dealing with classical waves, so we won't use Dirac notation. Though, I will still use our famous Greek letter to represent functions. The total water wave function is:

$$\Psi_{\text{total}} \tag{4.1}$$

It describes the amplitudes of the water waves at each point in space. My quest for the rest of this section is to convince you that the decomposition of this function into numerous functions is rather arbitrary.

To start, we might as well decompose Ψ_{total} in a simple manner. Let's say the superposition is:

$$\Psi_{\text{total}} = \Psi_{\text{L}} + \Psi_{\text{R}} \tag{4.2}$$

I show these two parts in the next figure (Fig. 4.3). *If* one were to block the left (L) and right (R) halves of the wide slit in turn, these are the waves one would see emerging from the new smaller opening. Take note that the last sentence refers to scenarios different from the present one. The waves in these two images don't have an independent physical existence in our *present* scenario. They are part of our mathematical imagination. You can't look at the waves in the tank and see these oscillations.

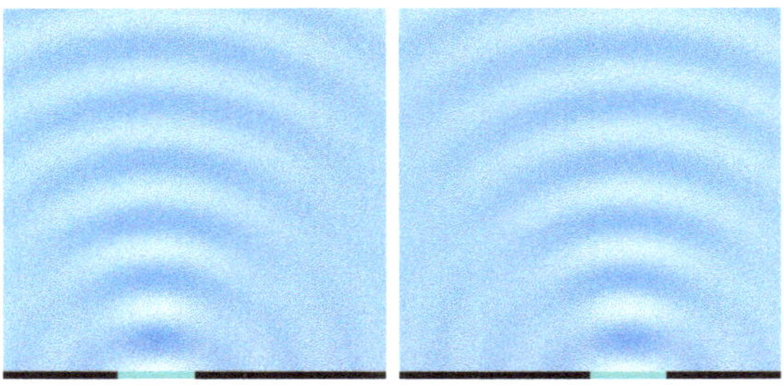

Figure 4.3: Total wave function decomposed into two parts. The shading of the crests and troughs is more subdued than before. They aren't illustrated as intensely white or blue because the amplitudes of the waves are smaller. Adding the intensities of the two images up will result in the image of Fig. 4.2.

By the way, the words **decompose** and **decomposition** I have used a few times now both refer to a simple mathematical concept. Here, you should think of the concept as the reverse process of superposition. Whereas superposition makes a single, new function out of many functions, decomposition takes a single function and breaks it into many functions. I bring this term to your attention because it has the connotation of human choice in it. It makes you more aware of the arbitrariness of how many and which functions are used to compose a superposition.

Next, let's break the ripple tank's wave function into three. I've already used L and R, so let's go back to the beginning of the alphabet:

$$\Psi_{\text{total}} = \Psi_{\text{A}} + \Psi_{\text{B}} + \Psi_{\text{C}} \tag{4.3}$$

Fig. 4.4 shows these three parts of the total wave function. Similar to the last decomposition, each image appears as if we blocked two-thirds of the original slit (but, of course, the present scenario has the slit unblocked).

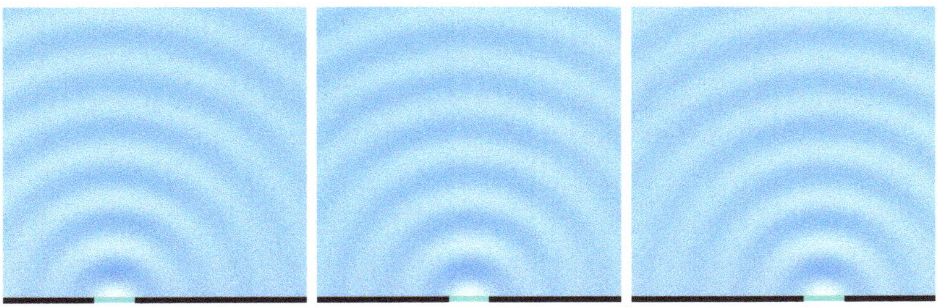

Figure 4.4: The total wave function divided into three. The shading is more subdued as compared to the previous images. The amplitudes of these waves are slightly less intense than the previous images because we are dealing now with three parts.

I'm sure you have figured that I could continue chopping up the total wave function into finer and finer parts. This completes the illustration that an arbitrary number of functions can go into a superposition. The next level of arbitrariness that I want to show you involves picking functions that don't correspond to an imagined physical modification of the present scenario, such as blocking portions of the present slit.

In my final decomposition (Fig. 4.5), I break the total wave function into two parts called *chosen* and *what?*:

$$\Psi_{\text{total}} = \Psi_{\text{chosen}} + \Psi_{\text{what?}} \tag{4.4}$$

I constructed the function Ψ_{chosen} so that no amount of creativity in blocking the wide slit will get you this mathematical function. In Ψ_{chosen}, it looks as if waves 180° out of phase come from outside the length of the gap. Once I chose this first part of the superposition, I had no choice in the second part. That lopsided function is forced upon me because the two parts must add to give the original image.

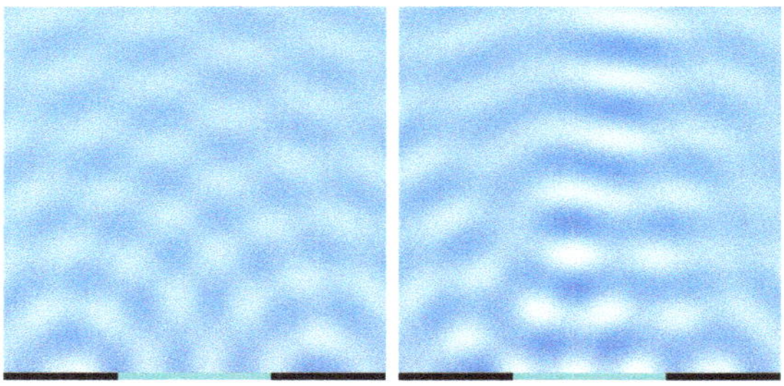

Figure 4.5: Two wave functions that add to make our real-world example. The function I chose is on the left.

Notice that, away from the central portion of the original image (Fig. 4.2), the troughs and crests are less intense than in Fig. 4.5. That means, in the latest images, the wave amplitudes of the parts are more intense than the whole. This can help you understand that superpositions are mathematical. None of the underlying waves in the parts of the superpositions are "really there". Moreover, if any single Ψ on the right-hand side of any of this chapter's equations had an independent physical existence, then they would all have it. But that's clearly more parts than is needed to describe the whole. In reality, what is happening on the right-hand side of each equation is the demonstration of our choice in how we choose to imagine (i.e., choose to decompose) the total wave function.

It is important to accept the explanation in the previous paragraph. We need to know that the composing functions do not have an independent physical existence because that knowledge constrains our thinking of what is possible and what happens in quantum mechanics. In particular, this knowledge will also assist you in accepting what occurs in spin superpositions when we examine spins in a later chapter.

Overlap and approximations

Have a look at a new scenario in the following figure (Fig. 4.6). The bottom image is the total wave function, and the top two images are the functions I used to create the total wave function.

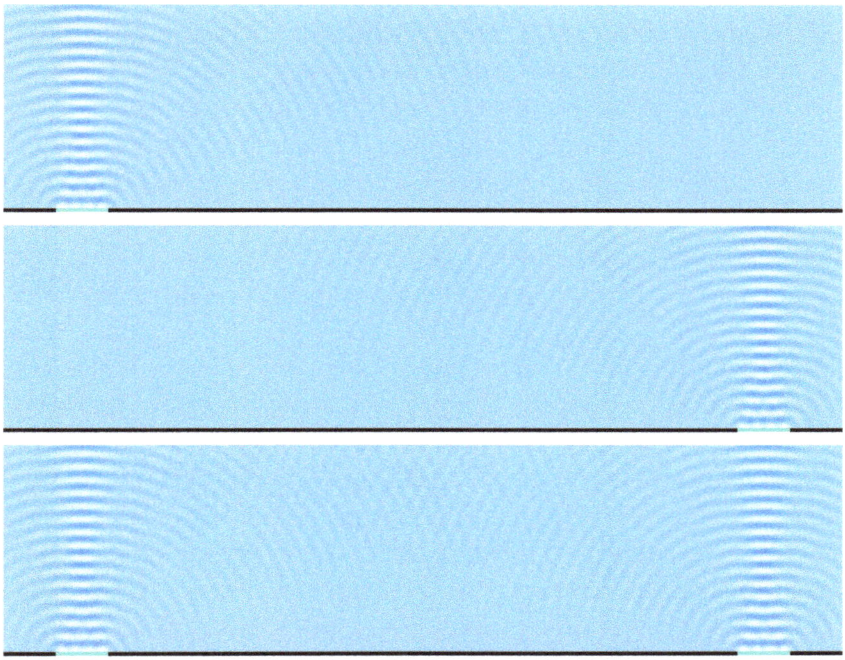

Figure 4.6: A new scenario where there isn't significant overlap of the functions in a chosen decomposition. To ensure there wasn't significant overlap, I had to 1) set the wavelength so it wasn't large compared to the opening in the slit and 2) separate the slits so that the water waves dissipate their energy as they spread, which reduces the amplitude of the waves.

Coming vertically out from each slit, we see regions of the most significant amplitudes in each of the two functions of the superposition. This scenario is different from the previous one because those regions of significant amplitude do not overlap. Had I only given you the chosen functions, you could have roughly imagined or sketched the total wave function. After all, if we cut the top two images in half and paste together those pieces with significant amplitude, the result will look somewhat similar to the total wave function in the bottom image. Using "looking somewhat similar" as our metric, we would say that there isn't significant interference between the two contributing functions.

The same line of thought won't work for the scenario in the previous section. Cutting the images of the decomposed functions into parts and pasting them together won't get you something that looks like the original ripple tank image. Moreover, if I gave you any one set of images from the previous example, I think you would have difficulty sketching the total wave function. By just about any metric we choose, we would say there is significant interference between the functions in any one of the chosen decompositions in the first half of the chapter.

What I am trying to illustrate with this new example is that sometimes it is okay to treat the functions in a decomposition/superposition as separate entities that are not part of a whole. In the strictest sense, they are never truly separate from the physical situation, the whole. But there are plenty of situations when the approximation works fine. All that is needed is for those functions' amplitudes to decay away before they overlap significantly.

For example, suppose we want to be sticklers and say the truest things. In that case, we must admit that the atoms and molecules in your body and your friend's body belong to the same total wave function because all wave functions overlap, at least to the tiniest degree. However, this is not practical, and any space larger than, say, a few lengths of water molecules that separates you and your friend allows us to approximate you two macroscopic objects as being individuals. If we couldn't do this, physics, chemistry, biology, and all the physical sciences would be far more complicated than they presently are.

However, whether an approximation is adequate depends on the needs of the researcher. At the top center of the bottom image in the latest figure, there is a weak cross-hatch pattern due to interference between the two chosen functions. If the interference matters to the researcher, then considering only one or the other function is inappropriate. Only a consideration of the total wave function would be adequate.

The meaning of it all

From now on, you should understand that superposition, or interference, of functions refers to addition at all points of the space under consideration. Therefore, on the left side of the equations above is the symbol that represents the whole thing that is "out there" in the real world, and on the right side of a particular equation is still the whole thing that is "out there" in the real world, but the symbols on this right side are one way that we have chosen to represent how we currently imagine mathematically decomposing the thing that is "out there". You must keep in mind that quantum mechanics is about quantized things, whole things. You must know which things are whole and physical and which are the notional and mathematical parts. Strictly speaking, the individual parts of a superposition have individuality in the math on the page but not in the real world. However, there are plenty of situations when the parts can be approximated as having their own identity. But when and whether an approximation is sufficient depends on the details of the situation and the needs of the researcher.

5 The cat, the keg, and the cut

> *In theory, theory and practice match. In practice, they don't.*
>
> —Unknown.

The cat

As a term **Schrödinger's cat** is part of our cultural lexicon. Everyone knows it refers to a cat that is both alive and dead, and I'm sure that most people think it was meant to show an interesting aspect of quantum mechanics. You may have heard that it shows how quantum measurements work or enlightens us on how something can be in a superposition of two states.

However, Schrödinger's cat must be understood in the historical context. The truth is that Erwin Schrödinger was not comfortable saying that quantum theory was complete. He felt Heisenberg, Bohr, and others were too cavalier by advocating against reality.

Leading up to the cat paper were the famous, multiyear Bohr-Einstein debates over the completeness of quantum mechanics. Einstein was on the side of reality and against the idea that quantum mechanics was complete. In 1935, Einstein, along with Boris Podolsky and Nathan Rosen, wrote the famous Einstein-Podoldky-Rosen (EPR) paper: *Can Quantum-Mechanical Description of Physical Reality Be Considered Complete?* (The missing "the" is one piece of evidence that Boris Podolsky, a native Russian speaker, was the primary contributor to the work.) With this work, they felt they demonstrated the incompleteness via the need for extra variables (aka hidden variables).

This was Schrödinger's inspiration to start a months-long letter exchange with Einstein on the completeness of quantum mechanics.

A byproduct of this correspondence was Schrödinger's so-called cat paper. In it, he used the accepted quantum mechanics mixed with the Copenhagen interpretation to lead to an absurdity. The intent was to have a laughable example which everyone agrees cannot happen so that the understanding of quantum mechanics could progress. The cat paper is a long review article, and the description of his cat thought experiment is surprisingly brief:

> One can even set up quite ridiculous cases. A cat is penned up in a steel chamber, along with the following diabolical device (which must be secured against direct interference by the cat): in a Geiger counter there is a tiny bit of radioactive substance, so small that perhaps in the course of one hour one of the atoms decays, but also, with equal probability, perhaps none; if it happens, the counter tube discharges and through a relay releases a hammer which shatters a small flask of hydrocyanic acid. If one has left this entire system to itself for an hour, one would say that the cat still lives provided meanwhile no atom has decayed. The first atomic decay would have poisoned it. The ψ-function of the entire system would express this by having in it the living and the dead cat (pardon the expression) mixed or smeared out in equal parts.

To get the cat in a superposition of both alive and dead (Fig. 5.1), he started with one of the atoms in the radioactive substance and applied the logic of the Copenhagen interpretation. In a simple model of the atomic nucleus (Fig. 5.2), the wave function of a soon-to-be-detected particle slowly leaks out from the barrier that binds the nucleus. This is the often-mentioned **quantum tunneling**.

In his mind, as the wave function slowly escapes, the atom transforms into a superposition of decayed and not decayed. The spreading wave function will then mix with and induce the Geiger counter to be in a superposition of two states. Next, the hammer will be in a superposition of two states. Then, the hydrocyanic acid will be in a superposition of being released and unreleased. Finally, the cat goes into its famous neither-dead-nor-alive superposition.

Figure 5.1: Illustration of Schrödinger's "diabolical device" that might or might not kill a cat within an hour. There is a chunk of radioactive material inside the Geiger-Muller tube, which is connected by wire to its counter. This is connected by a thicker wire to a coil of wire and armature. Together, they are the electrical relay. The triangle on the armature is the head of the hammer. The flask and cat are obvious. All of this is inside a steel chamber. (FYI: HCN is colorless, but I had to make it some color.)

Since its publication, many people have tried to explain away the problems of this enigmatic cat. Though, they consistently miss Schrödinger's point. Some say the decoherence interpretation resolves all problems because, in this interpretation, the universe is always interacting with itself. Some say that the cat collapses into a particular state when it becomes aware or unaware of the poison gas (how is unawareness of the gas a point in time?). Most commonly, it is said that the state collapse happens when the box is opened and the cat is observed by a person.

In Schrödinger's next paragraph, which is always left out of the most superficial descriptions of his thought experiment, he comes to his conclusion. Because a cat that is both alive and dead is an obvious absurdity – most people would agree we never see **macroscopic**

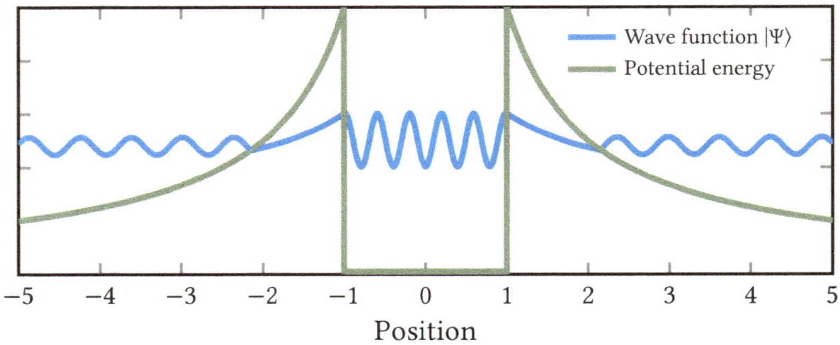

Figure 5.2: Idealization of quantum tunneling out of a generic radioactive nucleus. The potential energy curve arises from the interaction of the nuclear and electric forces with a positively charged quantum object that will leak across the barrier. The curves that decrease to the left and right are due to electrical repulsion. The rectangular region represents the "inside" of the nucleus (more appropriately: represents the region of the majority of the density of the nucleus) and is due to the nuclear forces being more strongly attractive than the electrical force is repulsive. The vertical axis is not labeled because potential energy and the wave function don't have the same units. I placed the wave function at this height to show how the potential energy changes its behavior. If you could see an animation of this, you would see that the amplitude of the wave function inside the nucleus decreases (unbelievably slowly) as the wave function leaks out and travels away from the center. I can't emphasize enough how schematic this is and how it does not represent well an actual nucleus. The potential energy curve around the nucleus should be smoother. (We don't know the exact form of the potential energy, though). The wave function inside a nucleus is not a pure sine wave, nor is it a sine wave outside the nucleus. The relative amplitudes of the sine waves should also be vastly different. The *Hyperphysics* website has a rough calculation for the alpha decay of polonium-212 (half-life: 0.3 microseconds). They found that the sine wave outside the nucleus should be approximately 10^{15} times smaller than the one inside. If I plotted that, the wave function outside the nucleus would look like a flat line.

superpositions — the mathematical description of the alive and dead superposition is not a description of reality:

> It is typical of these cases that an indeterminacy originally restricted to the atomic domain becomes transformed into macroscopic indeterminacy, which can then be resolved by direct observation. That prevents

> us from so naively accepting as valid a "blurred model" for representing reality. In itself it would not embody anything unclear or contradictory. There is a difference between a shaky or out-of-focus photograph and a snapshot of clouds and fog banks.

He is saying the wave function is like an out-of-focus photograph, but reality isn't out of focus, so a measurement is what reveals the reality that was already there. The argument relies on the reader accepting the fact that we do not see macroscopic wave functions. Later in the article, he illustrates again how current theory says the wave function is a mathematical tool. The final sentence of his article tells us this too:

> Perhaps, the simple procedure of the nonrelativistic [quantum] theory is only a convenient calculational trick, but it has at the moment attained a tremendous influence over our basic view of Nature.

If this view of the wave function seems at odds with his reality interpretation in the first chapter, know that it is absolutely at odds with his earlier view. Over the decade between the two works, his views on the wave function evolve. It is a commendable trait to have one's view develop with how one sees the evidence. A good scientist must accept that there is a possibility that their favorite theory may be modified or surpassed with new evidence.

With his final sentence in the paper, you now have a proper account of Schrodinger's intended message when he introduced his cat. His point was that the wave function is only a rough tool for calculations and reality does exist.

As for the decoherence interpretation I mentioned earlier, due to its popularity, it deserves a brief comment. I don't think it is a bad explanation for why we could not see a cat smeared between states of being alive and dead, and I think it is a very intriguing interpretation of quantum mechanics. The decoherence interpretation will rightly tell you that Schrödinger's description ignores other processes that could interfere with the wave function in the chain of objects on the way to killing the cat.

No one can deny that every object emits a spectrum of infrared light at room temperature, air molecules jostle up against anything in the atmosphere, cosmic rays are sources of low-level radiation that penetrate everything on the surface of the Earth, and all matter is above zero Kelvin. All of these can change a wave function. Matter can't stay in the exact same state from one moment to the next if it isn't isolated. All of these disrupt the clean connection of events from the radioactive atom to the cat.

Saying that decoherence is the answer does miss Schrödinger's point, though. He was specifically against the ideas of the Copenhagenists, who said their quantum mechanics is a complete theory. Decoherencists then implicitly agree that the device Schrödinger invented out of the Copenhagenist's way of doing things leads to states that no one expects to occur.

The keg

Those who assert that opening the box collapses the state would have a more challenging time arguing against Einstein. During his correspondence with Schrödinger, Einstein created an example that would have been more suitable than a cat to demonstrate that an absurdity results from the Copenhagen interpretation. He imagined a gunpowder keg that is so unstable that it might or might not explode within the next year. Initially, it is in a comprehensible state, but as time goes on, it evolves into a superposition of quiet and violent.

Einstein believed his gunpowder example nicely showed the problem with the supposed completeness of quantum mechanics. In a letter to Schrödinger, he wrote:

> Through no art of interpretation can this psi-function be turned into an adequate description of a real state of affairs; in reality there is just no intermediary between exploded and not-exploded.

Combining Einstein and Schrödinger's reasoning, if we place the gunpowder keg inside a box, it would be unobserved according to the Copenhagen interpretation and thus can be in a superposition of exploded and unexploded. However, the box will be unable to contain the exploded state. So, what does that say about the superposition? As soon as the superposition is created, one portion can no longer be contained by the box and is thus observed.

It's quite amusing trying to picture a superposition of the gunpowder keg, where half of the superposition is not as subtle as a dead cat in a box. I think you will agree that there is a problem with either the argument or the quantum mechanics that led us here. Einstein and Schrödinger think the problem is with quantum mechanics, more specifically, the Copenhagen interpretation of quantum mechanics. As you know, I think wave functions are real things, so I must think there is a problem with the details of the cat and the keg. I'll detail my thoughts on this at the end of the chapter.

Entanglement - the briefest of remarks

When writing his cat paper, Schrödinger was largely motivated by entanglement. In fact, he was the one who coined the term in quantum mechanics. **Entanglement**, in brief, occurs when one has a single quantum object composed of what are normally considered separate particles. We say the normally separate particles are entangled. The confusing bit happens when we make a measurement on one of the entangled particles and force the other one to adopt a particular state because certain quantities, like spin, need to be conserved.

In interesting experiments, two photons or electrons travel so far apart that no signal traveling at the speed of light could communicate between the measurement apparatuses. Yet, the values we measure seem to have a connection across the vast spaces. A less talked-about example occurs in atoms. All the electrons in an atom are entangled. However, they aren't spatially separated, so they aren't as exciting.

Recall that in the paragraph where he introduces his cat experiment, Schrödinger said: "The ψ-function of the entire system..." You can see that his setup is a series of quantum objects entangled because the wave function is leaking (tunneling) out of the radioactive nucleus and blending with other wave functions. Instead of considering just the cat, the two superposed states should be characterized as states A) and B) like this: A) atom-decayed, Geiger counter triggered, hammer swung, poison released, and cat dead; and B) atom not decayed, Geiger counter not triggered, poison still in vial, and cat alive.

In the Copenhagen interpretation, the whole setup goes into one of those two states when the lid is opened and the setup is observed. Assuming we observe the dead-cat series will lead to some interesting thoughts about the temporal order of things. If we see a dead cat, that means a Geiger counter clicked, the solenoid created a magnetic field, the hammer swung and went *thwack*, the glass shattered, and the cat wailed. Ask your nearest Copenhagenist when all of those happened. Did they happen before or after the box was opened? From the hammer swing, could the vibration or the change of the center of mass be felt outside the box? When? What happened to the live cat that was happily going about its business being a cat in a box before the lid was open? Do we hear the cat retching from the poison and collapsing? Could we detect the magnetic field produced by the relay's solenoid?

The Copenhagen interpretation does not provide good answers to these questions, which explains why people naturally gravitate towards decoherence, multiverse, or other interpretations. If one accepts his setup, for the past nine decades, Schrödinger's cat-killing box has shown that the Copenhagen interpretation has serious problems with regard to entanglement from some of its assumptions.

The cut

In response to the EPR paper, Heisenberg tried to formalize how he dealt with the measurement problem and Einstein's extra unknown variables. He wrote up a manuscript that he shared with Bohr, but because it didn't convince Bohr, it was never submitted for peer-reviewed publication in their lifetimes. Heisenberg's solution to the measurement problem – which should also solve the cat and the keg conundrums – takes up the first half of the manuscript. The solution requires an arbitrary dividing line between classical and quantum realms. Often, Heisenberg called this a *schnitt*, which translates to *cut*. According to John Stewart Bell, of Bell's theorem fame, this is an unacceptable "shifty split". Importantly, Heisenberg admitted that this cut is essential to make his version of quantum mechanics internally consistent.

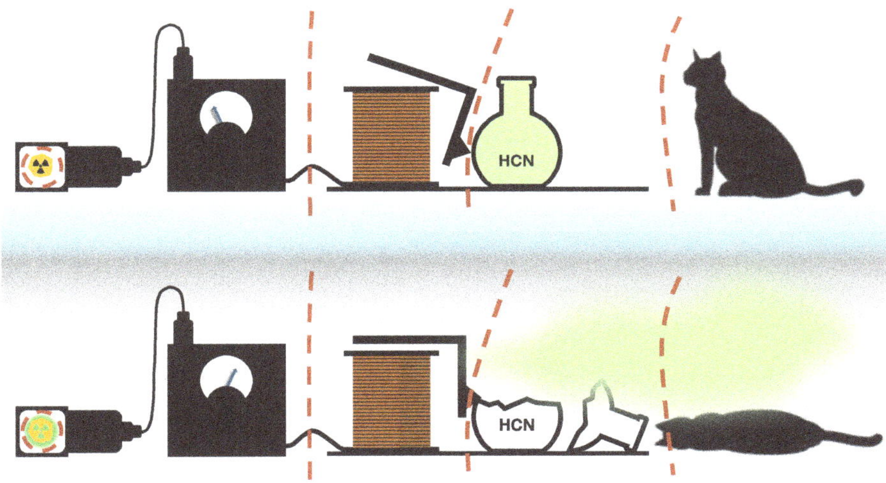

Figure 5.3: Heisenberg's cut applied to Schrödinger's cat experiment. I've put several potential cuts between the items Schrödinger considered as separate devices. Here, the cat is treated as a measurement device. Interestingly and quite accidentally (I didn't consider the cuts when making Fig. 5.1), for the movable devices, the *schnitt* cuts through them after the radioactive atom decays. Are the cat's ears quantum once it has died because we made an arbitrary choice of where to put the cut? This alone shows how ill-considered Heisenberg's cut idea is. He couldn't have put serious time into considering what he claimed. My guess is that it was more of a desperate attempt to force his quantum mechanics to be complete.

According to Heisenberg, the placement of the cut in the chain of events is rather arbitrary, with plenty of choice given to the scientist.

Going from micro to macro, it must come after the quantum system of interest but before both the person observing the quantum system and the final apparatus, which is observed by that person.

In addition, one must be able to group the other objects in the chain of events into systems that would count as measuring devices in classical physics (which is hardly precise). The cut can be on either side of any of those classical devices. (He never says how to deal with any cords connecting them.) If a part of the system is clearly quantum mechanical and can't be considered a classical physics apparatus, then in the entire chain of events, the cut must come somewhere between that part and the observer. In Fig. 5.3, I show the possible locations for the cuts in Schrödinger's thought experiment.

We can now critically examine some cases. First, we could consider the cat as the final link in a (psychopath's) desired way to measure whether a radioactive atom has emitted a particle or not, much like a canary in a coal mine. The cut would be before the cat. Everything quantum happens earlier in the chain, and the cat is never in a superposition. Easy-peasy. No problem with the cat. Nevertheless, we still have all the conceptual problems we had with hammers causing vibrations, glass shattering, the solenoid creating a magnetic field, and when they would be noticed. Even more confusing, do we now potentially have a live cat (which is our final measuring device) in a box full of things in a superposition?

As another case, we can consider the cat in a superposition. To follow Heisenberg's rules exactly, this will require us to add something in the box like a camera between the cat and the observer.

Okay, now, we add a camera, and the cat is in a superposition of alive and dead until a measurement is made. We are still left with the same conceptual difficulties as before of moving hammers causing vibrations, but we also have sentient beings disappearing again. Drawing lines on paper and calling them cuts clarifies exactly nothing in these cases.

The cut does worse with Einstein's gunpowder keg. According to Heisenberg, there *must* be this dividing line in the Copenhagen interpretation. But there is no measuring device involved with the keg, and a cut needs at least one measuring device. (If I put a camera up to my eye and take a picture, can I cause the keg to explode?)

Since observers are in the classical realm and there is no cut between them and the gunpowder keg, the gunpowder and its chemical bonds must be classical things, right? Therefore, everything you learned in chemistry about quantum mechanics must be incorrect, is it not? Even if we completely ignore the vibration and noise problem from Schrödinger's experiment, the cut has this fatal problem remaining. It is silent on how to deal with situations in which the observer does not receive information from a measuring device. This is why it can't deal with Einstein's keg and Schrödinger's cat when they are in superpositions. The unavoidable conclusion is that Heisenberg's Copenhagen interpretation does not completely explain quantum mechanics.

The justification for the cut

What he says

It is quite informative to study how Heisenberg tried to formalize his *schnitt*, and the best translation of his unpublished manuscript is by the scholars Crull and Bacciagaluppi. I will use their translation to examine his thoughts.

Before Heisenberg dives into some rather questionable mathematics, he starts with two foundational assumptions about there being two realms. Worry not. We will not get to the math. He trips over his laces with his first two steps. He begins by saying,

> The quantum mechanical way of describing nature thus begins with a peculiar rift: on the one hand, one proceeds from the assumption that the task of physics is the description and synthesis in terms of laws of [physical/vivid], objective processes in space and time; ...

Here, by using "task", "description", and "synthesis", all words that apply to humans, he says that a physicist's job is to describe reality. On its own, this assumption is rather reasonable. In fact, Crull and Bacciagluppi pass over this section rather quickly, thinking they are mundane introductory remarks. However, everyone should recognize that physics and the task of (the field of) physics are two different things. Most of us would accept physical laws and objects exist irrespective of whether someone studies them. Still, Heisenberg has just nested the human endeavor of the study of physics inside his argument concerning the reality of physics. He does this because his brand of philosophical positivism demands it.

I can't emphasize enough that this step in his thinking sets the stage for much of the viewpoint that the quantum realm is unreal and unpicturable, which has confused physicists, students, and the public for a century now.

Once one has read has read much of Heisenberg's writings, they can recognize that "laws of [physical/vivid], objective processes in space and time" is reserved for laboratory measurements and thus assumes a classical realm exists.

Moving on, his second misstep is in the remainder of the paragraph:

> ... on the other hand, one uses for the mathematical description of physical processes these wavefunctions in multi-dimensional configuration spaces, which in no way can be seen simply as representatives of the objective happening in space and time, as can, say, the coordinates of a point-mass in classical mechanics.

In this second part, he creates the unreal quantum realm. His description makes it seem like this half of the description of nature is not much more than math on a page.

He has thus set the stage to have people (recall: "task", "synthesis", "description") and measurements ("[vivid], objective processes") conjure the remaining reality from the quantum realm (recall his "the possible becomes actual" mentioned in the first chapter).

Analysis of his second misstep

The words "multi-dimensional **configuration space**" signal to the expert physicists his reasoning for rejecting the reality of the wave function. By "multi-dimensional", he means the number of position variables needed to describe the state of the spatial wave function. This number is found by taking the three space dimensions and multiplying by the number of quantum objects. Configuration space does not have a dimension for time. It is considered to be a parameter that allows the wave function in configuration space to change.

In other places, he explains his critique of the realist's view further by saying that there are too many dimensions (too many variables) in configuration space to describe reality. Therefore, wave functions don't correspond to objects. (Full disclosure: a cautious Schrödinger said something similar when he first introduced wave mechanics to the world.)

That is not great reasoning for such a strong conclusion. First, the idea of a multi-dimensional configuration space came from classical physics, and the answers that classical physics spits out describe a world of three spatial dimensions.

For example, suppose you are having a party with ten people in a room. Suppose you can also mathematically describe each of their positions for the whole duration of the party. The number of variables is ten people times three space variables, for a total of 30 variables in the configuration space of classical mechanics. Despite how cool it sounds, you don't think you are having a 30-dimensional party. You know the partygoers exist in three dimensions.

Secondly, single-electron wave functions, such as in the hydrogen atom, hint at the fact that wave functions will give matter densities. In hydrogen, the matter density is simply $|\Psi|^2$ because that function is already in three dimensions, but in more general cases, there is another step. With his attitude towards using very loose arguments, had he been motivated to do it, he could have attempted to reduce the $|\Psi|^2$ of a multi-electron wave function down to three dimensions and found a matter density function.

He would not need to have proved it, as this was a time of great liberty to attempt things mathematically without proof. Herr Professor Doktor, why not play with the math of the wave function to see if your philosophy is foolproof? Experts who know calculus and quantum mechanics can guess the process easily: every student of calculus knows that a single integration reduces the number of variables by one. It's informally called *integrating out a variable.* So, you keep integrating over the excess variables in $|\Psi|^2$ until you get down to three spatial dimensions.

Nowadays, it's commonplace to take the squared magnitude of the wave function and reduce it to three dimensions to get the electron density. Numerous variables can go into figuring out a wave function, yet a computational physicist can still tell you the matter density at every point in three-dimensional space. I would like to show you an example now of how it's done. Integration of $|\Psi|^2$ is like addition and is easy to describe. But setting up $|\Psi|^2$ for that requires you to learn more about wave functions first. So, I will delay showing you an example of the matter density of two electrons in a one-dimensional universe until the *Spring cleaning* chapter.

Summarizing my view on his second misstep, I claim that the quantum realm is not empty as Heisenberg suggests but instead has matter density in it. (I also think that a three-dimensional wave function might exist, but I can't conclusively show that they exist in general. However, I will also outline an idea regarding this at the end of the *Spring cleaning* chapter.)

Macroscopic superpositions

I chose to write this chapter because Schrödinger's cat is hardly ever characterized well by popular science writers. Even though I disagree with the Copenhagen interpretation, I don't think Schrödinger nor Einstein are correct. Their examples are thought-provoking but not compelling. They are missing some convincing and rigorous details.

The objects of focus in this chapter are macroscopic wave functions. In the following, I take two different views of what we mean by the term *macroscopic wave function* and give you an analysis according to each meaning. Both of these analyses will reduce the mystery further surrounding ideas that come from Schrödinger's cat.

Macroscopic wave functions are everywhere

As I said, I think the wave function is a real object, so I clearly disagree with Schrödinger's conclusion about macroscopic wave functions representing our fuzzy knowledge. At the time he wrote his cat paper, he thought there was something that physics could not yet resolve, and wave functions were imperfect reflections of reality, like out-of-focus photographs. I instead see wave functions like the clouds and fog banks he mentions: as very real things. In addition, as a solid-state physicist, I often see orderly, macroscopic wave functions and superpositions all around us.

They are in materials that have their atoms in a regular arrangement where the pattern of atoms repeats regularly whether we move left, right, up, down, in, or out. This describes a crystalline structure, and we say these materials are crystals. Certain insulators have crystalline structures, such as diamond, sapphire, quartz, fool's gold, silicon, salt, and ice crystals. The list is endless.

Metals can also be crystals. If made carefully enough, ingots of gold, copper, nickel, and silver could each be a single crystal. As a solid-state physicist, I feel compelled to put that statement into context: the metals you see are most often composed of tiny and randomly arranged crystalline grains. In those cases, pieces of metal are said to be polycrystalline. In an alternate case, if you are even less careful and cool a molten metal rapidly, it may not be polycrystalline or a single crystal but have an amorphous or quasi-crystalline arrangement of atoms instead. With that out of the way, let's stick with picturing a metal as a single crystal so that I don't have to include more caveats and teach you all of solid-state physics in the explanation that follows.

Most of the electron wave function for an isolated atom is located in a compact volume near the nucleus. When atoms come together in a crystalline solid, one could reasonably guess that the electron wave functions would, more or less, stay near their original atoms. This, however, is not true. Rather, we know that crystalline metals and insulators have a macroscopic wave function because there is a theorem

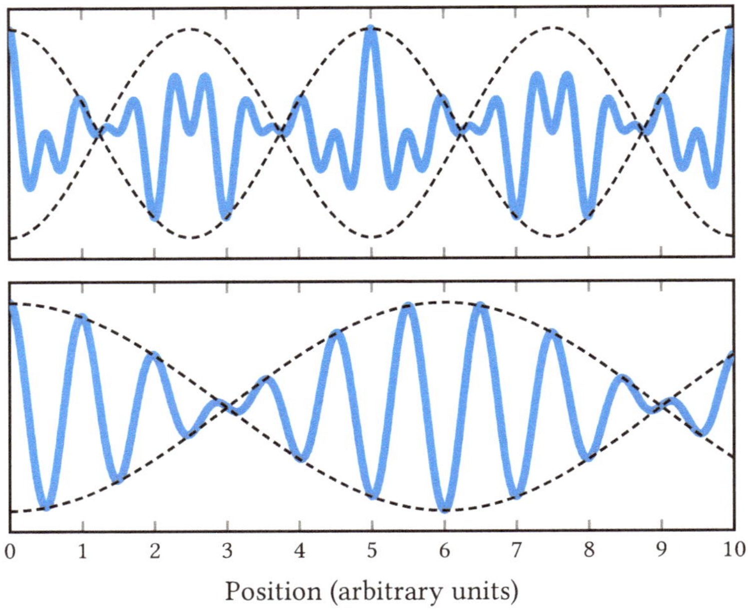

Figure 5.4: Wave functions (solid blue lines) in solids with a crystal lattice. The horizontal axis shows the position in space. The atomic nuclei of the crystal are at the integer values of position. The vertical axis shows the amplitude of the wave function. The black dashed lines are guides to the eye. An incomprehensible number of waves of different patterns superpose to give the overall wave function of the solid. The number is found via: # of atoms in the solid × # of electrons per atom.

(i.e., Bloch's theorem) that tells us exactly that. It tells us, because of the regular arrangement of atoms, each electron's wave function runs from one end of the solid to the other. The wave function's pattern repeats regularly in the solid. Each electron occupies the three-dimensional volume of the diamond, metal, or table salt grain you see with your own eyes (see Fig. 5.4).

The reflectivity and colors of metals and the index of refraction of diamonds are direct consequences of the details of the wave functions. We can see how different colors are reflected by copper, gold, or silver. Or we can see how diamonds bend light. The details confirm that there are macroscopic wave functions in each of them. Thus, I say, we see macroscopic wave functions all the time. And, they are not like out-of-focus photographs.

Do not be in doubt. Everything you see is quantum; it's just that crystals are the orderly macroscopic quantum objects that must be treated as macroscopic superpositions. When you view or touch a crystal, such as a diamond, you are viewing or touching a single quantum object. Any one and all electrons are distributed across the crystal in a massive superposition.

Because of your previous exposure to descriptions of quantum mechanics by other physicists, you may consciously or unconsciously feel that wave functions should be ethereal and have unexpected behaviors. With that one could wonder why the diamond isn't ethereal and why your hand doesn't penetrate the diamond. The reason is that wave functions are not ethereal; each wave function will have some interaction with some other objects. The atoms of your body and the diamond are positively charged nuclei surrounded by negatively charged electron clouds. You don't penetrate the diamond because the electron clouds repel each other. That's what touching is: the repulsion between electron clouds.

We will never manifest an alive and dead cat

Yet, I have missed something with that meaning of *macroscopic wave function.* I didn't mention what I think about some hypothesized *interesting* macroscopic wave functions, those macroscopic wave functions that excite our interest and make us imagine things that seem impossible. So, in the following analysis, we move away from crystalline metals and insulators.

The final three pages of Griffiths' textbook are devoted to Schrödinger's cat. In that section, he ponders if we will ever see something in two distinct states. He gives the example of a baseball in the superposition of being in Seattle and Toronto.

In the poll of researchers studying the foundations of quantum mechanics, which I mentioned in the first chapter, they were also asked a question about whether we will ever observe superpositions of distinct macroscopic wave functions, and one-third of the researchers believe they "will eventually be realized experimentally". So, the possibility of a superposition that is both macroscopic and in two distinct positions is also on researchers' minds. If we get an answer to this, then we could have a better idea of why we don't see cats, gunpowder kegs, and other objects existing in two distinct states.

Before I begin my analysis, I must mention that there are journal articles in the scientific literature that talk of experimental realizations of *cat states* and distinct macroscopic superpositions. Those aren't the *interesting* superpositions that you want. They don't have states in distinctly different positions, and, as such, those experiments have created states that are weak imitations of what you imagine when you wonder about Schrodinger's cat.

We desire an everyday object we can see, like a baseball, that is both here and over there. This would require the wave function for each of the innumerable atoms to split into two halves (two localized charge distributions for each atom), which travel apart some macroscopic distance (say 10 cm).

This scenario is obviously complicated. We should first examine whether we can get a single atom to split in two and have each half stay put. Let's say it is a hydrogen atom in its ground state (a 1*s* orbital). Ultimately, we want half of the wave function over here and half over there. Additionally, we want the character of the atom not to change. Each half must still be recognizable as a 1*s* orbital of hydrogen.

For clarity, here is how to understand the terminology I will use going forward. The single hydrogen wave function will split in half, and I will say the two halves are *atomic blobs*. Each blob carries half the mass and half the charges of the atom. Additionally, I will soon enough talk about the parts of the atomic blobs. Each atomic blob is composed of a tiny *proton blob* surrounded by an *electron blob*.

Separation of an atom into two parts (two atomic blobs) is feasible. We could imagine that we would make an electromagnetic trap of crossed laser beams. It is called an *optical trap*. There would be two wells, or traps, that would contain the atom. We put one hydrogen atom in, and via quantum tunneling, the wave function is equally distributed between the two wells. We thus have an atomic blob in each well. We then slowly separate the wells until they are 10 cm apart. In theory, this should be possible. In practice, moving the atomic blobs 10 cm apart has many difficulties associated with it. Nevertheless, we will assume we have a great and persistent team of experimentalists.

With all that said, going forward, don't forget the lesson from the last chapter. These blobs do not have an independent existence. There is only one object in our scenario, and it is the whole atomic wave function. The matter density and wave function smoothly change from a region of higher density through an intervening region of imperceptibly low density to a second region of higher density. We are only mathematically decomposing them into blobs.

Recall that we need to make a baseball or some other object out of this process. Thinking practically, we could never do it in a reasonable amount of time. Putting numbers to it, a single gram of hydrogen atoms — which hardly approaches the mass of a baseball — is about 6×10^{23} atoms. If it takes one second to separate one atom into two blobs, it will take 19 zillion years to split a gram of hydrogen. Even though I pretended a baseball is made out of only hydrogen, you get the point. Had I used the proper mass and a realistic ratio of hydrogen and carbon atoms in the analysis, I would have still ignored how we would get the atomic blobs to bond, which would take even more time. So, it already seems practically impossible to achieve our goal in a reasonable time.

Moving away from practicality, we also can't achieve what we want in theory. The electron blob on each proton blob will very likely look similar to a 1*s* orbital if the optical trap doesn't distort them too much (non-distortion would be difficult in practice). However, there is a big difference. The difference is that each electron blob is not the same size as a hydrogen 1*s*-orbital. Because there is half as much electron charge, half as much proton charge, and half as much electron mass in each atomic blob, if I were using the equation for the 1s orbital, I would have to put a factor of one-half at three spots in that equation. This would result in the 1s orbital expanding by a factor of eight.

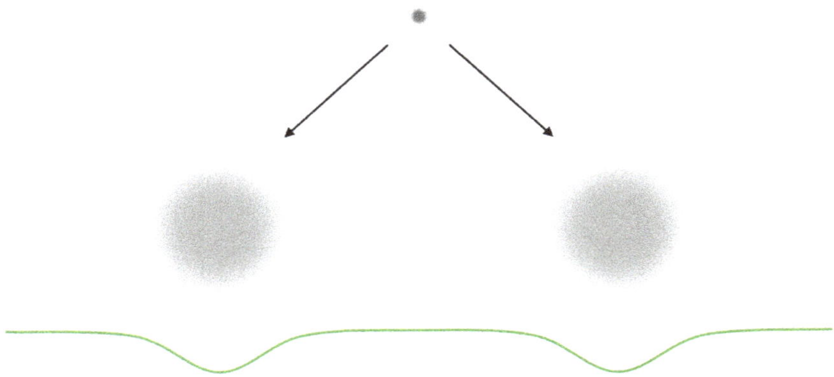

Figure 5.5: Schematic picture of splitting a hydrogen atom into two. The green line gives the idea that the atomic blobs are in the potential of an optical trap. The traps should be much wider to not perturb the blob shape we imagine. Because the electron density has been split in two and the wave function has expanded eight times, the density of each blob will be 1024 times less than the original atom.

For the expert readers, this should seem reasonable without even having an expression for the electron's wave function written on the page in front of you. If we had a dial on the magnitude of the electron's charge and dialed it down to the halfway point, the attraction to the proton would be dialed down too, and we would expect the electron density to spread out because it is less bound. The same effect occurs when dialing down the proton's charge. Regarding the electron's mass, if we dialed it down, the momentum distribution would become thinner, and the position distribution of the electron density would correspondingly widen.

For completeness, I must say the proton is unimaginably compact, and the extent of its wave function comes into only the most precise analyses of the electron wave function, so I ignore how it expands.

The proton's purpose in this discussion is to contribute its charge and mass.

The electron blob expanding to eight times that of the atomic wave function is a fatal error. Our distinct macroscopic states do not look as we expected. They are much larger and have a lower matter density than a hydrogen atom. This problem would continue to exist if we scaled it up to an object the size of a cat or a baseball because each of the two blobs of those objects would still have half as much electron charge, proton charge, electron mass, and proton mass.

No one talks about superposed cats becoming lightweight cougars or baseballs becoming beach balls, so the expanded blobs do not meet our original intentions. More importantly, since chemistry is the science of interactions between electron wave functions, the expanded orbitals will change the chemistry of gunpowder kegs, cat physiology, and the bonds that hold baseballs together. Beyond not being the right size, none of the real objects that we have mentioned will be like the objects that we get in a superposition.

Ignoring completely the issues above and bashing on, we could turn off the EM trap and also allow the environment to interact with the two atomic blobs. I don't know what is going to happen. But I suspect one of two things. The first possibility is the two blobs will, within a short amount of time, rejoin due to the inexplicable collapse of the wave function. Alternatively, the proton blob on the right and the electron blob on the left will attract each other via the electric force. The same holds for the proton blob on the left and the electron blob on the right. The Schrödinger equation makes things more complicated than Newtonian mechanics, so I can't say for sure that they would come together without modeling it in detail. However, modeling realistic quantum phenomena is not my specialty.

In summary, we have an impractically long time scale needed to create the interesting macroscopic superposition, wave functions expanding like balloons, and the unknown effects of turning off our EM trap. From this, we see that we aren't going to get our interesting and superposed macroscopic states, whether they are cats, kegs, or baseballs. My analysis signals that Schrödinger and Einstein didn't rigorously justify how to achieve their superpositions. If they had, they would have discussed the problem of expanding wave functions. And, without a rigorous first step, the rest of their descriptions fall apart.

6 Mathematics vs. measurements

> *If you try and take a cat apart to see how it works, the first thing you have in your hands is a non-working cat.*
>
> — Douglas Adams

The beginning of the problem

A dubious claim

Regarding measurements, physicists try to make axiomatic quantum mechanics say and do things which are inconsistent with experiments. The two-part claim I refute in this chapter is the following:

1. The measurement process forces a quantum object into a new state or possibly leaves it in the same state. The state after a measurement has a precise correspondence with the value measured.

2. Immediately after this process, we can repeat the measurement and find the same measurement result.

The first part is a cryptic statement for the uninitiated. I must write vaguely because you don't yet know the required math. To get a feel for both parts, let's look at an imaginary experiment. Portions of this example will be relevant in the next chapter during the discussion of the double-slit experiment.

Imagine a wave function emerging from a tiny, micrometer-sized slit and spreading as it crosses a vacuum chamber, much like water waves emerging from a slit. To be concrete, we will say it is a helium atom's wave function. Someone wants to measure the position of the helium atom, so they fire another quantum object, most likely a photon, in the general direction of the wave function. Many believe and the assumption I use when not invoking the postulates is that the photon somehow collapses the helium wave function into a compact volume. The atom can be said to be, more or less, around a particular position. On the other hand, a mathematical version of the first part of the two-part claim above, says the wave function somehow becomes a geometric point. I show these two divergent hypotheses in Fig. 6.1.

In a previous chapter, I have already shown that quantum mechanics doesn't deal with point particles, so I have already show the first part of the two-part claim above is doesn't work with the current scenario. So that I can illustrate the second part of the claim, in the current scenario, we will continue forward using the compact–wave-function hypothesis.

We now return to visualizing the experiment. After localizing the helium atom, the probing photon is randomly scattered into a detector array that determines the direction from which the photon came. From the geometry of the situation and the timing, the position of the helium atom is determined. (The details don't matter. We'll just say it can be done.) Luckily, we have good experimentalists, and this detection and calculation happens so fast that they are ready to measure again before the atom has had a chance to move much.

They do the measurement a second time. Another photon scatters off the compact helium wave function, and we get essentially the same answer. We can say immediately after the measurement we found the quantum system in a position state that is associated with the first position measured. The second part of the two-part claim works in this example. However, axiomatic quantum mechanics says

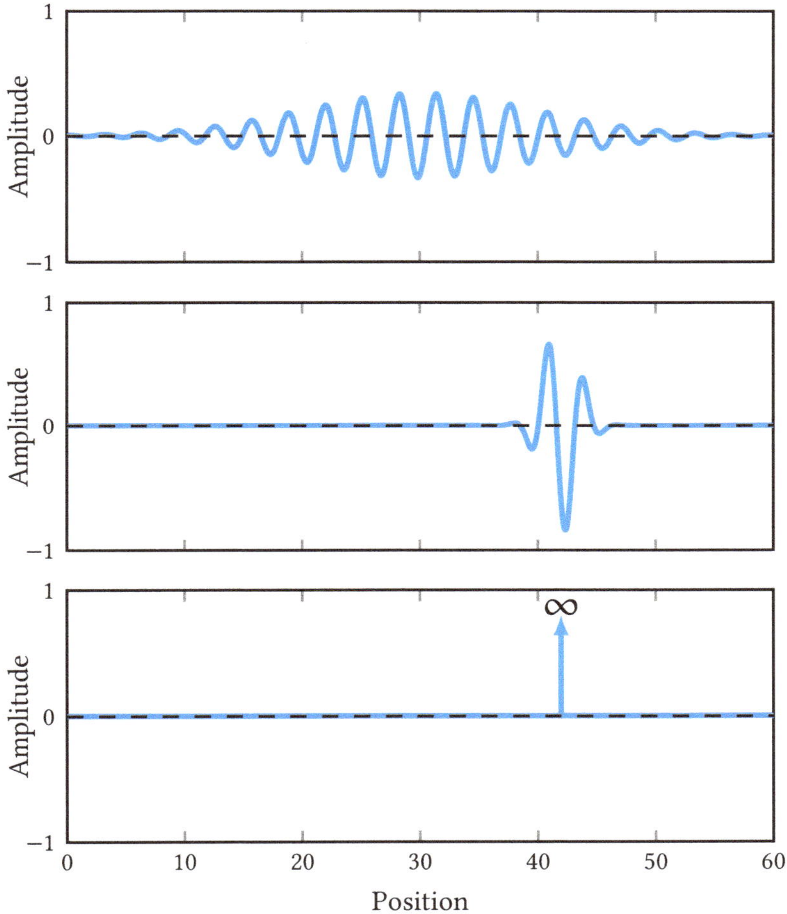

Figure 6.1: Two hypotheses of how a wave function localizes due to a measurement Top image: a spread-out wave function before measurement. Middle image (collapse of the wave function): the hypothesis of a compact wave function after a position measurement. Bottom image (projection postulate): the hypothesis of the wave function as a geometric point after a position measurement. The wave function in the middle image could be wider or slimmer. We don't understand the collapse of the wave function, so we don't know how thin it will be. It could be so narrow that it *seems* to be a point in the image. However, the amplitude will still be finite because it isn't actually infinitely thin (a geometric point). On the other hand, in the bottom image, the wave function is infinitely thin, and the amplitude is infinite (which is not a number).

that the second part of the claim always works because the first part always works.

As you will recall, there is a list of allegedly true statements in axiomatic quantum mechanics — the postulates — which are the foundation for quantum mechanics. I haven't mentioned them since the first chapter, but a physicist could go back through the topics I wrote and point out every time something could be mathematically abstracted to one of the postulates. In this chapter, I will detail five of the postulates of quantum mechanics and mention a sixth one in passing that has to do with spin.

Strictly speaking, the postulates are all intertwined, and rarely can we speak of only one at work. But, for simplicity, I will do just that for a few pages before I introduce the other postulates. This brings me to finally giving a name to the first sentence in the two-part claim that opened the chapter. Due to the minimal math in this book, it is a vaguely worded version of the **projection postulate**. The problem is that this rule is violated in plenty of situations.

A supposedly universal postulate

Some professors tell their students the projection postulate works for all measurements. They say that when we make a measurement, the quantum object is projected into a particular state that has a precise correspondence with the value measured. Your measurement device says the helium atom is at $x = 42$ cm from some reference point? Well, your helium atom was projected, or forced, completely into that position. Your hydrogen atom was in its ground state, and now your measurement process says that it has a ground-state energy of 13.6 electronvolts (eV)? Then, the projection keeps it in the ground state. Your silver atom was in a superposition of spin-up and spin-down and now your device measures it to be spin-up? Then it has been projected into a spin-up state. They double down on this by claiming that if you're quick enough to make another measurement before things change, you'll get practically the same measurement, thus demonstrating that the atom was projected into that state.

You will see at the end of the chapter that their plans to make a second measurement go awry in those energy and spin measurements. The reason for this is because this postulate ignores the **observer effect**.

The Observer Effect

What is it?

To observe an object, we must either interact with it in some way or interact with something it emitted. However, the interaction or emission changes the state of the object. In short, that's all there is to the observer effect.

In our macroscopic world, the observer effect is often too small to be significant. To measure the current in a circuit, we insert an ammeter in the circuit, which slightly changes the current. Measuring the pressure in your car's tire releases some air and slightly lowers the pressure. A light microscope bathes critters from a pond in bright light and heats up the water, which changes their behavior slightly.

Your vision relies on photons that bounce off or are emitted from the object you are viewing. Because those photons carry linear momentum, conservation of momentum says that the observed object gains a recoil momentum. In the macroscopic world, this recoil is insignificant because the objects we see with our eyes are so massive compared to single atoms. But it's still there. The state of the massive object changed because of the photons.

In contrast, our helium-scattering position measurement is of a tiny mass. I neglected to mention that the photon will also change the state of helium by imparting momentum to it. This is in addition to the projection/collapse of the wave function into a localized volume. The observer effect caused no problems in that case because, even though the photon gave momentum to the helium atom, our measurement was concerned with position and the second measurement happened before the atom moved significantly. Had we also been concerned with a momentum measurement, it might have complicated things.

A possible point of confusion

Since there is a measurement and a disturbance, the observer effect rightly sounds similar to Heisenberg's measurement-disturbance relation: $\Delta x \cdot \Delta p_x \approx h$ (commonly and incorrectly called the uncertainty principle). However, the observer effect is more general than Heisenberg's relation.

The observer effect would even occur in a non-quantum-mechanical world of billiard-ball physics. If you want to observe an atom that is like a microscopic billiard ball, you need to throw another billiard ball at it and collect one or both scattered balls in some measurement apparatus. By doing that, you have altered the original state of the first ball. This is still the observer effect. Whereas, Heisenberg's measurement-disturbance relation is a *mathematical estimate of a particular case* of the observer effect in a quantum-mechanical world. I won't mention it again here. But, in a later chapter, I'll say a bit more on Heisenberg's relation.

A second possible point of confusion

Despite the well-behaved scattering example above, you may get the observer effect and the projection postulate (or the collapse of the wave function) confused in other cases if you are not careful because there is a measurement and a change of state in both ideas. Allowing myself some hyperbole to contrast them for you, I would say that the observer effect's change of state is characterized by foreseeable violent change, and the projection postulate's change of state is governed by the mysteries of quantum mechanics. (Less hyperbolic: substitute *randomness* in the place of *mysteries.*)

Where we go from here

In the next section, I introduce the mathematical notation that I use throughout the rest of the book. Understanding the math makes the confusing things you have been told about quantum mechanics so much, well, less confusing. The patterns will be the same over and over again, so don't stress about memorizing anything. I also describe four other postulates (i.e., the state postulate, the observable postulate, the single-measurement postulate, and the probability postulate) because they naturally arise. In the final section, we can see how the math and measurements disagree because of the unavoidable observer effect.

Dirac's notation: a crash course

Despite the stereotype, physicists aren't madly in love with the grind that comes with doing math and don't want to copy long, complicated formulas from line to line. It's tedious. Paul Dirac came up with a notation that is quite powerful and hides the long, ugly math. I introduce it to you because, along with hiding much of the math, it makes it clear to the readers the functions of certain symbols.

There are four types of mathematical objects in Dirac's notation you need to know in this book. In this chapter, I explain three of them: **kets**, **operators**, and **numerical values**. The fourth, *the quantum mechanical amplitude*, is explained in the *Spin* chapter. Though, you've already seen the idea multiple times. It is the amplitude of the wave function (e.g., Fig. 6.1).

Kets

Kets represent the state of a particle, whether it is the whole state (total wave function) or part of the state (either the spatial wave function or the spin state). At the end of the third chapter, I introduced you to the ket of the spin state:

$$|\chi\rangle \tag{6.1}$$

As written, this is an ambiguous ket. You don't know if the particle is spin-0, spin-$\frac{1}{2}$, spin-1, etc. You also don't know the direction of the spin. If we want to represent a spin-$\frac{1}{2}$ particle, we could be more specific and write a ket for spin-up or spin-down (i.e., $|{+z}\rangle$ or $|{-z}\rangle$), spin-left or spin-right (i.e., $|-x\rangle$ or $|+x\rangle$), or any other direction in three-dimensional space. Note that these symbols still hide the matrix math of spins. In the third chapter, I also wrote the ket for the spatial part of a generic wave function as follows:

$$|\Psi_{\text{spatial}}\rangle \tag{6.2}$$

We can also be more specific when referring to which wave function the ket represents. We can represent the ground state (GS) of the hydrogen atom as:

$$|\text{GS}\rangle \tag{6.3}$$

Of course, this still hides the algebraic formula of the wave function.

Of the three symbols, the ket is always the center of attention. Without it, the other two symbols are nothing. Experts will recognize I am alluding to the postulate that always appears first in the lists of the postulates (even though there is no order for the rest of the postulates). It is the **state postulate**. There are many different ways to word it that mean slightly different things, each of which may reveal some of the authors' leanings to those familiar with the field:

- All the information about the quantum system is contained in the ket/wave function.
- The ket/wave function corresponds to the quantum object.
- The quantum system is fully specified by the ket/wave function.
- The properties of the quantum system are fully specified by the ket/wave function.

Those statements may be confusing. I have been using the word *wave function* to refer to a quantum object, but here, it's more like the wave function refers to the math on the page and there is a connection between the two. Since they are so closely linked, I say you might as well think there is little difference between the math of the wave function on the page and the quantum object in reality.

Finally, be warned. An author, including myself, might verbally point at a ket on the page and call it a *state*, an *eigenstate*, a *vector*, an *eigenvector*, a *wave function*, a *state function*, an *eigenfunction*, a *spin state*, or whatever. They all mean approximately the same thing: a description of a quantum object. The terms with *eigen-* out front refer to states in special situations, and I'll explain what they mean before this section ends.

Operators and numerical values

The next type of symbol is an operator, as in the thing that tells you to do a mathematical operation. Operators tell the state (tells the ket) to either:

- stay in the same state (don't change the ket), or
- change into a new state (become a different ket).

Simultaneously, the operator also instructs the state to:

- return a single numerical value, or
- return many discrete value (there is a caveat to this), or
- return many values in the form of a continuum (i.e., a continuous function; so it is considered to be one mathematical object).

If you are programming savvy, maybe you can think of the initial state as input data and the operator as an algorithm that returns output data. Another analogy might involve the operator as a manager telling an employee (the quantum state) to sit up straight (change state) and get that report in (return a value).

When an operator works on a single ket, either a single value or a continuous function (a continuum of values) appears in front of the ket symbol in place of the operator. The first example I'll show you will have a single energy value. You'll see a function of values in the chapter after the next.

Often, when a single value is returned, it is the hidden number “1”, which is not considered a value of interest. In that case, the operator (e.g., a rotation operator) is interesting because it only changes the state. In this book, you will only see one example of an operator that both changes the ket and returns a value that isn’t a “1”; it is used to show that we made a physics mistake, but not a math mistake (Eq. C.29 on p. 363). If you are interested in seeing an example of an operator that both changes the ket and returns a value that isn’t a “1” where no physics mistake is made, you can search online for *ladder operators*.

Here is the caveat I alluded to in a bullet point: sometimes an operator is applied to a single ket, but it is really meant to work on a mathematically equivalent superposition of discrete kets. In that case, each ket in the superposition will return a discrete numerical value. You’ll see that with spin later in the chapter. And, by *discrete*, I mean things that are isolated like the set of integers and that are not part of a continuum, like the real numbers. Were we to apply these operators to the single ket rather than the equivalent superposition, we would be making the physics mistake of the previous paragraph.

Our first operator

The symbol for the operator that asks for the total energy (kinetic and potential combined) of some system is:

$$\hat{E} \tag{6.4}$$

This one is often called the **energy operator**. Physicists call the circumflex on top a *hat*, and it is used to tell us that it is an operator in its most abstract form. Unfortunately, not all physicists use the hats on top, so their work is harder to read. There’s a small controversy with calling this an operator that I will briefly touch on in the first of the chapters on Schrödinger’s equation. For the curious, this abstract operator represents $i \cdot \hbar \cdot \partial_t$. I don’t want to get into fine details too soon, so I will hold off explaining those symbols until the second of the chapters on Schrödinger’s equation.

To illustrate another failure of the projection postulate, I will use the energy operator to tell us the hydrogen atom's ionization energy, which is the minimum energy needed to make the electron and proton of the hydrogen atom no longer bound together. While my use of $\hat{E}$ as revealing the ionization energy is fine for atoms, it is not appropriate for unbound systems because those systems, by definition, don't have an ionization energy. So, please, don't leave this chapter thinking there is an ionization-energy operator.

This operator raises an important point. In many cases, operators have a connection to quantities that can be measured. I think the postulate for this should have been called the *physical-quantity postulate*, but I wasn't around one hundred years ago when they named it. Instead, it is called the **observable postulate**. It states that for every observable quantity (aka physical quantity) in classical physics, there is a corresponding observable quantity in quantum mechanics, and that observable quantity has an associated operator. To be clear, this means some operators correspond to physical quantities, and some operators do not.

Interestingly, the observable postulate misses out on spin angular momentum because it relies on the physics we knew at some point in time (i.e., classical physics). So, spin's existence and its operator need a postulate of their own. Or, they need to be included in the observable postulate. For now, let's say the observable postulate has spin covered.

That an observable postulate is needed is uncontroversial. Because the macroscopic world is built up from quantum mechanics (well, to a realist, it is), we had better be able to trace the physical quantities we see in our macroscopic world back to the quantum realm. And since the math describes that quantum realm, there should be a mathematical thing associated with the physical quantities.

The energy operator, therefore, corresponds to an observable because we can measure the total energy of a quantum object. In contrast, a *time-evolution operator* is an example of an operator that isn't an observable. This operator tells the state function written on the page to fast forward into a new state function. There is no measurable quantity that a quantum or classical object possesses associated with this operator, so it can't be called an observable.

To have the energy operator ask the hydrogen atom's ground state for the value of energy, we put the operator to the left of the ket ("out front") like this:

$$\hat{E} \cdot |\mathrm{GS}\rangle \tag{6.5}$$

I will include central dots between many mathematical objects to make the expressions easier to read.

The numerical value that is returned

I said that the operator asks the ket for a response, so we must get a value in return when we apply the operator to the ket. Every physicist knows that the ground state of hydrogen has an ionization energy of 13.6 eV. The equation that shows how kets, operators, and values come together in our example is:

$$\hat{E} \cdot |\mathrm{GS}\rangle \;=\; -13.6\,\mathrm{eV} \cdot |\mathrm{GS}\rangle \tag{6.6}$$

You can always read an equation as a sentence, and short equations are the easiest to read. This equation says, "Applying the operator to the ground-state ket of hydrogen (and doing the hidden math) will tell us the ionization energy of the ground state. The value returned is −13.6 eV."

Unfortunately, we write the energy in a somewhat confusing convention. Zero electron volts or greater would represent a free electron, so the negative value tells us that some energy needs to be added to ionize the atom. We can see the added amount must be at least 13.6 eV to ionize the atom.

If a physicist doesn't know yet or doesn't want to show the specific value, they may write this equation as:

$$\hat{E} \cdot |\mathrm{GS}\rangle \;=\; E \cdot |\mathrm{GS}\rangle\,, \tag{6.7}$$

where the E is a single numerical value in this case. This can get confusing because the $\hat{E}$ and the E look so similar. A careful physicist will make sure everything is clearly defined.

The **measurement postulate** says that if we were to conduct a measurement of an observable, the result of the measurement would be one of the numerical values returned by the application of the observable's corresponding operator. The numerical values will be real valued because measurement devices display real values. If the operator has been applied to the appropriate kets, none of the kets are transformed; if the kets are transformed, then the physics mistake I mentioned on p. 92 has been made. By the way, I will break with convention and call this the **single-measurement postulate** because there is another postulate that describes repeated measurements.

In quantum mechanics, the results of measurements are generally random, and the probability of finding a particular value is determined by the weighting given by $|\Psi|^2$. This is the essence of the **probability postulate**. Further details depend on the state and observable in question. Regarding position measurements, in the matter/probability density images or plots, the brightest pixels or highest points of the curves are the most likely locations to register a measurement in some idealized experiment. I never give you a full accounting of this postulate where you can walk away and do general calculations because that is not needed. However, as we proceed through the book, there will be different examples where you incidentally learn more about the probability postulate.

Eigenkets, eigenoperators, and eigenvalues

Notice in the previous two equations we get one value returned. This situation has no randomness because there is no choice the universe can make between various numerical values. Notice too that the ket is repeated on both sides of the equation. In this case, when the operator asked for a value, it didn't change the state.

When a state doesn't change by application of the operator, and we get one value back, we change the words we use for ket, operator, and value. We call them **eigenstates**, **eigenoperators**, and **eigenvalues**. It might help you to know in German that "eigen-" means "peculiar", as in a special case of something. A sentence that uses these goes like this: "the eigenoperator acting on its eigenstate returns the eigenvalue without changing the eigenstate." Cases like this are of interest because we know exactly the quantity that would be found in a measurement if the operator corresponds to an observable.

You'll notice I didn't write *eigenket* above. There's nothing wrong with that word. It's just rare to see written. Most often, we use *eigenstate* or *eigenfunction*. In addition, it is also rare to read of an eigenoperator. More often, we simply write *operator*. So, a more common sentence to read is: "the operator acting on its corresponding eigenfunction returns an eigenvalue."

The most important thing to know about the "eigens" is that equations with the form of the previous two equations are called **eigenvalue equations** and signal that we have a situation with no randomness.

Physicists who say "when a second measurement is done rapidly enough, we get the same value" are relying on an eigenvalue equation. More accurately, in general, they first rely on the projection postulate and then on an eigenvalue equation (with the other postulates implicitly used). Let me show you what I mean.

The problematic projection postulate

Bringing ideas together

At the beginning of the chapter, I wrote the following claim, which contains the projection postulate in its first sentence:

1. The measurement process forces a quantum object into a new state or possibly leaves it in the same state. The state after a measurement has a precise correspondence with the value measured.

2. Immediately after this process, we can repeat the measurement and find the same measurement result.

To illustrate those statements, I gave the example of a helium atom's wave function localizing due to a measurement. Now is a good time to illustrate the projection postulate and the second claim by relating the words and graphs to some math.

The helium atom's wave function $|\Psi\rangle$ is spread through all space, but, again, I pretend it is one-dimensional for simplicity. The two equations that follow are representations of the upper and lower images in Fig. 6.1. The standard textbook math I will show is limited and cannot represent the middle image.

We can describe the wave function by all the positions along the x-axis (essentially a number line) and the height, or amplitude, of the wave function. For simplicity, as I said, I will ignore the amplitudes in the math of this chapter. The top image of Fig. 6.1 is then described mathematically as a sum of all the individual position kets:

$$|\Psi\rangle \quad \text{“=”} \quad \cdots |-9.2\rangle + \cdots + |\sqrt{2}\,\rangle + \cdots + |\pi\rangle + \cdots + |42\rangle + \cdots \tag{6.8}$$

The equal sign in quotation marks tells you that this isn't really how I should write the wave function because I'm missing the amplitudes for each ket. But it's good enough for what I want to show. Units are implied inside the kets (i.e., -9.2 cm, $\sqrt{2}$ cm, π cm, and 42 cm). The dots tell us that the whole x-axis is implied. Therefore, there is an unlimited number of kets.

The projection postulate says that we can start with a particle in a state written as one, many, or an infinitude of kets, then a measurement is made and, randomly, the particle will be projected into only one of those kets. The images of Fig. 6.1 show that we randomly found the helium atom at 42 cm from the entrance to the vacuum chamber. Mathematically, we think of the spread-out wave function becoming projected into a state like the following.

$$|\Psi_{\text{before}}\rangle \quad \text{``=''} \quad \cdots|-9.2\rangle + \cdots + |\sqrt{2}\,\rangle + \cdots + |\pi\rangle + \cdots + |42\rangle + \cdots$$
$$\downarrow \tag{6.9}$$
$$|\Psi_{\text{after}}\rangle \quad \text{``=''} \quad |42\rangle$$

Think of the arrow as the measurement process. The quotation marks hide the unphysical infinity that this illogical process produces. However, we are ignoring that problem to learn the math of both parts of the two-part claim, so I'll stop using the quotation marks in the rest of this example.

Many physicists say if we make a second measurement essentially instantaneously, we will be guaranteed to get the same result because there is now an eigenvalue equation for the position operator $\hat{x}$:

$$\hat{x} \cdot |42\rangle \quad = \quad 42\,\text{cm} \cdot |42\rangle \tag{6.10}$$

Because of this eigenvalue equation, they picture the state evolution during the second measurement like this:

$$|\Psi_{\text{before 2}^{\text{nd}}\text{ measurement}}\rangle \quad = \quad |42\rangle$$
$$\downarrow \tag{6.11}$$
$$|\Psi_{\text{after 2}^{\text{nd}}\text{ measurement}}\rangle \quad = \quad |42\rangle$$

As I said before, this example ignores the observer effect, but there is no practical problem with the idea of a second position measurement if we pretend the experimentalists are quick enough.

However, again, to be perfectly clear, I do have a problem with the implication of a geometric-point particle that this math shows. Near the end of the Schrödinger equation chapter, I will give you another reason to wholly reject the idea of a geometric point particle, which involves the fact that the bottom image in Fig. 6.1 is not smoothly changing. We say that it is discontinuous. For now, we'll ignore that and say this claim is right-ish in this particular case.

The primary purpose of this chapter is to illustrate that the projection postulate and measurements are in conflict. Part of the reason for the confusion among physicists is that they are loose with math, like that of infinities. Another reason is that when they talk about the math of an observable quantity, they may not have any clue how an experiment could be conducted. There could be multiple ways to conduct the measurement, each of which could do varying levels of violence to the state.

Finally, now that you have been introduced to some of the necessary math, you are ready to see how it work with other types of measurements. We turn to two illustrative cases that show that the observer effect and the projection postulate clash. The first case is the ionization-energy experiment that we have already examined. The second case is a random spin measurement.

Ionization-energy measurement

When a hydrogen atom is in the 1*s* state, the ground state, it takes 13.6 eV to ionize the atom. We know this because Equation 6.6 told us so, which I repeat here:

$$\hat{E} \cdot |\mathrm{GS}\rangle \quad = \quad -13.6\,\mathrm{eV} \cdot |\mathrm{GS}\rangle \tag{6.12}$$

You can now recognize this as an eigenvalue equation. Let's say we measured a hydrogen atom and found that its ionization energy is 13.6 eV.

We can now consider the projection postulate. It says the state had a do-nothing projection:

$$\begin{array}{rcl} |\Psi_{\text{before}}\rangle & = & |\mathrm{GS}\rangle \\ & \downarrow & \\ |\Psi_{\text{after}}\rangle & = & |\mathrm{GS}\rangle \end{array} \tag{6.13}$$

If we repeat the measurement on the same atom, we should again get the same 13.6 eV because it should still be in the ground state.

But this couldn't be more wrong. To measure this energy, we could fire photons at the atom and tune to higher frequencies (i.e. higher energies) until an electron is ejected and swept into a detector by an electric field. (The photons will need to be widely spaced in time to allow the atom to relax after it gets excited to higher bound states.) Ionizing an atom means that we have removed an electron from its nucleus and put it in an unbound state.

The problem is evident. It is the observer effect. A measurement that rips apart a hydrogen atom will obviously not leave the atom in its ground state. The projection postulate is wrong in this case. It would be more accurate to think of the state changing like this because of the first measurement:

$$\begin{array}{rcl} |\Psi_{\text{before}}\rangle & = & |\text{GS}\rangle \\ & \downarrow & \\ |\Psi_{\text{after}}\rangle & = & |\text{✸ KABLOOEY! ✸}\rangle \end{array} \tag{6.14}$$

We can’t conduct a second measurement to confirm the ionization energy of the ground state because there is no ground state left. (By the way, we really can write whatever we want inside the ket, as long as the reader knows what we mean. It’s just a label.)

Spin measurements

Dirac notation for spin

Suppose we want to measure the spin of a particle along the z-direction (spin-up or down), but the spin is currently in the positive x-direction (spin-right). Quantum mechanics says that this state is equivalent to an equally weighted superposition of spins in the positive and negative z-directions:

$$|+x\rangle \;\; \text{“=”} \;\; |+z\rangle + |-z\rangle \tag{6.15}$$

Again, the quotation marks indicate that I am leaving out the quantum amplitudes.

The result of a spin measurement along the z-direction will be determined randomly. It'll be either one of the discrete values of $+\hbar/2$ or $-\hbar/2$. But how could you find these numerical values? First, you need to know the $\hat{S}_z$ operator corresponds to spin along the z-direction. This is specified by the observable or spin postulate. We don't call this symbol the *spin-z operator*, but, if it helps, feel free to say it to yourself. According to the single-measurement postulate, when applied to the equation, this operator returns the values that correspond to a measurement:

$$\hat{S}_z \cdot |+z\rangle + \hat{S}_z \cdot |-z\rangle = \left(\frac{+\hbar}{2}\right) \cdot |+z\rangle + \left(\frac{-\hbar}{2}\right) \cdot |-z\rangle \qquad (6.16)$$

Of course, I'm still hiding the math that gets these answers (it's in the appendix on matrix mechanics).

Now, let me say more about the $+\hbar/2$ and $-\hbar/2$. Here, the plus and minus signs tell you the directions of spin angular momentum along whichever axis is under consideration. The $\hbar$ is rather special. It is Planck's constant h divided by 2π and is called either *h-bar* or *Planck's reduced constant.* Whenever you see $\hbar$ or h, you know the equation is definitely quantum mechanical. The units of $\hbar$ and h are joules-seconds. In the case at hand, these are the units of angular momentum, but they are often combined with other variables to give different quantities. Planck's constant is also said to have units of *action*, but I won't cover that ground. When you see Planck's constant in other places in this book, like in the uncertainty principle, you should think of it as a number and not worry what the units mean.

When discussing spin, physicists often drop the $\hbar$ and say that the amounts are +1/2 and −1/2. Technically, this change turns them into **quantum numbers**, which are more like labels and much less like physical quantities. Specifically, they are the possible *spin-magnetic quantum numbers* (m_s) for a spin-$\frac{1}{2}$ particle. We'll be using spin quantum numbers s and m_s as labels for states other than spin-1/2 states in the chapter on spin.

Naive math

With that out of the way, the naive description from the projection postulate goes as follows. If we measured the spin to be pointing up, the original state of the particle $|+x\rangle$ has been projected into the $|+z\rangle$ state:

$$\begin{array}{rcl} |\chi_{\text{before}}\rangle & = & |+x\rangle \text{ “=” } |+z\rangle + |-z\rangle \\ & & \downarrow \\ |\chi_{\text{after}}\rangle & = & |+z\rangle \end{array} \tag{6.17}$$

That means, if we so wished, we could conduct a second, identical experiment immediately after the first, before the quantum object has a chance to interact with anything. We should get the same answer. *If* the particle is now in the $|+z\rangle$ state, we will measure the same thing again because of the following eigenvalue equation:

$$\hat{S}_z \cdot |+z\rangle = \frac{+\hbar}{2} \cdot |+z\rangle \tag{6.18}$$

Examining an actual spin experiment will show us why Eq. 6.17 isn't true, which, in turn, means we don't get to the point of having an eigenvalue equation (Eq. 6.18). We then can't say that we can make a second measurement and get the same spin result.

An actual spin experiment

Back in 1921, before spin was even hypothesized, Otto Stern and Walther Gerlach conducted an experiment on silver atoms to measure the atom's orbital angular momentum. A few years later, it was realized that they made a mistake in their assumptions about which orbitals occur in the atom. It turns out that, when the contributions from each orbital are added in atomic silver, it has zero overall orbital angular momentum. With this realization and hints of the existence of a two-valued and quantum quantity from various atomic spectra, physicists realized that the *Stern-Gerlach (SG) experiment* showed that spin angular momentum exists.

I will make a simplification that most texts do when talking about the SG experiment. In the real experiment, the oven that created the silver-atom beam made the spins in random directions. I will instead start with the spins in a particular state, namely the $|+x\rangle$ state of the example above.

With that in mind, let us begin by looking at Fig. 6.2. In the SG experiment, we start with a beam of silver atoms moving along the y-axis with their spins aligned along the positive x-axis. The beam then encounters an inhomogeneous magnetic field created by an oddly shaped electromagnet. A beam is what is used in the actual experiment, but conceptually, you should think of one atom going through the apparatus at a time. The atom has no overall electric charge, so the magnetic field does not act on any electric charge to move the beam. Rather, the inhomogeneous magnetic field acts on the magnetic dipoles of the $|+z\rangle$ and $|-z\rangle$ kets to split the wave function of a silver atom into two wave packets. The wave packets then separate vertically in space.

Incidentally, this is the second time I get to show you the spatial wave function and spin state in the same equation. The wave function of the silver atom $|\Psi_{\text{atom}}\rangle$ is broken down into two separated wave packets $|\Psi_{\text{wp1}}\rangle$ and $|\Psi_{\text{wp2}}\rangle$, each of which carries half of the $|+x\rangle$ spin superposition:

$$|\Psi_{\text{atom}}\rangle \quad \text{“=”} \quad |\Psi_{\text{wp1}}\rangle \otimes |+z\rangle \ + \ |\Psi_{\text{wp2}}\rangle \otimes |-z\rangle \tag{6.19}$$

Because the tensor product symbol $\otimes$ tells us that a spin state is distributed over a wave packet, we can see that the electromagnet acting on the spin causes the wave packet to move.

Eventually, the spatially separated wave packets reach the imaging detector. Somehow, when the wave function encounters the detector, it collapses to one of two regions on the detector. It's important to recognize that the spatial separation of the two wave packets is a critical part of the experiment. We don't have devices that can measure spin angular momentum directly. We must use a position measurement to deduce the spin state that the silver atom had for a brief moment

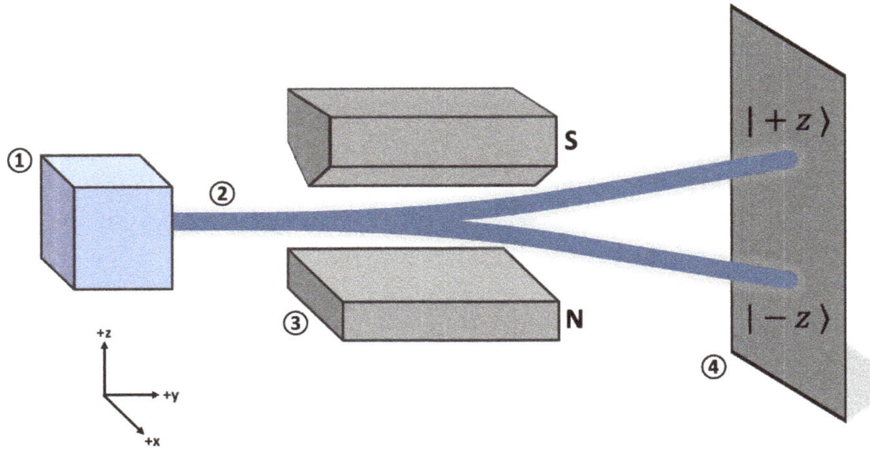

Figure 6.2: A Stern-Gerlach experiment measures quantum angular momentum. ① Oven and collimators produce an atomic beam of silver atoms. (Neutrally charged particles are needed because moving electric charge will also respond to the external magnetic field.) ② Beam of spin-$\frac{1}{2}$ atoms. ③ Electromagnet creates an inhomogeneous magnetic field that splits wave functions into two parts. ④ The imaging detector collapses the wave function down to one of two spots for every atom impinging on the screen.

after the collapse of the wave function.

The wave function of the silver atom pinballs its way through the semiconductor of the detector, giving up energy and unleashing a cascade of electrons as it goes. The spin of the silver atom has a chance of changing with each collision. The detector registers an atom in a pixel (either in the upper or lower spot but not both) by sweeping the energetic electrons into a detecting circuit and counting the current pulse as a detection.

This is the observer effect at work: to detect the silver atom's position, which is our proxy for spin, we must put a massive object in the atom's path to interact violently with the atom's wave function.

The silver atom eventually ends up at rest in the semiconductor detector. Good luck finding the silver atom in the silicon haystack for a second measurement. Even if we could find it, the detected spin is not guaranteed to match the first measurement. It's unlikely to be exactly spin-up or spin-down. The spin could be pointing in any direction in space. Obviously, again, this violates the projection postulate, which says the silver atom will be spin-up immediately after the measurement. At best, I guess we could say that the atom is in the same pixel it was measured at, but that's not what was claimed. Therefore, it's more accurate to say a measurement of the spin changed the state as follows:

$$\begin{array}{c} |\Psi_{\text{before}}\rangle = |+x\rangle \text{ ``="} |+z\rangle + |-z\rangle \\ \downarrow \\ |\Psi_{\text{collapse}}\rangle = |+z\rangle \\ \downarrow \\ |\Psi_{\text{after}}\rangle = |???\rangle \end{array} \tag{6.20}$$

These two real-world experiments also show that the projection postulate does not work in all cases. As such a failure, we can conclude that it shouldn't be a postulate of an axiomatic system that is expected to apply universally to all systems. Therefore, we don't have a true axiomatic system in quantum mechanics.

Some clarifying remarks

So that you don't get lost when you leave this book behind I must mention that three terms I used in this chapter are often synonyms outside of this book. Many authors would say that there is as much difference between collapse and projection as there is between a violin and a fiddle: they differ in name only. In contrast, in the way that I use them, they are functionally different. Though, their Venn diagrams overlap more than not. They both involve a state suddenly changing by an unknown mechanism. When I use *collapse* I mean that the wave function is forced into a different configuration. Most often this means the wave function has become more localized and the exact form is unknown. When I use *projection* I mean that the wave function is forced into a state specified by a well-defined ket. So, be aware of this subtle point as I occasionally use the terms in the rest of the book.

Many would also say that both of those terms fall under the umbrella of the observer effect. However, lacking better terms established by the community of physicists, I think it is more useful to separate the fairly predictable exchange of quantities, like energy and momentum, during a measurement from processes which we don't understand and are apparently random. The first is what I call the *observer effect* and the second are what I call a *collapse* or a *projection.*

7 Wave-particle duality and the double-slit experiment

> *'And when you know all the names of that bird in every language, you know nothing, but absolutely nothing, about the bird.' So I had learned already that names don't constitute knowledge.*
>
> — Richard Feynman quoting his father

On wave-particle duality

One of my main efforts is to show that wave functions are at the center of quantum mechanics. To accomplish my goal, I must correct the record on two of the most damaging delusions pushed by pop-sci physicists: mystical **wave-particle duality** and misleading explanations of the double-slit experiment. Starting with wave-particle duality will be best.

The problem with the term *wave-particle duality* is that it can mean different things to different people. The problematic version insists that quantum objects are sometimes corpuscular particles. In an early chapter, I already dismissed these ideas, so there is only a little more that I should say. Authors of misguiding statements, such as "sometimes it is a particle, and sometimes it is a wave", want us to picture quantum-mechanical objects as *alternating* between a binary of conditions. They say: it is or is not. Ones or zeroes. One moment it is a wave; the next it is a particle. They unconsciously yearn for classical mechanics and cannot embrace all that quantum mechanics reveals.

It is common to tell the public that physicists say a particle has **wave-like** and **particle-like** properties rather than say it is a wave and a particle. We supposedly do this to avoid the traditional, pre-quantum way of using the words *particle* and *wave*. They say it does a better job of getting the truth of the matter across. I believe these claims are rather dubious. From everything I have seen, appending the word *like* onto the ends of *wave* and *particle* has not helped many.

In my experience, the terms *wave-particle duality*, *wave-like*, and *particle-like* aren't even commonly used after a second-year survey course on modern physics. If someone had used one of those terms with me after I had completed my first course in quantum mechanics and I was attentive, I would have asked them to list or define the properties because I am unsure what each person believes they refer to. (For example, do they think *particle-like* means a ball, a geometric point, or something else?) These terms aren't normally used by physicists in our day-to-day research because we have more precise vocabulary and mathematics to describe the quantum realm than *particle-like* and *wave-like*.

As a quick illustration of what is taught to physics students about these dueling ideas, I turn to two of the most recognizable textbook authors in quantum physics. In his undergraduate textbook, Griffiths makes one reference to wave-particle duality. He mentions and dismisses it within three sentences of a footnote in his Afterword. In J.J. Sakurai's graduate-level *Advanced Quantum Mechanics*, his first general graduate textbook, he mentions wave-particle duality once in a passing reference to history. In Sakurai and Napolitano's *Modern Quantum Mechanics* text, which is a standard graduate textbook, they make no mention of wave-particle duality. These are far from endorsements for the importance of wave-particle duality by physicists who are writing to educate the next generation of physicists. You can rightly surmise that, in the modern age, it is rare for students in a class devoted to quantum mechanics to talk about wave-particle duality.

Despite the rare use of *wave-particle duality*, do not think the word **particle** must be banished from our vocabulary. Physicists regularly use the word, but not in the way pop-sci authors lead you to believe.

There is a common concept in mind when popularizers say a thing is sometimes a wave and sometimes a particle. They are referring to a thing that keeps its identity in both of the forms they imagine. That thing, that concept, which I have been calling a *quantum object*, is what working physicists call a *particle*. It has particular amounts, or chunks, of invariant quantities, such as rest mass and electric charge, at all times. These never-changing quantities serve as labels for the particles. With the labels in hand, we can say it's an electron, a proton, or a neutrino. Importantly, a physicist also recognizes that a particle has an oscillatory nature, a wave function, at all times. In summary, when a physicist says "particle" in a professional setting, they do not mean something like a tiny tennis ball. They instead refer to a particular class of quantum object (i.e., electrons) or a particular quantum object (i.e., that electron in particular).

This version of the word *particle* is the correct one to use in *wave-particle duality*. Though, my position is that we should not promote the use of the term *wave-particle duality* because it is so misleading.

From this point onward, the common understanding you and I will have when I use the word *particle* (with no explicit modification of the meaning) is that it refers to a quantum object or a class of quantum objects. I will not use it to mean a classical particle, as some other authors do.

With this new understanding of particles, some may misguidedly say that, when measured, particles can no longer have definite positions in my scheme and this is at odds with what we know about the **double-slit experiment**. I see where this comes from. Explanations of the double-slit experiment show images of a detecting screen with minuscule dots representing the measurement of particle location. And you, the public, are misled to think that we can measure the spatial extent of the particle with infinite precision and that they are geometric points upon measurement. In place of the point-particle notion, I want you to understand that, instead, the measurement is forcing the wave function to be localized into a region that is small when compared to the experimental apparatus's resolution. Should a physicist ever try to convince you a particle is a geometric point, know that, in their scenario, the wave function is actually compressed into a tiny volume. In the situation they imagine, it is an infinitesimal object.

The ultimate goal of the rest of the chapter is to illustrate how explanations of the double-slit experiment – which is held up as a perfect example of wave-particle duality – have deceived you. The depth of the deception will surely surprise you. The public is told multiple things that are undeniably contrary to facts. Before you learn those details, I will briefly introduce two well-known equations that have connections to wave functions.

Equations of light and matter

Let there be light...that comes in wave packets

With a bit of injustice to history, we can say that Max Plank and Albert Einstein revolutionized physics by showing that light comes in discrete packets, which were called light quanta and are now called **photons**. They tell us the photon's wavelength λ_{photon} (Greek lowercase lambda) can be related to the magnitude of its momentum $|p|$ or related to its kinetic energy KE:

$$\lambda_{\text{photon}} = \frac{h}{|p|} = \frac{h \cdot c}{KE} \tag{7.1}$$

The c is the speed of light and h is Planck's constant. You don't need to understand this equation as it is. You can ignore the constants in the numerator of the fractions. I will focus on energy, and I want you to see that there is a relationship like this:

$$\lambda_{\text{photon}} \propto \frac{1}{KE} \tag{7.2}$$

I had to use the proportional-to symbol $\propto$ because the two sides are no longer equal. You can now more clearly see that a light of a specific wavelength is related to a specific amount of kinetic energy. The figure below shows how physicists view the relationship between the two quantities λ and *KE* for photons.

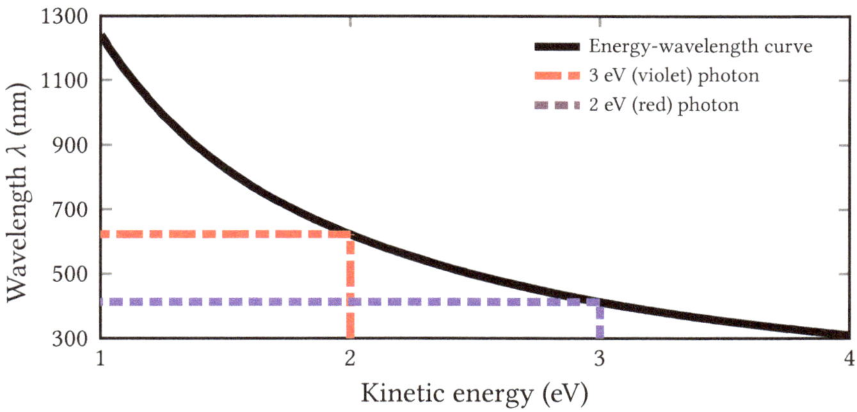

Figure 7.1: Photon wavelength vs. photon kinetic energy. The vertical axis is the wavelength of a photon. The horizontal axis is the kinetic energy of the photon. Energy has units of joules (J) in the metric system, but a joule is too large in quantum mechanics, so we use electron volts (eV), which is the amount of energy an electron acquires when it is accelerated across an electric potential of 1 volt. The black line represents Eq. 7.1. The dashed lines show a rough rule of thumb I use to remember the approximate range of the visible spectrum in eV. Red light is around 2 eV, and violet light is around 3 eV. You can use the lines to see that my rule of thumb says that the wavelengths are 414 nm and 620 nm. Our vision doesn't have a sharp cutoff, but in your day-to-day life, the visible spectrum limits are close to 420 nm and 680 nm.

It's common to use this equation in introductory classes and as approximations in experimental situations. However, this equation has a limitation as is. It doesn't correspond to a finite photon wave function. It would be the wavelength of an infinitely long sinusoidal wave. An actual photon will be a wave packet (recall the images on p. 45) composed of a weighted superposition of an infinitude of sinusoids of different wavelengths. The photon wave packet will then have a characteristic, or average, wavelength. It would also have a KE from a weighted average of the superposition.

Matter waves

Two decades after the discovery of Einstein's light quanta, Louis de Broglie derived an equation that showed objects with rest mass have a characteristic wavelength. These quantum objects are said to be **matter waves**. His equation that tells us the wavelength of a matter wave λ_{matter} relates to momentum in exactly the same way as a photon's wavelength does:

$$\lambda_{\text{matter}} = \frac{h}{|p|} \tag{7.3}$$

But the p is the momentum of a matter wave, and I want to make a plot with kinetic energy, so I'll rewrite that to use *KE*:

$$\lambda_{\text{matter}} \approx \frac{h}{\sqrt{2 \cdot m \cdot KE}} \tag{7.4}$$

The m is the particle's mass, so it tells us this equation only deals with matter waves. I use the approximately-equal-to sign $\approx$ because I used a non-relativistic formula to go from momentum to kinetic energy. The approximation is nearly perfect in the energy range I use. To show the similarities and differences with the photon's equation, I plot this equation in Fig. 7.2, using the electron's mass, over the same energy range that I used in Fig. 7.1.

The de Broglie equation also refers to an infinite sinusoid. Just like in the photon case, an actual matter wave function is composed of an infinite, weighted superposition of wavelengths. You can preview a superposition of three sinusoids by looking at Fig. 13.2 on p. 289.

Despite the fact that there is a superposition of wavelengths in wave functions, during the explanation of the double-slit experiment below, I will approximate the wave function as having a single wavelength. If I desired to be more accurate, I could say that the super-

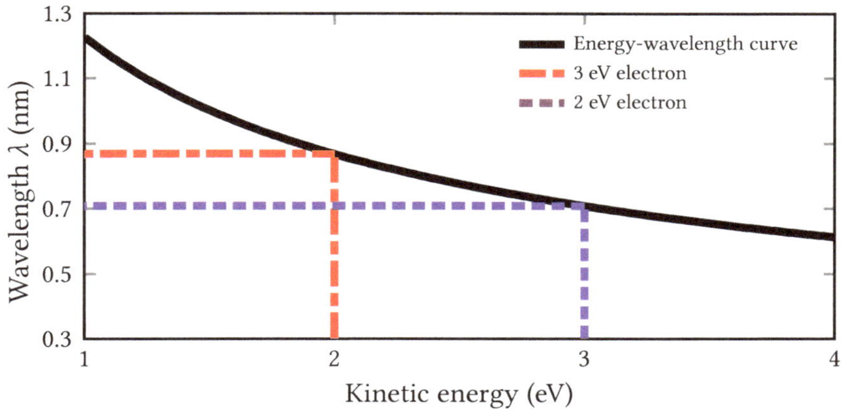

Figure 7.2: Electron wavelength vs. electron kinetic energy. Notice the wavelength has been scaled down 1000 times smaller than the previous figure and the black lines, at 1 eV, seem to be at a similar height along the axis. There is no deep connection here. These are numerical coincidences. Had I chosen a proton, you wouldn't think anything of the numbers. The black line represents Eq. 7.2. The line doesn't fall as fast as the previous figure because of the square root in Eq. 7.2. Electrons don't have color, but I still included dash lines at the same energies as in the previous figure.

position is weighted so heavily in a small range around a particular wavelength that it appears to have a single wavelength.

The double-slit: incorrect explanations

Everyone wants to be like the great Richard Feynman

Looking at the topics in popular books on quantum mechanics, one might think that it is almost mandatory to include a discussion of the double-slit experiment (Fig. 7.3), which purports to demonstrate wave-particle duality. Since you now understand that quantum objects are always waves, we can investigate dubious explanations of this experiment. As best as I can figure, nearly all descriptions are inspired by Feynman's 1962 lecture on the double-slit experiment, which can be found in *Feynman's Lectures on Physics*.

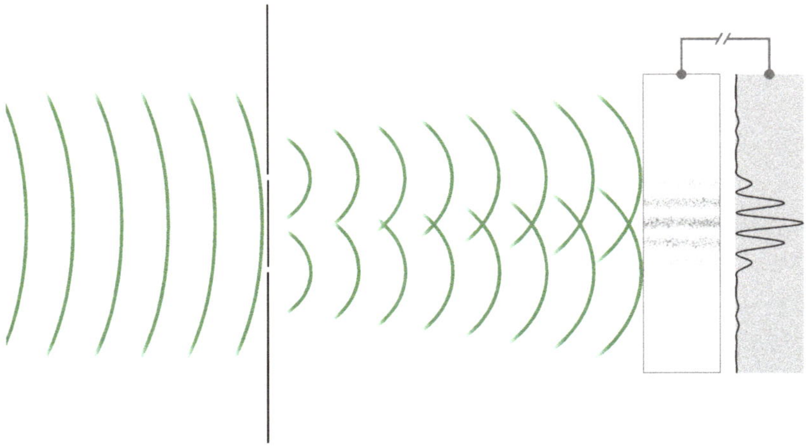

Figure 7.3: A schematic overview of a double-slit experiment with light or matter waves. This represents an overhead view of an experiment with waves coming from the left and traveling to the right at a snapshot in time. The green crescents represent when photon or matter density is the greatest. The vertical black line is an absorbing screen with two thin slits cut into it. The waves on the downstream side of the slits superpose to cause constructive and destructive interference. The wave function of each individual particle collapses somewhere on the imaging screen with a probability proportional to the intensity of the light on the screen. The imaging detector registers a count in one of its pixels. Afterward, the experimentalist sums the two-dimensional image data along one dimension to give a smooth curve representing the photon or matter density or, equivalently, the probability of wave function collapse.

Feynman said that the essential mystery of quantum mechanics is wrapped up in the double-slit experiment. For a newcomer to truly understand all the parts of this experiment takes a great deal of time. Because of this, I call on all physicists, whether they are professors writing books or grad students giving physics outreach presentations, to only tell the public about Feynman's version of the double-slit thought experiment if there is sufficient time to do it without short-cuts. With limited time, can one not make do with a simpler topic?

I checked, and it would be rare for a public speaker to be in the same situation as Feynman. For his students, he took 67 minutes to explain his double-slit thought experiment. Seven and eight lectures before this, he had already covered the mathematics of single-slit diffraction and two-source interference. The students also had completed homework and a take-home quiz on those topics one month before his double-slit thought experiments. That's quite a different preparation for his lecture than the general public has.

With that finger-wagging done, I am not going to pretend I will give you a complete explanation of the multitude of physics topics necessary to fully comprehend the double-slit experiment. I suggest you read or watch a few explanations of the double-slit experiment before proceeding because I will write as if you are already familiar with the gist of the experiment. It will be even better if you understand constructive and destructive interference and why the waves emerging from the slit must be characterized narrowly around a single wavelength and must be in phase across the slit.

Many video and written explanations are available. I should pick one to discuss, though. The one I have chosen is Prof. Jim Al-Khalili's talk on YouTube, where he is giving a speech to the Royal Institution of Great Britain and mentions he is following in Feynman's steps. It is called *Double Slit Experiment explained! by Jim Al-Khalili.* Note that Feynman uses electrons, and Al-Khalili uses atoms, but the choices make little difference. You can follow the video during my deconstruction of it. You will find a similar explanation and images in the first chapter of his 2003 book *Quantum: A Guide for the Perplexed.*

Al-Khalili is not alone in leading his audience astray. In nearly all explanations, popularizers give some significant falsehoods. I will step through the standard description often given and make reference to Al-Khalili's explanations. As we go through them, you will see how inaccuracies in single-slit diffraction and how the "which way" flimflam pretty conveniently allows someone to bring the conversation to the "sometimes it's a wave, and sometimes it's a particle" delusion.

Single-slit diffraction is often presented incorrectly

Imagine yourself and a friend at the exterior corner of a building. Each of you leans against a wall of the building with the corner between you. You don't have a direct line of sight to each other, and there is nothing else nearby for about 200 m. You want to communicate discretely, like in a bad spy movie.

How is it possible for you to hear the other? It's not like the sound is traveling 200 m, reflecting off a tree, traveling 200 m back, and entering your ear. The sound would be incredibly diminished, and there would be more than a full-second delay between the time you speak and your friend hearing you. The reason you can hear your friend is because sound waves, like all waves, spread around corners and barriers. This spreading is called **diffraction** and is an essential component of all double-slit experiments, no matter whether they involve water, sound, light, or electrons. Back in Fig. 7.3, I showed a conceptual drawing of waves diffracting after the slits.

An interesting property is the degree of diffraction depends on the wavelength. Short-wavelength waves will spread less than long-wavelength waves. This is partly why you can hear the bass outside a nightclub very well, but the higher-frequency sounds are muffled. The long-wavelength bass travels more easily through all the twists and turns of the building's halls and doors to get to you (see Fig. 7.4).

Before getting to two slits, explanations typically have waves going through a single slit. A proper explanation of single-slit diffraction will emphasize that interference causes an interesting pattern in the measured data, somewhat similar to the one you saw in the double-slit apparatus above. This happens because each point along the slit width can be considered a source of waves, all with the same phase, and those waves interfere with each other as they propagate downstream from the slit. In Fig. 7.5, I show how increasing the slit width causes more destructive interference and causes weak side bands to appear at angles away from the central axis.

A proper explanation hardly ever happens. Instead, Al-Khalili and others say that the particles go through the slit, and a pattern similar

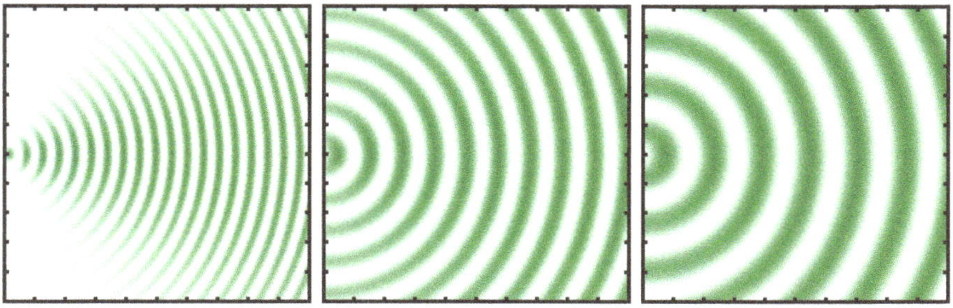

Figure 7.4: Various wavelengths diffracting through a slit. From left to right, the wavelengths scale as follows: λ, 2λ, and 3λ. The slit is 8% the width of the image and is located at the center of the left edge of each image. The darkest region corresponds to zero photon or matter density. (A major caveat is due: in this and the next figure, to make the intensity more visible to you, I didn't let the waves decay with distance from the source. Among other violations, the lack of decay violates the conservation of energy.) In the λ image, the waves are less intense at the wings. As the wavelength increases, the intensity does not decrease as much toward the edges. Hence, we say that longer wavelengths get around corners more easily.

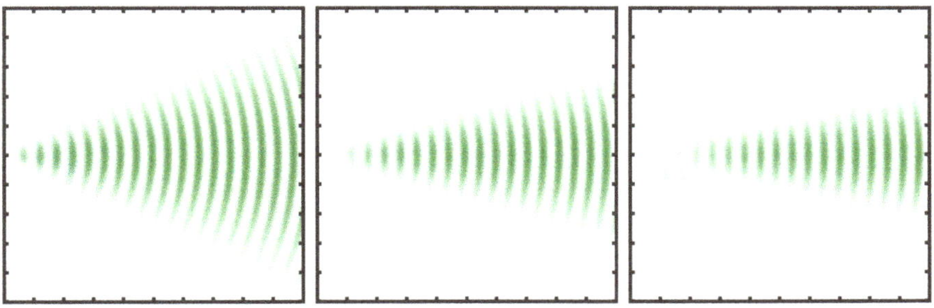

Figure 7.5: Single-slit diffraction with wavelength λ and increasing slit width. The slit widths are 16%, 24%, and 32% of the image width. If I had extended the plots a considerable distance farther, the arcs of green would approximate straight lines. In those places, we can place flat detectors and find a particular pattern. Otherwise, we have to use a more complicated mathematical expression than the sinc-squared function I mention below. (Note again that to make the waves visible the intensity is not decaying with distance.)

to the one in Fig. 7.6 is observed. If this pattern were true, the particles certainly would seem to be acting like classical particles since they arrive at the detecting screen in a single band that imitates a blurred image of the slit. Whoever says it, to sell more mystery, they need the implication that wave interference does not occur during single-slit diffraction. Al-Khalili specifically says in his presentation, "So far so good — nothing strange here", with regards to his incorrect diagram of atoms through a slit. But the blurred image of the slit is not the full truth. It has features missing.

In Fig. 7.7, I show what should have been shown. It is the *Fraunhofer limit* of single-slit diffraction, which is what is taught to undergraduate students. This limit occurs under a few conditions. The wave hitting the slit needs to be uniform and characterized by a single wavelength. And the detector needs to be far from the slits.

We strive for this limit because it makes the data analysis easier. For your information, the pattern in the figure has the shape of a sinc-squared function. The sinc function, aka the sine cardinal function, is the sine of x divided by x (with a 1 at $x = 0$ because we can't have division by zero).

The true detector pattern may surprise you if you are familiar with popular explanations of the double-slit experiment. You probably expected one smooth, narrow bump now and multiple bumps to show up later when I eventually include a second slit. Your memory is not failing you. You have been previously taught a simplification so that they can sell you their story of mystery.

Multiple bumps in a double-slit pattern will undoubtedly appear more impressive if multiple bumps do not appear in the single-slit pattern. Nevertheless, the figures I show you are what would occur in an ideal experiment. To be clear, if there is a *narrow* central bump in the single-slit pattern, the pattern will also have multiple, smaller bumps whether it is in the Fraunhoffer limit or not.

Single-slit diffraction is not an esoteric result. Ripple tank demonstrations are standard in many first-year physics laboratory demonstrations. Searching the internet for images of a ripple tank with a single slit, or looking back at Fig. 4.6 on p. 57, will show alternating bands where water waves exist and cancel out due to interference. It's difficult to conceive of a professor of physics who is unaware of the proper pattern of single-slit diffraction when the central bump is narrow.

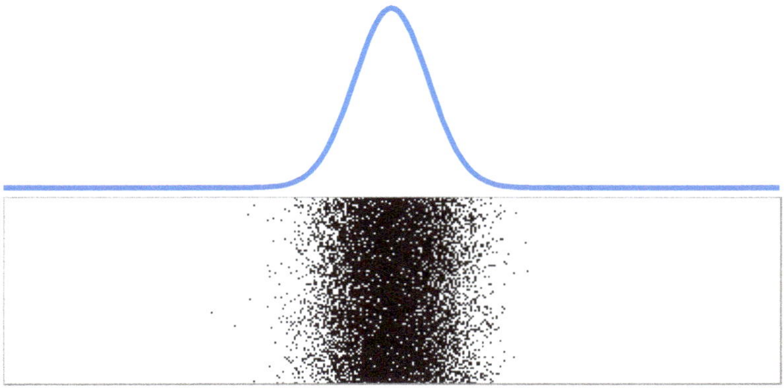

Figure 7.6: My reproduction of a blurred-image-of-the-slit pattern supposedly created by atoms going through a single-slit. Beware: This figure does not reflect any single-slit diffraction pattern. Take note that there is only a single narrow bump in the probability (top) and a single narrow band in the imaging detector (bottom), which is like a blurred image of the barrier slit.

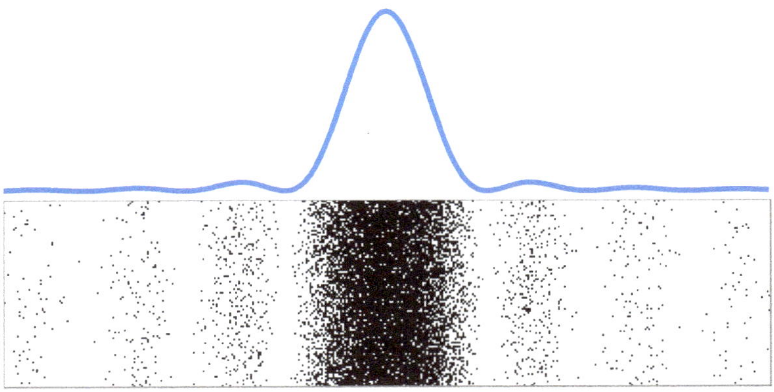

Figure 7.7: The pattern due to single-slit diffraction. We are guaranteed to see multiple bumps if there is a narrow central band on the imaging detector. The bands with few pixel spots in the detector and the minor bumps on the blue curves correspond to the green, off-axis bands in Fig. 7.5. We won't see multiple bumps if we are under the conditions that produce the images in Fig. 7.4. In those cases, there would be a single broad bump that does not go to zero.

The double-slit pattern

The next step in every explanation is to open a second slit, mention the waves from each slit interfere to give a new pattern, and show the interference pattern. I have already shown you this with a schematic of a double-slit experiment and its pattern (Fig. 7.3).

The revealing of the double-slit pattern does not have any problems. The only minor wrongdoing our expositors of quantum mechanics have done concerning the double-slit pattern is to not tell you the pattern we usually try to produce is a sinc-squared function multiplied by a cosine-squared function.

For those curious to see the math, the probability density at the imaging detector is:

$$|\Psi|^2 = \left(\operatorname{sinc}\frac{\pi \cdot b \cdot x}{D \cdot \lambda} \cdot \cos\frac{\pi \cdot d \cdot x}{D \cdot \lambda}\right)^2 \tag{7.5}$$

The λ is the wavelength of the light or matter wave. The width of the opening in each of the slits is b. The separation of the slits measured from the center of each opening is d. The distance from the screen with the double slits to the imaging detector is D. The position on the screen is represented by x, and $x = 0$ is where the central bump occurs. To ensure this equation is the most accurate, D should be a couple hundred times larger than d or more.

Changing the variables in that equation will change the pattern. Setting D, the distance to the screen, or λ, the wavelength, to a higher number will make the pattern wider. The opposite will happen if these variables become smaller. Making b, the slit openings, wider will make the sinc-squared envelope narrower while keeping the cosine oscillations under the envelope peaking at the same positions. In contrast, making the distance d wider will increase the number of oscillations under the sinc-squared envelope.

In Figs. 7.8, 7.9, and 7.10, you will see scenarios with different scalings of the variables. You should compare the variable and pattern changes among the images.

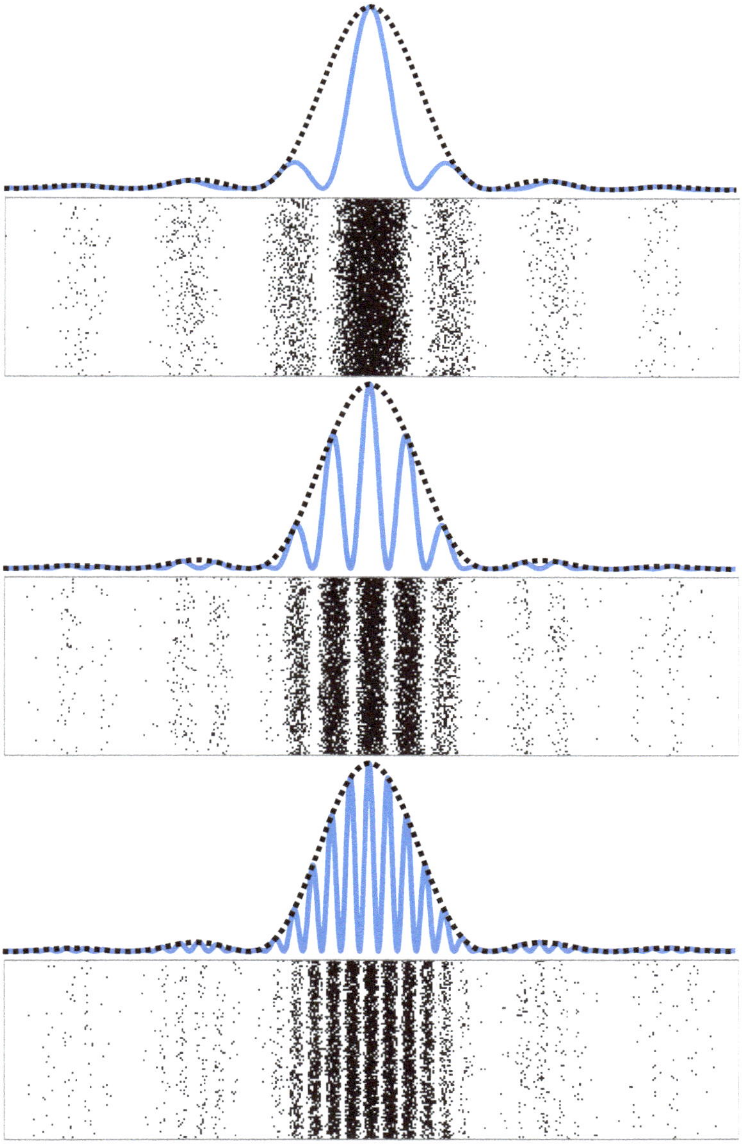

Figure 7.8: Various-double slit patterns with a dotted line showing the enveloping sinc-squared function. The sinc-squared envelope doesn't change, so this means the individual slit widths did not change. Going down the image, the slit separation distance is scaled from d to $2.5d$ to $5d$.

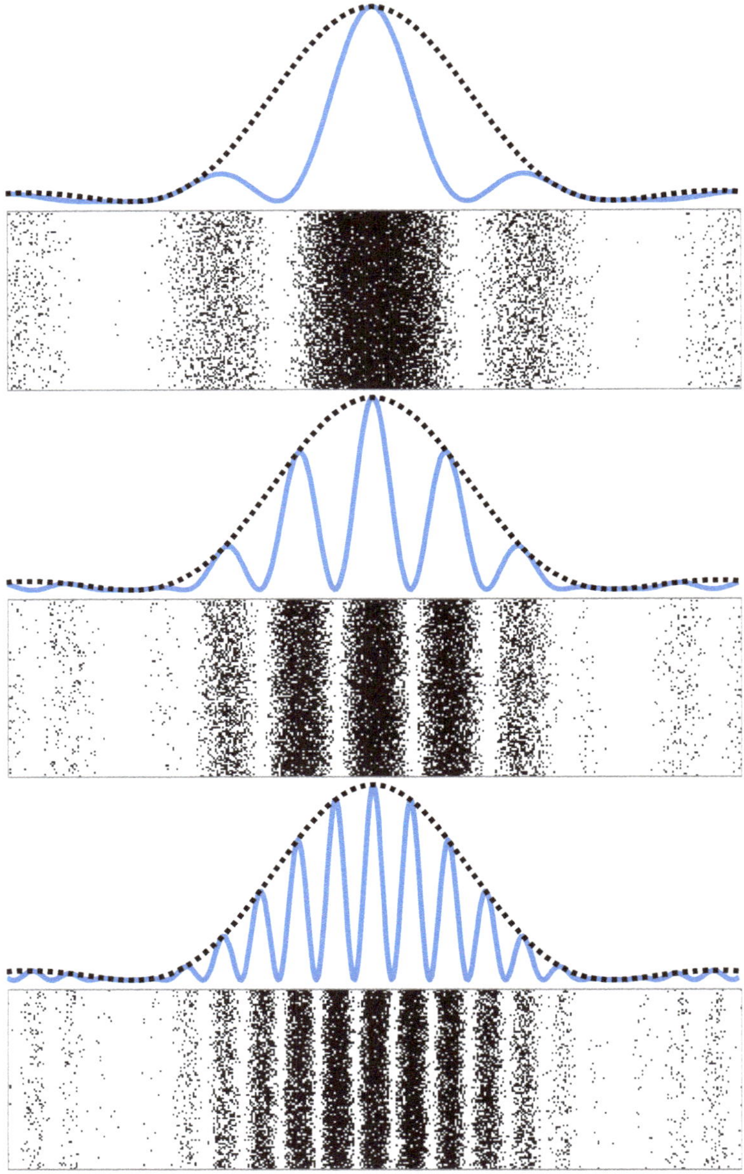

Figure 7.9: Same as previous figure with one change. You can consider that, compared to the previous figure, the wavelength has doubled (λ to 2λ) or the distance from the two slits to the detector has doubled (D to $2D$). The result is that the entire pattern stretches horizontally outward. In a sense, it is the same as zooming in on Fig. 7.8.

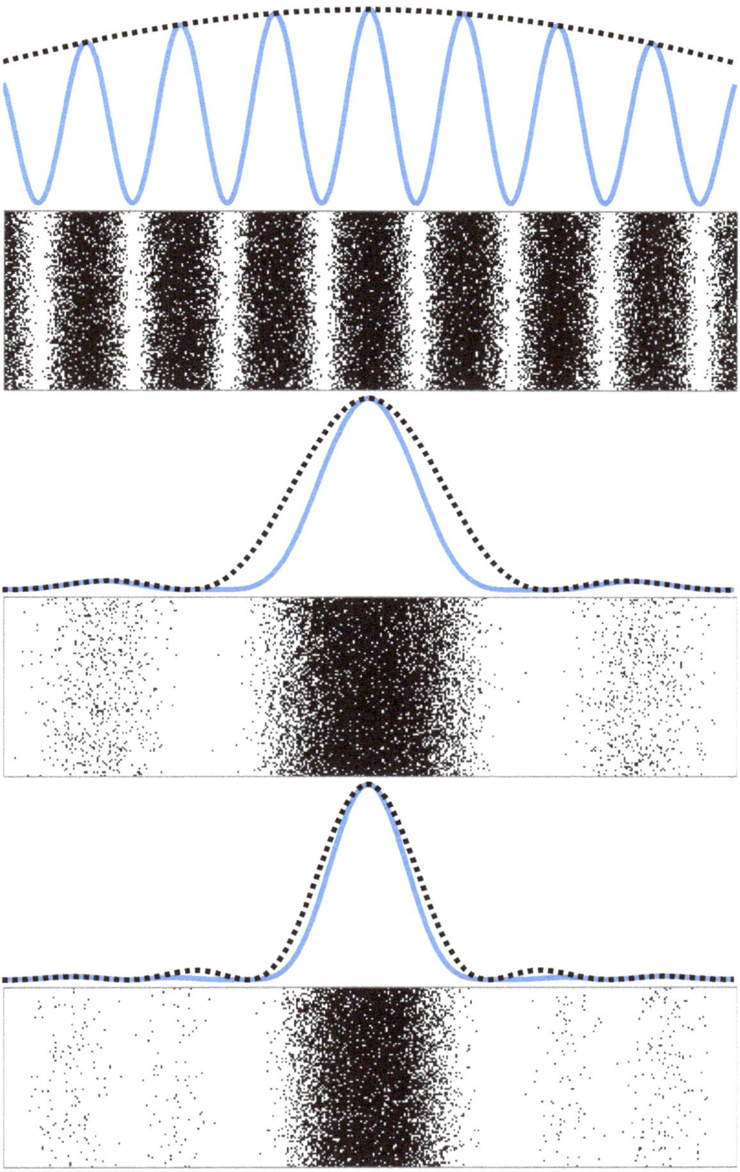

Figure 7.10: Extreme double-slit patterns. 1st scenario: D, d, and $b/10$; smaller slit widths make the envelope expand outwards. 2nd scenario: $3D$, d, $2b$; also the slits are not small compared to the separation distance, so the pattern is approaching that of a single slit. 3rd scenario: $3D$, d, $3b$; even closer to having overlapping slits, so closer to resembling a single-slit pattern.

The "which way" flim-flam

What is said

Some can't resist making quantum mechanics as different and mysterious as possible. Often a question is posed: "Which way did the atom go, through the right or left slit?" Beware, this question assumes that an atom is a classic particle that only goes through one slit or the other. The question rejects the atomic wave function as a quantum object where at some point during the measurement process it collapses to a small region downstream of the slit.

Nevertheless, they mention placing a detector next to one of the slits to determine which one it went through. It's just a casual thing out of the way, right? They make it seem as unintrusive as taking a photograph of a horse race to see which crossed the finish line first.

They say that the simple act of observing a quantum particle forces it to decide to go through one of the slits. And when we measure the atoms passing through one of the two slits, the observed pattern is no longer the double-slit pattern but instead is two blurred-images-of-the-slit patterns overlaid, giving a pattern with two bumps. Apparently, we have forced it to be a classical particle by simply looking at it. This conclusion is, of course, wrong.

Popularizers need the implication that observation is passive and unobtrusive to sell the marvel of their tale. Indeed, Al-Khalili told the Royal Institution audience that observation is gentle and even showed a cartoon detector that is like a passive CCTV camera.

The truth is much different and much more complicated. We can contrast Al-Khalili's explanation with one Feynman told his class. Feynman warned his students that he needed to fire photons at the electrons to see them and that the

> ... light exerts a big influence on the electrons. By trying to 'watch' the electrons we have changed their motions.

This is because light is made of oscillating electric and magnetic fields, which an electrically charged particle, like the electron, will readily respond to.

Al-Khalili was talking about atoms and not electrons, but he still needs something to interact with each atom, which will then be collected in some detector. Otherwise, how is he going to know when an atom passes by? His CCTV camera needs photons, and those photons would have interacted with his neutral atoms because the photon's electric field interacts with the electric charges of the atom.

Do we know what happens?

This leads me to mention again the undeniable observer effect and the hypothesized collapse of the wave function. Simply put, we cannot make a measurement without disturbing the particle. Recall that I characterized the observer effect as a predictable violent change to the particle. In contrast, the collapse of the wave function is a century-old mystery whose details we don't yet comprehend. We, therefore, know the photons used to detect atoms will do two things simultaneously in a "which-way" measurement.

First, the photons will impart momentum to the atoms, and because this is quantum mechanics, the scattering angle and momentum change will be random (so the violence done to the state is not completely predictable). In most explanations, the photons are seen coming at more or less 90° to the direction of travel of the waves. The random scattering of the photons would then cause random shifts in the direction of the waves. This would blur the pattern seen. However, as long as the momentum of the photon is much smaller than the momentum in the atoms, Al-Khalili can safely ignore this.

In a footnote in his book, he mentions a real-life experiment with atoms but doesn't cite it. Assuming that I found the correct publication, the experiment used a 1 μm (1 micrometer = one-thousandth of a millimeter) slit width and helium atoms with de Broglie wavelengths around 0.1 nm. Looking at Eq. 7.1 and Eq. 7.3, we can conclude that when light and matter have the same wavelength, they will have the same momentum. Additionally, we can see that wavelength and momentum have an inverse relationship. Therefore, as long as the probing photons' wavelength is much greater than 0.1 nm, the momentum will be small enough. The atoms' wavelength of 0.1 nm corresponds to a photon energy of about 12.4 keV, which is the energy of X-rays. So which wavelength is needed? In the following paragraphs, you'll see that neither I nor anyone can be sure which photon wavelength is needed. However, I *feel* that wavelengths much greater than 0.1 nm photons are acceptable. So, we will say Al-Khalili can safely ignore the observer effect, and we can move on.

The other effect I need to analyze is collapse. The photons cause the atoms' wave functions to be more localized than they were before. Recall that I keep saying we don't understand the collapse of the wave function. The fable often told to students is that the photon will localize, or collapse, a particle to the size of the photon's (characteristic) wavelength. If this were true, we would need an infrared photon around 1 µm to localize the photon to about the width of the slits in order to find "which way" it went. This fable is an odd claim because the wavelength is a property associated with only one dimension. Additionally, it's just speculation. There is no known mechanism for this. Moreover, why would it work the same for every particle in every scenario?

I could hypothesize something different and suppose that some measure of the extent of a photon wave packet in space determines the degree to which the electron wave function is localized by collapse. This is appealing because it deals with three dimensions for both particles. Despite my hypothesis's attractiveness, localization could be determined by something else. The degree of localization could even have some randomness to it. As you can see, starting down the "which-way" rabbit hole can lead to an unknown wonderland of hypotheses. And it may surprise you to learn that we only have hypotheses because we haven't conducted the "which-way" experiments as they have been described to the public.

The reason for the "which way" flim-flam

Their purpose for asking what happens when you try to measure a particle's position at the exit of the slit is to amaze the audience by showing again the exact same pattern that was shown for single-slit diffraction. That pattern is the blurred-image-of-the-slit falsehood. But we have disposed of that and have no need to discuss it anymore.

Assuming you were shown the correct diagram for the particles' single-slit diffraction, would we see the correct pattern again in a "which way" measurement? The answer is no. It would require being able to control the collapse of the wave function so precisely that we could emulate the wavefront that emerges in a single-slit situation. And, by emulate, I don't mean to only get a wave function approximately the same size as the width of the slit. The single-slit pattern won't be seen unless the which-way wave function copies the single-slit wave function fairly precisely. You can't have any random set of amplitudes along the width of the slit. At the present, we can't control collapse and have no chance of doing this. Feynman also went beyond what is acceptable when he included this.

What is speakable and unspeakable?

To understand how confused people are by explanations in quantum mechanics, anyone can search their way through the comments of Al-Khalili's YouTube video or any physics forum. You will find plenty of people left with questions after hearing from experts.

I have been emphasizing the importance of the wave function to understanding quantum mechanics. Yet, in the video, Al-Khalili never even mentions the word *wave function* to the Royal Institution audience. I ask why should physicists get to know about the wave function and not the public? That's not a rhetorical question. I ask because the public desires to learn about the true nature of reality, and they are mature enough to handle a discussion of the wave function.

Behind this is a larger question: Why the gatekeeping by popularizers of physics? For example, I find that, in his books and documentaries, Brian Greene obscures much physics in inappropriately loose language. (In the prefaces of at least two of his books, he gives himself license to write like this by saying he will use metaphor and analogy.)

As another example, it is uncommon for popularizers to explain the true purpose Schrödinger had for introducing us to his cat.

No matter the reason, the obscure dumbing down is unnecessary. People can handle the truth. Even the Catholic church found that laypeople could handle their services in the local vernacular in 1963 when they dropped Latin. If they can drop their cryptic speeches, I think physicists can, too.

Chapter appendix

To avoid cluttering the main points with details, I've moved some commentary here. The first point involves why Al-Khalili chose to describe the experiment they way he did. The images in his presentation came from the first chapter of his book. He has the copyright to his entire book, but an illustrator created the images. The images naturally constrained what Al-Khalili could say, which likely contributed to his mistakes in the video. Nevertheless, Al-Khalili chose to use the images in his lecture and spoke the words he spoke when describing the images.

Second, expert readers will notice Al-Khalili's double-slit probability curve seems a little wrong because the troughs don't all go to the same minimum. In a footnote in Al-Khalili's book, he says that his description of the double-slit experiment was conducted in the 1990s on a gold foil with slits. It seems he is talking about a 1991 experiment by O. Carnal and J. Mylnek of the Universität Konstanz. Al-Khalili's wonky double-slit probability curve seems to vaguely echo the experimental data of Carnal and Mylnek's Fig. 5, which I surmise was the inspiration for the illustrator. Any scientist can tell you that it is common for experimental data to deviate from the ideal patterns of theory. So, this one image isn't quite as wrong as it appears. I will also mention that Carnal and Mylnek did not measure a single-slit diffraction pattern.

Third, Al-Khalili doesn't mention the word *wave function* to the Royal Institution audience. However, starting in the third chapter of his book, he does use the word hundreds of times. He also eventually updates his explanation of the double-slit experiment with an explanation of the wave function and probability density.

Finally, Al-Khalili mentioned he was following the example of Feynman. In Feynman's presentation, he too failed to draw the full sinc-squared shape. Though, there are some facts that help redeem Feynman. As I said, he previously taught his students about diffraction several lectures before. Additionally, he drew his single-slit bumps very broad, and they never went to zero, so he was using geometric conditions different from Al-Khalili's narrow bumps. It seems like Feynman was assuming conditions close to those used in one of the images in Fig. 7.4 (p. 117) or the first scenario in Fig. 7.10 (p. 123). Despite these redeeming features, I maintain that Feynman's or anyone's explanation of the double-slit experiment is unlikely to be as educational with regard to quantum mechanics as they think it will be.

8 The mysterious hieroglyphics of Schrödinger's equation —As an energy balance

> *"The Schrödinger equation has been one of the great triumphs of physics. By providing the key to the underlying machinery of atomic structure it has given an explanation for atomic spectra, for chemistry, and for the nature of matter."*
>
> — Richard Feynman

Never taught to read the hieroglyphics

From the title and the length of these two chapters, you can conclude it is not obvious how to read the Schrödinger equation. You might hear a physicist introduce the Schrödinger equation as amazing or beautiful. And from this, the implication is that you must be able to naturally appreciate the equation. I'd like you to know that physicists don't usually wax poetic like that in their day job. Normally, we are content with people assuming that we are clever. Using colorful adjectives to refer to physics usually only happens in a book or when the public is watching.

In its typical form, the Schrödinger equation is neither beautiful nor amazing to the average person. It is stupefying and scary. I'm sure when many physics students first saw it in class, a little piece of them said, or at least a little part of me said, "yuck, derivatives."

With its unexplained symbols and concepts, Schrödinger's equation is completely indecipherable to a lay reader. It is reasonable to be intimidated by these hieroglyphics. Would you find it acceptable if a book on Egyptian hieroglyphics only gave you the pronunciation of the glyphs in a phrase, then stated: "From this phrase, the whole *Book of the Dead* is put in context", and said no more about the phrase? Of course not. No, that's poor exposition.

Yet, physicists get away with doing something comparable with the Schrödinger equation. In several places, I've seen images with the Schrödinger equation written out in full. Arrows and blocks of text are doing their darnedest to teach readers the individual quantum-mechanical hieroglyphics. But a single image can never be sufficient. An adequate understanding of Schrödinger's equation requires will take two entire chapters in this book.

Though, after these two chapters, you certainly won't be in a position to solve or prove anything. Schrodinger didn't even include "the exact and rather tiresome proof" of the solution to the hydrogen wave functions in his review paper that introduced his work to the English-speaking world. (He first wrote several papers in German developing the equation and wave mechanics.)

Be comforted, though, with the knowledge you will get from understanding the symbols. After I have finished, you will be able to look at:

$$\hat{E} \cdot |\Psi\rangle = \hat{H} \cdot |\Psi\rangle \tag{8.1}$$

and know that it is the Schrödinger equation in a very compact form. Before we proceed, I should warn you that $\hat{E}$ does not equal $\hat{H}$. In the next chapter, you will see the non-hat forms written out. What is equal, or balanced, are the numerical values that those operators extract from the ket.

Each chapter covers a mathematical aspect of Schrödinger's equation. The first aspect is a high-level understanding of how the various energy terms balance. I will show you this via abstract operators and kets and the values they return. The second aspect is how the Schrödinger equation governs how wave functions change with time. This will be done with a symbolic and graphical explanation.

Balancing energies in classical physics

In introductory physics, we teach students how to solve certain types of problems using energy conservation. In the Schrödinger equation, there are factors associated with **kinetic energy** (KE), **potential energy** (PE), and **total energy** (E).

In classical physics, the total energy is found by adding the kinetic and potential energies. In a loose sense, this is the Newtonian physics version of Schrödinger's equation for a solid object:

$$E = KE + PE \tag{8.2}$$

Kinetic energy is the energy associated with a moving object. You likely know it is found using the object's mass (m) and the square of the speed of the center of mass (v): $KE = m \cdot v^2/2$.

In simple situations, potential energy is associated with position. Whenever magnetic fields are involved, orientation matters too. It's also common for potentials to change with time. However, to keep things simple, when discussing Schrödinger's equation, I will only consider electric potential energy, which depends on position alone. This means in these two chapters, I completely ignore spin and orbital angular momentum.

In your daily life, gravitational potential energy is the potential energy you are most aware of. Before it meets the ground, an object will gain a particular amount of kinetic energy by being dropped from a second-story window. Whereas an object dropped from a third-story building will gain more kinetic energy and speed because it has traveled through more of the gravitational potential.

When we can ignore certain effects, such as air resistance and friction, these two energies sum together to always balance with the total energy. If the kinetic energy increases by some specific amount, then the potential energy decreases by the same amount. I give you a pictorial understanding of this in Fig. 8.1.

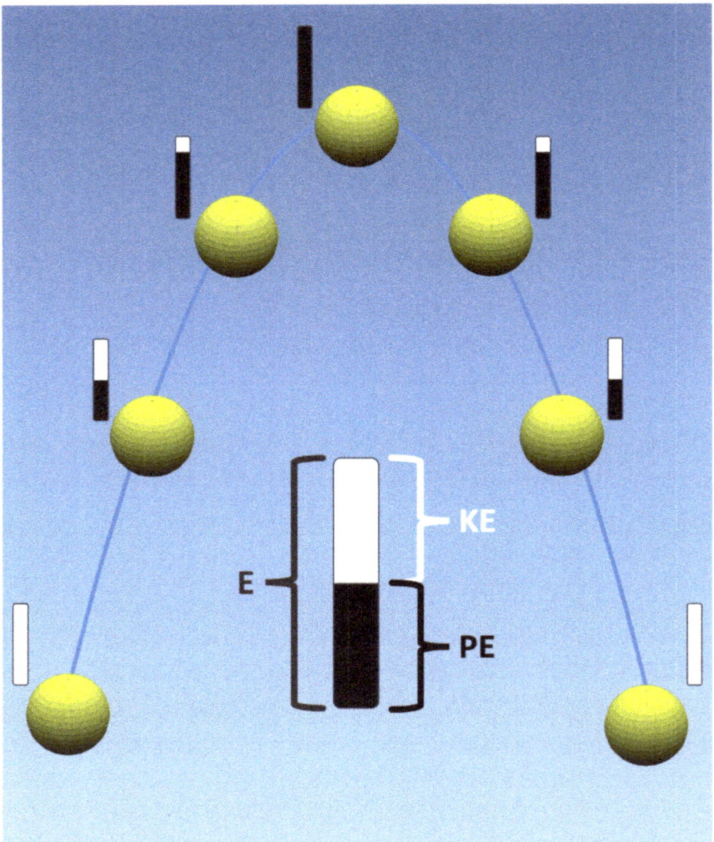

Figure 8.1: Energy conservation. If we throw a ball into the air, the total energy will be constant (if we ignore air resistance). The height of the black and white bar represents this. At the bottom of the arc, the potential energy is zero, and the energy is entirely kinetic. (Just like we had a convention for the zero of the ionization energy in Chapter 4, we conventionally set the zero of the PE to the lowest point in situations like this.) Note that we only consider motion and KE in the vertical direction; the horizontal motion is only for illustrative purposes. As the ball rises in the air, it loses speed and kinetic energy because energy is transferred to potential energy. At the top of the curve, the ball momentarily stops, and the total energy is entirely gravitational potential energy. Because energy is conserved, the process is symmetric on the way back down.

The Schrödinger equation is about balancing energies in quantum waves

The central equation of the chapter

The crucial thing to recall here about the math of quantum mechanics is that it involves operators and kets (aka wave functions, states, state functions, etc.). An operator acts like an instruction for a ket to return numerical values, change the state, or do both.

In Schrödinger's case, the state of the particle is the spatial wave function. He didn't yet know about spin angular momentum. His equation is a reflection of a general law, though, and it can handle spin states.

A compact way of writing Schrödinger's equation, which is great for beginners, is:

$$\hat{E} \cdot |\Psi\rangle \;=\; \hat{KE} \cdot |\Psi\rangle \;+\; \hat{PE} \cdot |\Psi\rangle \tag{8.3}$$

The hats above each energy symbol remind you that they are operators and are put on the operators' more high-level and abstract form. When we get precise later and define what they mean algebraically, there are no hats on the mess of symbols.

The energy operators in Schrödinger's equation

The equation above should mean little to you because you don't know what math goes into the wave function or each of those operators. For now, know that the form of $\hat{PE}$ changes with each problem we try to solve. The $\hat{KE}$ operator always has the same mathematical operation, but depending on the dimensions and symmetry of the problem at hand, we might transform the explicit symbols a bit. The energy operator $\hat{E}$, on the other hand, is ol' reliable. When its symbols are written out, they are always the same for every problem.

Also, and this is very important to understand, none of us knows the form the wave function will take until we decide on the form of $\hat{PE}$ and spend much time doing complex math.

A caveat: what is in a name?

I showed you this compact form of Schrödinger's equation with these particular symbols because it looks similar to the classical equation of Eq. 8.2 and will feel familiar. However, there is a bit of a lie in saying that the Schrödinger equation corresponds to Eq. 8.2. Let me introduce a mostly benign controversy so that I can teach you a bit about these operators.

Whoever chose to name the energy operator and put the E under a hat was likely unaware that the explicit form of it (i.e., $i \cdot \hbar \cdot \partial_t$) does not meet the definition of an operator in axiomatic quantum mechanics. The actual definition includes a description of a linear map that takes a function and transforms it into a function (which could be the same function) that is in the same abstract space. The so-called energy operator doesn't do this for technical reasons. However, it works perfectly well as a generic mathematical operator in wave mechanics. Fortunately, using it as a quantum mechanical operator causes no issues in this book until the final appendix. The lack of problems is likely why many physicists use it as such.

For the most rigorous of professors, using the letter E in $\hat{E}$ and calling it the *energy operator* is high on the scale of being wrong. These professors tend not to name the operator and do not use the abstract $\hat{E}$ when teaching quantum mechanics. They instead prefer to write out all the symbols explicitly.

I am between the two extremes. I am aware of the issue but will keep the name because it is so common and because it asks a question of the wave function and the answer returned is the total energy of the system, which follows how I introduced operators to you.

What these rigorous professors usually want people to additionally understand is that, despite the name and answer returned, $i \cdot \hbar \cdot \partial_t$ is not the group of symbols that *defines* the total energy in Schrödinger's equation. If you recognize that $\hat{E}$ is ol' reliable and its mathematical form never changes from one situation to the next, you could figure that it couldn't be the operator that defines each wave function's energy for all the myriad of different scenarios out there.

Instead, $\hat{KE}$ and $\hat{PE}$ together define the total energy. When considering both operators as a single unit, we call it the **Hamiltonian** or, less commonly, the **Hamiltonian operator**. In symbols, it is $\hat{H} = \hat{KE} + \hat{PE}$, which now allows you to see how to go from Eq. 8.1 to Eq. 8.3.

If some professors are getting in a mild huff because only one of the two operators defines the energy of a quantum object, how can the operators ask the same question of the ket and get the same energy? Well, it's because they ask "What is the energy?" to two different parts of the ket. The energy operator poses the question to the part of the ket's math that depends on time, and the Hamiltonian operator poses it to the math part that deals with position. (Recall, I'm restricting PE to be electrical potential energy, which only depends on position. General Hamiltonians aren't restricted to just position.)

On my choice of words

I anticipate some careful readers may have some pointed questions. I just said that the Hamiltonian operator *defines* the wave function's energy. I also said that the operators *ask* for the wave function's energy values. Aren't those at odds? It does seem like that, but I am imagining different scenarios.

When I say the Hamiltonian defines the wave function's energy, I am referring to the common scenario of a theoretical physicist wanting to examine a particular Hamiltonian (i.e., a particular situation) and see what the results will be. In this case, they must work from the Hamiltonian towards discovering or approximating the wave function. (Honestly, they don't always find the wave function because there are some shortcuts.) The Hamiltonian is the only operator that is special like this. Other operators don't get to define the system.

When I say the operators ask for a value, I imagine that a student has been given or a researcher has discovered a wave function, and they are now using the operators on it to discover something.

That being said, the $\hat{PE}$ operator never asks for the potential energy function from the ket. It's more like it always tells the ket what it is. If we want to stick with the asking-a-question-of-the-wave-function bit, it would be like asking the question "What color is the orange cat?" The answer is in the question.

The hydrogen atom can illustrate the balance

Once we have a wave function in hand, we can concentrate on what the operators are asking of the wave function. This is where we will see that the answers will all balance in the end.

I will reuse the example of hydrogen's ground state (GS) in this chapter. Writing Schrödinger's equation for the ground state requires only a slight notation change inside the kets to know we are discussing a particular state:

$$\hat{E} \cdot |\mathrm{GS}\rangle \;=\; \hat{KE} \cdot |\mathrm{GS}\rangle \;+\; \hat{PE} \cdot |\mathrm{GS}\rangle \tag{8.4}$$

The energy operator

Of those three operators, the only one that will return an eigenvalue is the energy operator $\hat{E}$. You've already seen this in the *Math vs Measurement* chapter, but I will repeat it here:

$$\hat{E} \cdot |\mathrm{GS}\rangle \;=\; -13.6\ \mathrm{eV} \cdot |\mathrm{GS}\rangle \tag{8.5}$$

Again, I remind you that I have hidden the math to get that answer. While there's no definite quantity as a line of math, I'll say that I hid a line or two of math from you. There are so few lines of math because the time part of the hydrogen wave function is so simple. Later, I'll show you how simple the time part is (Eq. 10.5 on p. 195).

We can substitute that eigenvalue into the Schrödinger equation for the ground state of the hydrogen atom. Two chapters back, I used a downward arrow to abstractly represent the measurement process.

Now, the arrows will represent substituting math in:

$$\begin{array}{ccccc} \hat{E}\cdot|\mathrm{GS}\rangle & = & \hat{KE}\cdot|\mathrm{GS}\rangle & + & \hat{PE}\cdot|\mathrm{GS}\rangle \\ \downarrow & & \downarrow & & \downarrow \\ -13.6\ \mathrm{eV}\cdot|\mathrm{GS}\rangle & = & \boxed{}\cdot|\mathrm{GS}\rangle & + & \boxed{}\cdot|\mathrm{GS}\rangle \end{array} \tag{8.6}$$

Because there is an equal sign in Schrödinger's equation, we need to get the same results on the right-hand side. When we apply $\hat{KE}$ or $\hat{PE}$ individually, the results aren't so simple, which you likely anticipate because there are two large empty boxes in the equation above.

The kinetic-energy operator

Applying $\hat{KE}$ to the ground state wave function, after several lines of math, returns a not-so-simple function:

$$\hat{KE}\cdot|\mathrm{GS}\rangle = \frac{C}{2}\left(\frac{2}{r} - \frac{1}{a_0}\right)\cdot|\mathrm{GS}\rangle \tag{8.7}$$

The entire function that has appeared out front of the state in place of the operator are the numerical values of the kinetic-energy operator. In this case, the operator assigns a continuum of kinetic energies to a continuum of positions.

At this moment, the point of this is not to understand the function's details but to see that an abomination has appeared in the place where we'd prefer nice, neat, single-valued numbers. I will show you a plot of this function and others once we have finished with the operator algebra.

Some will want to know what the symbols mean, so they get a paragraph of explanation. The C is a number that I use to hide some fundamental constants that would muddy up the equation. The a_0 is a distance called *the Bohr radius* with a value of approximately 0.05 nm. It is also found by combining a few fundamental constants. We tend to put zeros in the subscripts of constants, as you see here with a_0, when we think the constant is important or fundamental in some way. The r is a variable that represents the distance from the infinitesimal proton. This means that there will be a continuum of values possible for the infinitude of distances that could be input into that function (except $r = 0$ because the function is undefined there).

Before moving on to the next operator, we can update the Schrödinger equation:

$$\begin{array}{ccccc} \hat{E} \cdot |\mathrm{GS}\rangle & = & \hat{KE} \cdot |\mathrm{GS}\rangle & + & \hat{PE} \cdot |\mathrm{GS}\rangle \\ \downarrow & & \downarrow & & \downarrow \\ -13.6\ \mathrm{eV} \cdot |\mathrm{GS}\rangle & = & \frac{C}{2}\left(\frac{2}{r} - \frac{1}{a_0}\right) \cdot |\mathrm{GS}\rangle & + & \boxed{\phantom{-C \cdot \frac{1}{r}}} \cdot |\mathrm{GS}\rangle \end{array} \tag{8.8}$$

The potential-energy operator

Applying $\hat{PE}$ is easy for me to do, and I hide no math when I show you the result of using the operator. The non-abstract form of the operator is the same as the electrostatic potential energy from first-year college physics of an infinitesimal charge (the electron) in the field of another infinitesimal charge (the proton). The operator is a continuous function:

$$\hat{PE} \cdot |\mathrm{GS}\rangle = -C \cdot \frac{1}{r} \cdot |\mathrm{GS}\rangle \tag{8.9}$$

This function assigns a potential energy to each position according to its distance r from the infinitesimal proton. The C is the same constant as in the previous equation that hides several fundamental constants.

We can substitute in the final function to see what we have found:

$$\begin{array}{ccccc} \hat{E} \cdot |\mathrm{GS}\rangle & = & \hat{KE} \cdot |\mathrm{GS}\rangle & + & \hat{PE} \cdot |\mathrm{GS}\rangle \\ \downarrow & & \downarrow & & \downarrow \\ -13.6\ \mathrm{eV} \cdot |\mathrm{GS}\rangle & = & \frac{C}{2}\left(\frac{2}{r} - \frac{1}{a_0}\right) \cdot |\mathrm{GS}\rangle & + & \left(-C \cdot \frac{1}{r}\right) \cdot |\mathrm{GS}\rangle \end{array} \tag{8.10}$$

The Schrödinger equation is about energy balance, and it's difficult to see how the left and right sides are equal. We need to simplify it.

The Hamiltonian operator

By combining the two troublesome functions, the two terms with r's cancel each other out. The remaining fraction has only constants. And when I plug in the values for those constants and work out the fraction, I get the energy. So everything turns out well in the end:

$$\left(\frac{C}{r} - \frac{C}{2 \cdot a_0} - \frac{C}{r}\right) \cdot |\text{GS}\rangle = -\frac{C}{2 \cdot a_0} \cdot |\text{GS}\rangle = -13.6\ \text{eV} \cdot |\text{GS}\rangle \quad (8.11)$$

This is the right-hand side of our Schrödinger equation, and it gives the same eigenvalue as the left-hand side.

You can see that, when taken together, the potential and kinetic-energy operators return an eigenvalue. Since together these operators are the Hamiltonian operator $\hat{H}$, you know that when $\hat{E}$ has an eigenfunction that returns an eigenvalue (in this case, they are the eigenfunction $|\text{GS}\rangle$ and the eigenvalue -13.6 eV), you know that $\hat{H}$ will have the same eigenfunction and eigenvalue. It works vice versa as well. And there, the balance is seen between the operators $\hat{E}$ and $\hat{H}$ acting on the ket in:

$$\hat{E} \cdot |\Psi\rangle = \hat{H} \cdot |\Psi\rangle \quad (8.12)$$

Therefore, we see that Schrödinger's equation is a law that says the energies associated with the wave function must balance, even if the intermediate steps seem to suggest we will have difficulties.

The predicted energies

Have a look

As promised, I now show you, in Fig. 8.2, the plots of the values returned by the various energy operators. It is easy to get confused by

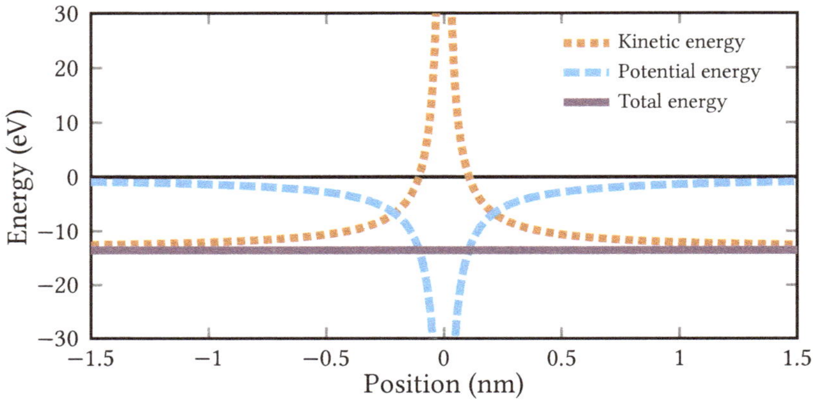

Figure 8.2: The numerical values returned by application of the energy operators when acting on the ground state of hydrogen.

all the curves that I plot throughout the book. I want to be clear that the curves in the figure here are not wave functions. Wave functions have quantum amplitudes on their vertical axes. The curves here are the energies at each point in space of the ground-state wave function of hydrogen. They are the expressions in front of the kets in Eq. 8.10.

Within axiomatic quantum mechanics, the numerical values returned by operators corresponding to observables have further significance. The single-measurement postulate says that they are the physical quantities that are predicted to be measured in an experiment. Looking at the predicted energies, we see the potential energy is always negative. There's no problem with that. We are allowed to shift the zero point of potential energy because we extract energy from differences in potential energy, and an overall shift won't affect that.

The set of values for the total energy is found by adding together the PE and KE functions. This set of energies is the constant value of −13.6 eV. There is also no problem with this being negative. With our convention, a negative value for total energy tells us that the ground state of hydrogen is a bound system (the system is an electron and proton). However, the KE values should give you pause. There is a conceptual difficulty. Why can the KE values be negative?

Negative kinetic energy?

In classical mechanics, kinetic energy depends on relative speeds v and can't be negative because of the v^2 in the equation for KE. I will tell you a story of a bullet to give an alternative way of seeing that negative KEs are not possible. If you stand on the ground and a bullet is fired at your leg, the bullet would get lodged in your leg and dissipate some large amount of kinetic energy, causing damage to your leg. If, however, you are fast like a superhero, you can run away from the bullet at nearly the same speed. Let's say you are 1 mm per second slower than the bullet. Eventually, the bullet will catch up to you and tap you very gently. A tiny amount of kinetic energy is dissipated.

If you match the bullet's speed, it will never catch you. To your eyes, the bullet is in the same spot relative to you and is at rest, while the whole world appears to fly by in the opposite direction to your travel. If something is at rest with respect to you, then it has no kinetic energy with respect to you.

Now, if you go faster than the bullet, its kinetic energy does not continue this downward trend in energy and *does not* become negative with respect to you. Let's say you weren't pushing your superpowers, but now you go ultrafast. You circle the pale blue dot that we call Earth. You come up behind the bullet and collide with it. Assuming the bullet is lodged in your leg, you will dissipate the kinetic energy inside you. That energy is, of course, positive. We can see, in the classical realm, there is no place for negative kinetic energies.

But quantum mechanics is weird, right?

The single-measurement postulate says the result of a measurement is one of the values returned by an observable's operator. Therefore, in our scenario, it should be possible to measure a particle with negative kinetic energy. That is obviously bonkers. How could an instrument register negative energy? How would an engineer use the math of classical physics — which doesn't allow for negative KE — to design such an instrument?

Few textbooks admit to the bonkers result, even though it is the inevitable conclusion. In a true axiomatic system, once you accept the postulates, you can only move forward via rigid logic. There is no room to maneuver. There are no opinions involved. New truths follow from the postulates. (Well, technically, there are no opinions only if we are all working with the same *unstated notions* in the axiomatic system. More on those in the final chapter.)

I reject some of the textbook postulates as they are presented because they don't make sense and it's obvious to me that measurements aren't a part of a quantum theory at the foundational level. My interpretation has a simple way of understanding the energies in a quantum system (simple if you know the math). I'd like to teach you further about the Schrödinger equation and wave functions first, so I'll wait until the *Spring cleaning* chapter to handle this. There, you will see that the set of numerical values returned by application of an operator are an intermediate result on the way to calculating an overall quantity.

Squaring a circle

It's not really necessary to precisely understand the projection postulate or the single-measurement postulate because they cause problems and aren't fully legitimate. Despite that statement, as is not uncommon in this book, I will explain more fully something I said earlier, namely the projection postulate.

At the beginning of the *Math vs. Measurement* chapter (p. 83), I described the projection postulate with the words "... into a new state or possibly leaves it in the same state. The state after a measurement has a precise correspondence with the value measured." Since you hadn't learned the Dirac notation at that point, the vagueness was necessary. Another description I used was "The projection postulate says that we can start with a particle in a state written as one, many, or an infinitude of kets, then a measurement is made and, randomly, the particle will be projected into only one of those kets." (p. 97). I then used mathematical examples of a position measurement projecting into a position state,a total-energy measurement projecting into a state with different total energy, and a spin measurement projecting into a spin state.

It's natural to not pay attention to the descriptions and get a feel for the rule from the examples. From them, you likely feel that the rule says that a negative kinetic energy measurement will project the quantum object into a state of negative kinetic energy, too. (Whatever that could possibly mean.) But that's not what the postulate says.

The postulate says that the quantum object is projected into the ket that is *at the location in the math where the numerical result was produced* that corresponds to the measurement result. When we find a continuum of values, the kets that made the continuum are labeled by the values of the horizontal axis when the continuum is plotted. In our case, those are position kets (recall the number line of kets in Eq. 6.8 on p. 97). Therefore, after a kinetic energy measurement that supposedly shows us a negative kinetic energy, the new state of the quantum object is described by a position ket that says that the electron is a point particle greater than 0.1 nm from the nucleus.

What, then, has the negative kinetic energy? I really don't know. The axiomatic system doesn't tell us how to make a measurement. Since it's not the quantum system of interest that has the negative KE, is it some quantum object we fired at the system of interest to make the measurement? That wouldn't make any more sense. Since it's obvious the axiomatic system won't give us a satisfactory answer, let's leave this behind.

There is even a further problem with the current axiomatic way of doing things. Notice that I plotted a continuous function for the total energy of the ground state of hydrogen, which I found via the addition of the numerical values that were returned after application of $\hat{KE}$ and $\hat{PE}$. But, I can find a supposed eigenvalue equation from the Hamiltonian operator, which is literally an identical operation:

$$\hat{H} \cdot |\mathrm{GS}\rangle \;=\; -13.6 \text{ eV} \cdot |\mathrm{GS}\rangle \tag{8.13}$$

A single value is not a continuous function, so it shouldn't be plotted as such if the eigenvalue equation is how we should view the situation. Nevertheless, everything leading up to the constant function in Fig. 8.2 legitimately used operator math, and when we add two functions, we get another function. Therefore, when we add the PE and KE functions, we find the total energy function, which is a constant function and not a single value. This then reveals the projection postulate has an ambiguity.

While this has no effect on the predicted total energy that is measured, because the prediction is −13.6 eV via both methods, should we choose one way over the other as the way to calculate results, it will affect the state the quantum object will be in after the measurement. In the eigenvalue case, the prediction is that the state after measurement will definitely be the ground state. In the continuum case, the prediction is that the new state will be a random position state (i.e., a point particle).

Although, what I have described here is never mentioned as an example of it, this is an example of what is known as the **preferred-basis problem**. It is part of the measurement problem. In summary, the problem is: how does a measurement know to pick between an energy ket and a position ket in this example?

Please don't ask me to have any of this make sense within the bounds of what is commonly taught to students. What I'm doing here is showing you the nonsense that arises from within those bounds. When the projection postulate or the single-measurement postulate are the mathematical tools with which we are working, eventually, we will find that we are trying to square a circle.

Finally, to be clear, when I look at the eigenvalue equation of the Hamiltonian operator in Eq. 8.13, the −13.6 eV is a constant-valued, continuous function in my mind, but I don't impart physicality to the function the way the single-measurement postulate does. I think some physicists are recklessly grafting their interpretation of quantum mechanics onto some math. That being said, so that you are comfortable with the common nomenclature outside of this book, I will continue to call the constant functions, such as −13.6 eV, eigenvalues.

9 The mysterious hieroglyphics of Schrödinger's equation —As a wave equation

> "*Historically, this achievement was considered to be one of the earliest signs that Schrodinger's wave mechanics was on the right track. We cannot get this kind of number by the purely algebraic (operator) method.*"
>
> — Sakurai and Napolitano on an approximation to helium's ground-state energy

The second mathematical aspect

In this chapter, we finalize our quest to expand the compact form of Schrodinger's equation and understand it. But, this time, we look to understand it as a wave equation and understand how it precisely specifies the time evolution of a wave function when uninterrupted by measurements/collapse of the wave function. It's essential to note both the equation's near universality and the restriction mentioned in the previous sentence. In most interpretations, Schrödinger's equation applies at all times except at the instant the state randomly collapses. However, all interpretations agree that when Schrödinger's equation is at work, there is no randomness. Hence, the state function is completely predictable when it is uninterrupted by collapse.

Schrödinger was inspired to find the equation after reading of the matter-wave hypothesis by de Broglie. We can't say Schrödinger's equation has been proven from more fundamental principles. Instead, it has been deduced as a law of the universe and represents a fundamental principle. In an axiomatic system, we say the time evolution of an uninterrupted state (whether it is a spatial wave function or a spin state) is governed by the **time-evolution postulate**, which in an introductory physics class could be informatively described as the *Schrödinger-equation postulate.* However, Feynman came up with a mathematically equivalent method, so, I think, that's why we have the more generic name of *time-evolution postulate.*

The previous chapter allowed you to understand that the operators show that energies balance in the Schrödinger equation:

$$\hat{E} \cdot |\Psi\rangle \;=\; \hat{H} \cdot |\Psi\rangle \tag{9.1}$$

But I hid the math of the operators. To understand how wave functions evolve, I must show you their non-abstract form and illustrate their meaning. Before we get to the quantum realm, it's best to use a classical equation as a stepping stone to understanding some of the math.

The classical wave equation

The classical counterpart to Schrödinger's equation is called the *Hamilton-Jacobi equation.* This classical equation does lead to solutions to wave problems. However, that equation is far too complicated to use as an example. I would guess that most physics majors don't even learn it.

Students instead learn about *the wave equation*, which has narrower applications. By learning this equation, students gain a first-approximation understanding of waves on a string, acoustic waves, water waves, etc. In the simplest situations, they all behave similarly.

We can use a guitar string as the object of our example. There will be energy losses to the air, the guitar body and the heating of the string, but we will ignore those and think of it as an idealized scenario. In broad terms, there will be a set amount of energy from plucking the string, which will constantly move between kinetic and elastic potential energy. The energy continuously changing its form is the reason the string vibrates.

Of course, even schoolchildren know a string will vibrate back and forth, but the wave equation can tell us the fine details. It describes how each position of the string changes or evolves with time.

A few ingredients go into the wave equation. Guitar players know that thinner strings and higher tension make higher notes. The wave equation has linear density (mass per length) and tension as set parameters to account for those. The character of the note emitted, the timbre, changes if you pluck at different locations along the string. This means that the initial waveform of the string matters.

The last and most important part is too difficult to motivate from your experience, so let's see the equation now. The wave equation for a guitar string is:

$$T \cdot \partial_x^2 \cdot A \;=\; \mu \cdot \partial_t^2 \cdot A \tag{9.2}$$

The tension in the resting string is represented by a T, and the μ (spelled: "mu"; pronounced: "mew") is the Greek version of lowercase m and here represents the linear density. Both of these are constant in a particular situation.

I've chosen to use A for the function that tells us the wave amplitude, as it is pretty standard to do that. It's how far each point of the string is from the rest position as it vibrates. Other common choices to represent the amplitude are f (for function), y (because the string moves in that direction), or u (I don't know why). Heck, it's not standard, but I could have used Ψ to emphasize its similarity to quantum wave functions.

Note the added complexity of considering a wave's behavior over calculating the motion of a single rigid object, such as the ball tossed in the air in the last chapter. To calculate the motion of a single non-rotating, rigid object (in environments with no air resistance or friction), we need to consider only one point at a time, the center of mass.

In contrast, we need to consider a continuum of points along a line when considering the motion of a guitar string because it is flexible.

I should also mention that physicists usually write the symbols for functions with more detail. Typically, they are written like $A(x,t)$ or $\Psi(x,t)$ to remind the reader that they represent many numbers. I generally don't write the symbols for functions like that to keep the equations looking simpler to you. Behind those symbols is a way to give a value (the string amplitude) at each point in time t and each point along the x-axis number line (location along the string).

A guitar pick plucking a string would make the initial waveform/amplitude look like a triangle before the pick releases the string. When the guitar pick loses contact, the string is momentarily at rest. Then, the wave equation takes over. In Fig. 9.1, I show the wave function at three different times: $A(x,t_0)$, $A(x,t_1)$, and $A(x,t_2)$.

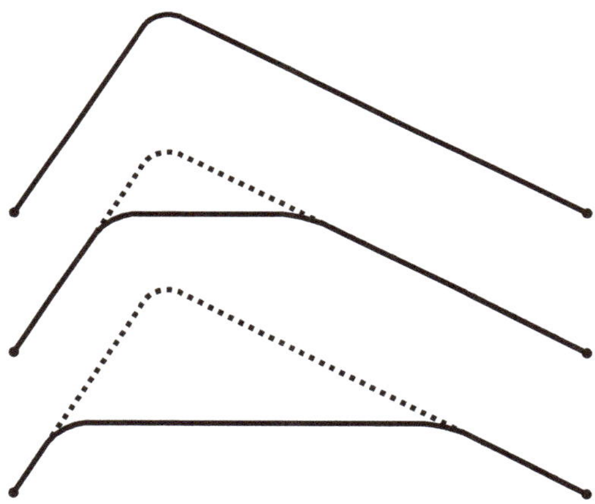

Figure 9.1: Curvature causes acceleration in a tense string. The guitar string's amplitude is greatly exaggerated. The top image is the plucked string at the moment of release. The two subsequent images show how it deforms at future times. Note that the string remains stationary until the curvature arrives at a point. Once the curvature leaves a point, the string continues traveling downwards at a constant speed.

Later in the chapter, you will see that the long and intimidating form of Schrödinger's equation has some similarities to the classical wave equation here. The most important commonality is that the symbols that look like lowercase curly *d*'s appear again. These are called *partial differential operators* or *partial derivative operators.*

Derivatives have a variable of interest as a subscript. The derivatives ask the question of the symbol to the right of it: "How much is this function changing when I increase this particular variable?" Put differently, it asks: "What is the rate of change of the function with respect to this variable?"

In the equation, you can see a "2" as a superscript, which has confused numerous generations of students. It does *not* mean to square the answer. The "2" means that it asks for the rate of a rate: "What is the rate of change of the rate of change with respect to this variable?"

That mathematical interpretation will be confusing to those uninitiated in calculus, so I will be specific and give you the physical interpretation for each derivative operator. The symbol ∂_x^2, when applied to the A, asks the question "At some point in time, what is the *curvature* of the string at each point along the string?" Whereas, the symbol ∂_t^2, when applied to the A, asks the question "At some point in time, what is the *acceleration* of the string at each point along the string?"

Let's make this easier to understand by ignoring the T and the μ in the equation. In that case, we have:

$$\partial_t^2 \cdot A \propto \partial_x^2 \cdot A. \tag{9.3}$$

You can always read a math expression out loud as English. In this case, it is quite simple. This expression is read as "the vertical acceleration of the string at a point is proportional to the curvature of the string at that same point." With this, the images of the guitar strings make sense.

If you want a feel for how strongly curved a function is at some point, first pretend that the function is a road map or an overhead view of a race track. Imagine traveling along the string at a constant speed of 100 km/h. Wherever you feel pushed strongly to one side or the other in your seat, you would say that is a tight corner. Tighter corners have higher curvature.

Using the constant-speed driving analogy as we look at Fig 9.1, you can see that we travel a straight road for a while and thus have no acceleration. Suddenly, we get to the kink in the string, which has a high curvature. That point of the string will have a high acceleration downwards. (By the way, technically, the string is not perfectly straight in the straight section.)

In Fig. 9.1, we jump forward in time to the second guitar-string image. It's not intuitive how the kink has split in two and leaves a flat portion, so let's just accept that. The important part to notice is that the string stays at rest until one of the two moving kinks reaches a point. The *curvature* of the kink causes *acceleration* of the string at that point, just like the guitar-string equation said it would.

Once the kink leaves that location, the string has been put into motion (but is not accelerating anymore) and will continue traveling downward at a constant speed.

With that brief introduction to the classical wave equation, we have seen how the symbols tell us how the wave will behave. With these useful ideas in hand, we are ready to move from the abstract symbols of the previous chapter to the symbols that appear in the intimidating form of Schrödinger's equation.

The explicit form of the operators

The "energy" operator

The operator that many think is inappropriately styled as the "energy operator" is:

$$\hat{E} = i \cdot \hbar \cdot \partial_t \tag{9.4}$$

There are two constants, i and $\hbar$. The imaginary unit is defined as $i = \sqrt{-1}$ (you could also say $i^2 = -1$ instead), which makes our math complex valued. You have seen Planck's reduced constant $\hbar$ before. It shows up pretty much everywhere in quantum mechanics.

The derivative symbol has a t that represents time. It does not have a "2" as a superscript, unlike the classical equation. When this derivative operates to the right, it finds the rate at which each point of the wave function changes with time.

You can see that there are no choices to be made here. For every physical situation, this operator will never change. But, because the wave functions differ for each new situation, this operator will return different results when it operates on them.

The kinetic-energy operator

We exist in a reality with three spatial dimensions, making the math three times as complicated as it needs to be when introducing someone to quantum mechanics. So, I will stick to one dimension along the x-direction for the following two operators.

The kinetic-energy operator is:

$$\hat{T} = \left(-\frac{\hbar^2}{2 \cdot m}\right) \cdot \partial_x^2 \tag{9.5}$$

I must come clean about the notation. Physicists rarely use the symbol $\hat{KE}$. I have been using it to make it easier for you to follow what I said. Physicists usually use $\hat{T}$ as the hieroglyphic for the kinetic-energy operator. The letter T has been used for centuries to represent kinetic energy, and it probably comes from *travail*, the French word for work. I show it to you here in case you read other people's work, but I will continue to use $\hat{KE}$ and $\hat{PE}$ in the majority of cases.

I also mentioned earlier the tension T of a string. There is no chance of confusing the two T symbols because tension doesn't exist in quantum mechanics and there is no hat on the one for tension. Reusing symbols is common in science, and one eventually gets accustomed to it.

Looking at the right-hand side, in the fraction, the only symbol I need to explain is m. It is the quantum object's mass. Therefore, as a whole, the factor in round brackets is a constant.

The most important part of the definition is the double derivative, which acts on the spatial part of the wave function. (The spin part of the wave function doesn't carry the kinetic energy. Nor does the spin part have the correct mathematical form for spatial derivatives.) The double derivative, like in the classical wave equation, asks what the curvature is at each point of the function. In the quantum case, the wave function has complex values for its amplitude, which makes the plots I show you later have two curves.

Once one has been doing math long enough, their brain starts chunking things for ease of processing. I guarantee that a physicist's brain sees only two chunks in this operator when their eyes glide over it. The first is "negative aitch-bar squared over two em". The constants get chunked together in our brains as a bit of a rhyme until we need to stick numbers into them. The second is the double derivative symbol: "del sub x squared". (Yes, I say "squared" in my head when I don't literally mean it.)

My tip to those of you who will go deeper into the math of quantum mechanics is to consciously make these rhymes as you go. No real effort is needed; just read it out loud. It eases the cognitive load when you re-encounter them.

The potential-energy operator

You are right if you think I will change the letters used for the potential-energy operator. For a physicist in a hurry, two uppercase letters in a symbol is one letter too many. The operator is usually written with a $\hat{U}$ or $\hat{V}$.

Like I said before, this operator is the special one. In introductions to quantum mechanics, the potential-energy operator is often restricted to the electrostatic potential energy, which are unchanging functions throughout space. (The V probably comes from voltage: when an electric charge is subject to a voltage difference, it has electrostatic PE.) When dealing with magnetism or time-varying potential energies, things are not so simple, so I won't bring those up.

In electrostatics, the potential energy depends only on the position. I have to give a vague definition for the operator because each situation is different:

$$\hat{V} = V(x) \tag{9.6}$$

That may seem rather redundant, but it's how we work. You should read this as "the details of the potential-energy operator $\hat{V}$, for the system we are thinking of, is some function $V(x)$ defined throughout all space." And, because we are working in one dimension for simplicity, all space in this case is along the x-axis. The details of $V(x)$, of course, need to be specified at some point.

You can now understand why most of us put hats on operators. It ensures no one gets confused by two Vs, two Ts, or two Es. Despite their usefulness, some physicists embrace chaos and want others to live in it, too, so they don't use the hats. They go to the same hell as theorists who set constants equal to one, like the speed of light and Plank's constant, so their calculations are easier, yet are too lazy to put them back in when they write academic papers. But I digress.

The Schrödinger equation in full

Finally, I can show you that in a compact form using a more traditional notation, the Schrödinger equation looks like:

$$\hat{E} \cdot |\Psi\rangle = \hat{T} \cdot |\Psi\rangle + \hat{V} \cdot |\Psi\rangle, \tag{9.7}$$

and in the intimidating form, in one dimension, it looks like:

$$i \cdot \hbar \cdot \partial_t \cdot |\Psi\rangle = \left(-\frac{\hbar^2}{2 \cdot m}\right) \cdot \partial_x^2 \cdot |\Psi\rangle + V(x) \cdot |\Psi\rangle \tag{9.8}$$

Remember that $|\Psi\rangle$ is some complicated function full of math symbols, and we only know the symbols once we crunch through the math. In fact, saying that is somewhat misleading. Only in a few cases can we find the exact symbols that represent the various wave functions. Often, we find written approximations of the wave function.

In other situations, we may have a computer find the values for the wave function, which often is a better approximation than can be found with written approximations.

As a final fun fact, I'll tell you that when physicists give conference talks where they've found some result for some quantum system, you won't see Schrödinger's equation. Everyone knows it and how it works, so they talk about the Hamiltonians they used to get their results.

A pause for some clarifying remarks

In these two chapters, I am showing you two mathematical aspects of the law discovered by Schrödinger. However, I'm doing that with one equation. You need to be of two minds when you think of the Schrödinger equation conceptually. You need to have the mathematical mind of calculus as well as the mathematical mind of kets, operators, and numerical values. The view depends on how we mentally group the mathematical symbols when we describe things.

Let's start with the energy operator $\hat{E}$ and an energy eigenstate $|\Psi\rangle$. We know that applying the operator will return an energy eigenvalue. However, its non-abstract form contains a partial derivative symbol ∂_t that also finds how quickly each point of the wave function changes. Looking at the following equation, you can see where each of those answers appears. The description above the equation relates to the previous chapter. Whereas the description below the equation will be what we examine in the rest of this chapter.

$$\hat{E} \cdot |\Psi\rangle \;=\; i \cdot \hbar \cdot \partial_t \cdot |\Psi\rangle \;=\; \overbrace{E}^{\text{energy operator's eigenvalue}} \cdot \underbrace{|\Psi\rangle}_{\substack{\text{how fast } |\Psi\rangle \text{ changes} \\ \text{at each point in space} \\ \text{multiplied by } i \text{ and } \hbar}} \tag{9.9}$$

By its nature, the $\hat{KE}$ operator will always return a non-constant, continuum of kinetic energy values. Additionally, inside the non-abstract operator is the symbol ∂_x^2 that finds the curvature of the wave function:

$$\hat{KE} \cdot |\Psi\rangle = \left(-\frac{\hbar^2}{2m}\right) \cdot \partial_x^2 \cdot |\Psi\rangle = \overbrace{\clubsuit}^{\text{kinetic-energy operator's continuum of values}} \cdot |\Psi\rangle \tag{9.10}$$

kinetic-energy operator's continuum of values: ♣ · |Ψ⟩ = curvature of $|\Psi\rangle$ at each point in space multiplied by $-\hbar^2/(2m)$

I used the ♣ symbol for the kinetic energy's values to make it stand out and be more memorable. In my opinion, using many different Latin letters for functions gets confusing for novices.

The interpretation of $\hat{PE}$ is the most straightforward. Because we declare the function $V(x)$, we don't need to use any more playing card suits. However, in a later chapter, I need to use the letter V for volume in the same equation as $V(x)$, so let's say $V(x)$ is sometimes instead represented by a heart.

$$\hat{PE} \cdot |\Psi\rangle = V(x) \cdot |\Psi\rangle = \overbrace{\heartsuit}^{\text{potential-energy operator's continuum of values}} \cdot |\Psi\rangle \tag{9.11}$$

It's just two functions multiplied together

With the different mathematical views clarified, I now show you how the time evolution of waves are calculated.

How does a wave change?

Ocean waves

Imagine a nice, gently rolling ocean wave in the deep water that a surfer might start on. Picture its cross-section traveling towards the shore at one moment and another cross-section a split second later.

We will consider how we might describe that mathematically. Instinctually, many of us probably feel that the mass of water moves horizontally as a car's mass does. As long as you are picturing gentle waves in deep water, water isn't like that. The mass of water is not getting pushed horizontally at the speed of the waves.

If the surfer is not trying to surf, they are bobbing up and down as the waves pass. What is happening from one moment to the next is that the wave's leading edge rises a bit and its trailing edge lowers a bit. That's like making a new wave from the original wave plus a little change (see Fig. 9.2). Using Ψ for the amplitude, or height, of the ocean wave, we can write that as:

$$\Psi_{\text{new}} = \Psi + \Delta\Psi \tag{9.12}$$

The Greek capital delta Δ is used in science to represent a change. The $\Delta\Psi$ represents one thing. The two Greek letters together are one symbol. It is also a function, so it will have different values at each point. The values of $\Delta\Psi$ will be small because we are considering a small change in time. They will be positive where the new leading edge will be and negative where the lagging edge will be.

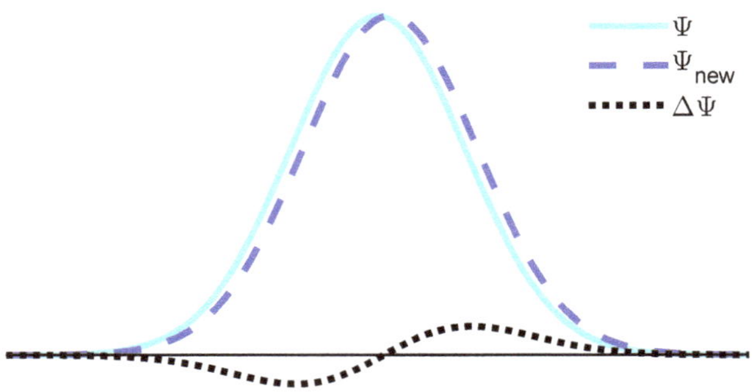

Figure 9.2: A cross-section of an ocean wave moving from left to right. The thin, horizontal line shows the sea level when it is undisturbed. The cyan, solid line shows the initial wave shape. The blue, dashed line shows the wave a fraction of a second later. The black, dotted line shows the difference between the two waves. Said differently, the dotted line shows how much the initial wave must change to become the new wave.

Quantum waves

Now, we switch to quantum mechanics and use Dirac's notation. We will say the starting quantum wave function $|\Psi\rangle$ is given. We will then find a later wave function $|\Psi_{\text{new}}\rangle$ by examining the parts of the Schrödinger equation. The idea of this section is similar to the ocean example. In this case:

$$|\Psi_{\text{new}}\rangle \ = \ |\Psi\rangle \ + \ \Delta\Psi. \tag{9.13}$$

The $\Delta\Psi$ is a function, but it is not a state, so it can't have the ket symbol around it.

We want to find the wave function on the left side of this equation. We have the starting wave function, so we only need to find $\Delta\Psi$. I will show we can find it approximately from Schrödinger's equation. (If you'd rather not get into the details, you can make a short quantum jump to Eq. 9.15 now.)

To find an equation for $\Delta\Psi$, we make two changes in the Schrödinger equation. First, we will work in computer-like time steps, so I will make an approximation. I can make the following substitution of symbols:

$$\partial_t \cdot |\Psi\rangle \ \approx \ \frac{\Delta\Psi}{\Delta t} \tag{9.14}$$

There's nothing quantum about the change; it's just a mathematical approximation. You can see the approximation is acceptable because the rate at which something changes in time (left side) is the amount of change per some duration of time (right side).

The left side of Schrödinger's equation becomes $\hat{E}{\cdot}|\Psi\rangle \approx i{\cdot}\hbar{\cdot}\Delta\Psi/\Delta t$. The symbol we are looking for is in the numerator. We additionally have the size of the time step Δt, which we can choose the value of. When the time step is smaller, we get a more accurate $|\Psi_{\text{new}}\rangle$.

The second change is to get $\Delta\Psi$ on one side of the Schrödinger equation. So, we move the i, $\hbar$, and Δt to the right side of Schrödinger's equation. And the math is done:

$$\Delta\Psi \ = \ \left(\frac{-i \cdot \Delta t}{\hbar}\right) \cdot \left[\hat{KE} \cdot |\Psi\rangle \ + \ \hat{PE} \cdot |\Psi\rangle\right] \tag{9.15}$$

If you have rigorously followed the algebra and are unaccustomed to imaginary numbers, you might think there is a mistake with the $-i$. I assure you it is correct because $-i = 1/i$.

The right-hand side is not too hard to understand. The three items in the fraction are constants. The two terms in the square brackets are the concepts labeled by the teal-colored lower braces in the previous section (i.e., curvature and the product of two functions).

To find the new wave function, we need to use the equations for $|\Psi_{new}\rangle$ and $\Delta\Psi$. (Eq. 9.13 and 9.15). Since I will show it graphically in two steps, we will consider this as a two-step process:

> **Step 1:** Find (a) the curvature of the current wave function and (b) the product of $V(x)$ and the wave function.
>
> **Step 2:** Add (a) and (b) from Step 1, and then multiply by some constants. This is how much the wave function changes. Add this change to the initial wave function to get the new wave function.

The steps are then repeated to find the function at the next point in time. That's enough algebra for now. Next, I will show you how this works.

Examining the equation graphically

The setup

Let's imagine that, by some clever process, we can make an electron in the shape of a Gaussian function and then set it free. There will also be no potential energy (i.e., $V(x) = 0$) and no initial motion of the center of mass. As always, we are going to work in a one-dimensional universe. We wonder, "what would the wave function look like as time goes on?"

Without potential energy, we are finding a simpler change:

$$\Delta\Psi = \left(\frac{-i \cdot \Delta t}{\hbar}\right) \cdot \hat{KE} \cdot |\Psi\rangle \tag{9.16}$$

The change at each point will be the curvature at that point scaled by some constants and made imaginary.

Before we get to the change, it's best to look at the wave function at the instant of release from whatever magneto-optical trap that had contained it. I finally must deal with what I have been avoiding for several chapters. I must show you a complex-valued wave function because we are dealing with time.

A mini-lesson on complex numbers

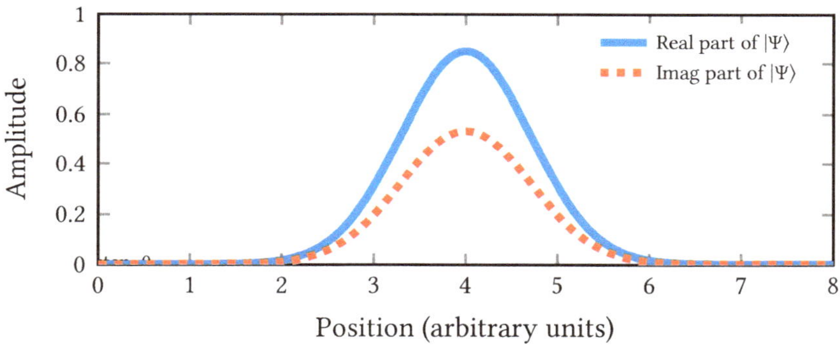

Figure 9.3: A Gaussian-shaped, starting wave function showing the real and imaginary values. The horizontal axis shows the position in space (in some arbitrary distance units). The vertical axis shows the quantum amplitude of the wave function. Note: The amplitudes on the vertical scale are not accurate for an actual quantum wave. Because of conservation laws, the matter density should be normalized, which would make the wave function amplitudes smaller in this case. The proper values are unimportant because we are concentrating on the shapes, so I use simple numbers.

In the figure above, I show you a wave function that has a Gaussian shape in space. You can see that to display the complex-valued nature, I had to plot both the real and imaginary parts. This is because, at every point in space, the wave function has a value represented by a complex number that can be written: $a + i \cdot b$. Both a and b are real numbers. We say a is the real part. And because the imaginary unit i

is beside it, we call b the imaginary part. Had I made a finer mesh of numbers on the vertical axis of Fig. 9.3, you could have figured that at $x = 4$ the wave function has a quantum mechanical amplitude of $0.85 + i \cdot 0.53$. Near $x = 7$, it is close to $0 + i \cdot 0$ but not exactly 0.

For simplicity, I gave it a constant phase function through all space, which means the real and imaginary parts have a constant ratio across the function. A constant spatial phase makes one of the curves look like a smaller version of the other and possibly flipped negative. In most situations, the real and imaginary parts don't look like little siblings, as you will soon see. (By the way, I'll say more about the phase of wave functions and the phase of complex numbers when we get to the next chapter on atomic orbitals.)

Soon, I will illustrate how the curvature and the constants (recall Eq. 9.16) determine the time evolution of the wave function. To understand that, you must know what happens when a complex number $a + i \cdot b$ is multiplied by i. When that occurs, you will get a new complex number where the original real part becomes imaginary and the original imaginary part becomes real (with its sign flipped because $i^2 = -1$). In other words, the new complex number is $-b + i \cdot a$. Keep it in your mind that multiplication by i switches the real and imaginary parts.

Start simple

Because this is a thought experiment, I don't need to settle for any random phase. I could choose the easiest phase to start with. It so happens that, twice in the cycle of the 360 degrees of temporal phase, the imaginary part of the wave function is zero, leaving the entire wave function to be real. Starting with a completely real wave function should make the first steps easy for you.

Be warned that I will play fast and loose with the constants' values in the Schrödinger equation, except for the negative signs and the i. The values I choose would belong to some fictitious universe. Additionally, my constants will cause large changes. To keep things accurate, a computational physicist normally wants the changes to be minuscule from one time step to the next. Instead, I want the large changes so that you can see the change and so that I don't have to make so many figures. I can do this because you are learning the principles and the exact values do not matter.

Wave evolution in computer-like time steps

In the figure that follows, I show you the two-step procedure to get from a starting wave function to the following snapshot of the wave function Δt seconds later. The steps are repeated to give a few more snapshots. Each row is a Δt later than the one before it.

The first column is the wave function, and the second is the complex curvature. We start at the first row with the purely real Gaussian wave function. Since it is purely real, the curvature will also be purely real.

The curvature in the top row will be scaled down and multiplied by the imaginary unit. It is then added to the original wave function. (In case you are wondering why the curvature isn't flipped upside down by the negative sign in Eq. 9.16, have a look at Eq. 9.10 for the other negative sign that cancels that flip.)

In the second row, you can see that the imaginary part that was initially zero everywhere now has the shape of the curvature of the real part. This is an interesting result. Time evolution guarantees that, even if one starts with a purely real wave function, a complex-valued wave function will exist in the next instantaneous moment.

Notice the real part decreased slightly going from the first to the second row. You should be wondering why that happened because the imaginary curvature is zero in the first row and contributes nothing to the real part in the second row. You should expect the real part to not change from what I have described so far.

Going from the first to the second row, I have done a *correcting normalization*. Doing things in computer-like time steps causes errors to accumulate. Assuming that the starting wave function was properly normalized (I was fast and loose with the constants, so it wasn't actually normalized in the quantum-mechanical sense), every

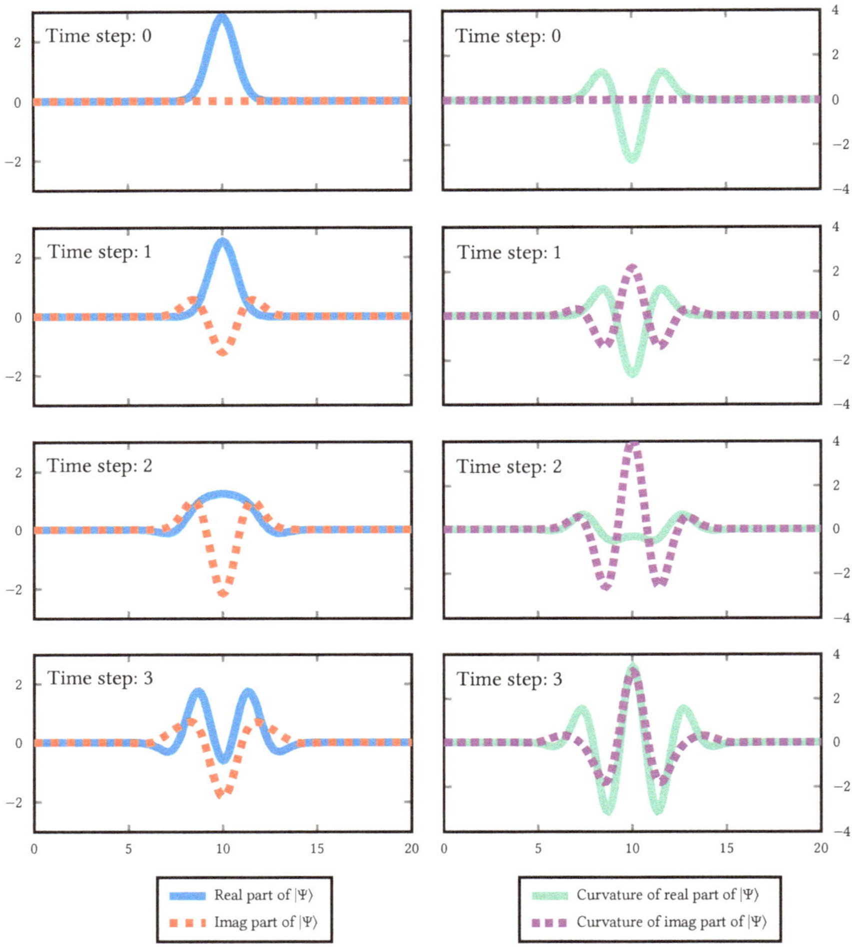

Figure 9.4: How the law of Schrödinger's equation tells the wave function to evolve. Start with the initial wave function in the top left plot. Its curvature is shown in the top right plot. Note how there is positive and negative curvature and that the sign depends on "which direction you accelerate towards while driving". The equation for the change in the wave function is found using the curvature. The first image in the second row is found by adding the equation for change to the initial wave function. This procedure is repeated to make the remainder of the plots.

time step, the wave function violates conservation laws by changing the mass, charge, and probability. And, to illustrate the idea of wave evolution to you, I am taking substantial time steps. These time steps cause large errors. So, I fix each step by multiplying the whole wave function by a correcting ratio.

Time evolution: a geometric point

When discussing the projection postulate, I promised to explain later why Schrödinger's equation says a geometric point can't be a wave function. I reproduce the pertinent image below, which represents a point in time immediately after the collapse of the wave function to a geometric point. The wave function is exactly zero everywhere except at one point where the quantum mechanical amplitude is infinite.

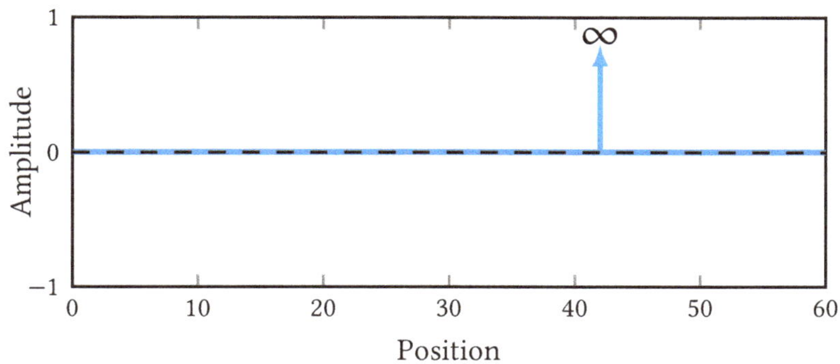

Figure 9.5: The textbook projection postulate. A measurement apparatus registered a position of 42 cm, so the wave function is purportedly a geometric point with an infinite quantum amplitude.

After learning about curvature in the Schrödinger equation, you have all you need to reason why this wave function cannot exist. I'll let you know it doesn't matter what the potential energy $V(x)$ is because the quantum amplitudes are zero and infinity. So, let's say it's

$V(x) = 0$ everywhere to keep things cognitively simple. After a time step of Δt, what will the wave function look like?

Stuck? Did you try to figure out the curvature using the road map analogy? The curvature is zero everywhere except at the one point, so the points where the wave function is zero cannot change the wave function. Additionally, because infinity is not a number (it is a concept), you can't calculate the curvature at the one point where this "function" is infinite. The curvature is undefined there. (If you know calculus, you'll also recognize even if it were finite at that one point, curvature doesn't exist at a discontinuity.)

After the supposed projection to a zero-dimensional point, how can the Schrödinger equation cause the wave function spike to spread out again to behave like a normal wave function? With no curvature, there is no time evolution. According to this analysis, a wave function can't be a geometric point.

This leads me to tell you that the "function" in the figure above does not satisfy the mathematical requirements to be a traditional function. We must input traditional functions into Schrödinger's equation to get a result. Therefore, the textbook projection postulate conflicts with the Schrödinger equation. Consequently, we must reject or modify at least one of them. My suggestion is to not bet against the Schrödinger equation.

Accurate time evolution of wave functions

A different path to the same result

With everything I said up to this point in the chapter, you now know how the Schrödinger equation causes wave evolution. However, you haven't seen what it accurately looks like because I was making the explanation simple. (Yes, that was as simple as I could make the math.) The simplicity of computer-like time steps cost us accuracy.

Going forward in this chapter, I will use a different mathematical formulation in the background because it gives more accurate graphs.

The math and the traditional pictures associated with it, a bunch of squiggly lines that start and end at the same two points, will remain hidden because I think they are not so useful in this book. Besides that, if I explained another equation, I would have to include another Nobel-prize-winning name in the title of the chapter.

This second math formulation still uses the same law of the universe that Schrödinger's equation represents. The math I will be using is known as **Feynman's sum-over-paths formulation**, **Feynman's path integral formulation**, or, less commonly, **Feynman's sum-over-histories formulation**. You could say that I am deviating from the hieroglyphics of Schrödinger and am now using the cuneiform of Feynman in the background. In other words, I don't show you new and (much more) complicated math; I just show you the results (i.e., the wave functions and matter densities) in some images.

In the context of Feynman's path integrals, you will hear people say "the particle takes all paths." No matter what it is called or how someone describes it, the results are equivalent to the results of Schrödinger's equation. It's the same law of the universe under a different name. What they identify as a point particle and a path are what we call a *dummy variable* and its integration over a path. A dummy variable is a temporary variable that disappears after calculating an integral. I want you to be aware that a path integral is an *intermediate* step on the way to calculating a single quantum mechanical amplitude for a single location in space.

Repeating the amplitude calculation for each point in space will build up a wave function. Repeating the wave function calculation for various times will give a series of snapshots of the wave function, which is what you will see below.

If you've heard before of "the particle takes all paths" nonsense, please don't think that there exists an infinitude of point particles that all have the premonitional ability to know and coordinate among themselves where to go in order to sample all the available space to calculate a number for a single point in space.

The pictures

In most situations, the path-integral formulation is a challenging calculation. However, I have chosen two simple situations to end the chapter. The first is a repeat of the Gaussian wave function that has been set free. In the second, I add a potential energy function. The accurate evolution of a freed Gaussian wave function is shown in Fig. 9.6 on p. 170. The matter density $|\Psi|^2$ for this same situation is in Fig. 9.7 on p. 171.

Dispersion

An important point to make is that time evolution makes the wave function **disperse** outward. In the matter density images, it's easy to see that the electron will eventually be smeared out into almost nothingness from its original, well-localized peak. Dispersion will make a few more appearances at important points in later chapters.

Some may think this dispersion of the wave packet into almost nothingness is the least appealing part of the reality interpretation. If they are to believe the wave function is a part of reality, they require it to always be a localized wave packet so that it matches their intuition about classical particles. However, I suggest that if one subscribes to the collapse of the wave function hypothesis, then dispersion shouldn't cause too much cognitive dissonance. Whenever or wherever collapse occurs, it brings the wave function back into a compact volume.

Time evolution: now with potential energy

To further your understanding of wave function evolution, I need to stop setting the potential-energy operator $V(x)$ to zero. The simplest case I can think of is a linear ramp. It is also a physical situation. This would occur in a vacuum between two metal plates with a voltage difference across them. To be specific, the ramp starts at zero at $x = 0$ and decreases (is negative) to the right. Stated symbolically, it is

$V(x) = -F \cdot x$. The potential energy represents a constant force F to the right on the wave function.

For a moment, let's return to thinking about $\Delta\Psi$ from the Schrödinger equation method of time evolution (Eq. 9.15). The mathematical contribution of $V(x) \cdot |\Psi\rangle$ to $\Delta\Psi$ is simple to understand. Considering $|\Psi\rangle$ to be real valued or complex valued, the contribution at each point in space is the height or depth of $V(x)$ multiplied by the amplitude of $|\Psi\rangle$. A linear ramp is lopsided, so the product will increase more on one side than on the other. Of course, we can't forget the curvature also contributes. The sum of the two is scaled by constants of nature and multiplied by the imaginary unit. The lopsided contribution leads to the wave function getting pushed in the direction of increasing wave function.

In Figs. 9.8 and 9.9 on p. 172 and 173, a Gaussian wave function is released at the same time a linear potential energy ramp is applied. The potential energy decreases from left to right. The wave function will gain energy and momentum, similar to a skier going down hill.

First, compare the matter densities in the two examples. They are the same, except the matter density travels to the right in the present example. You can also see that the constant force is causing the center of mass to accelerate (moving an increasing distance from one image to the next). Turning to a comparison of the wave functions, you can see that many more wiggles are accumulated in the present example than in the previous example. To be honest, I am in awe of how the wave functions look so different, but the matter densities are the same up to a shift in position.

The pictures in the next chapter

As I promised earlier in the book (p. 44), the next chapter will show you three-dimensional wave functions. As you might expect, there will be an emphasis on three-dimensional matter densities as well. Specifically, the wave functions and matter densities are of hydrogen's stationary states.

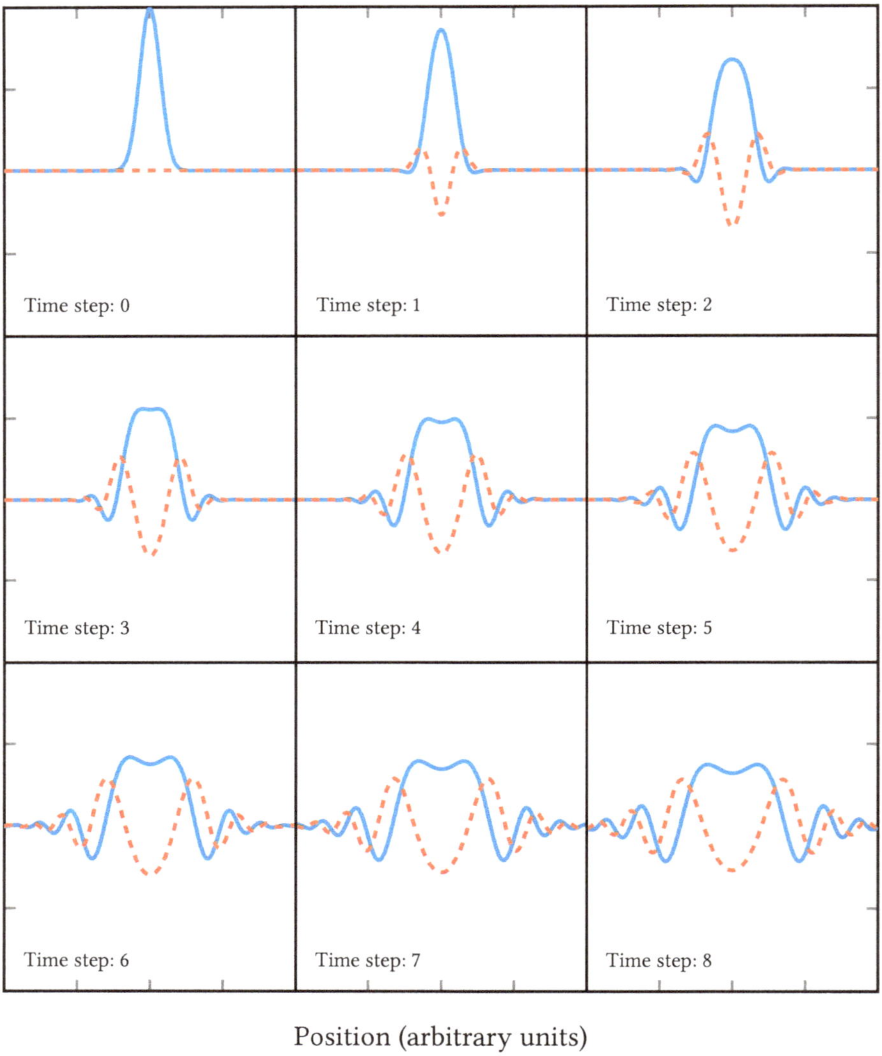

Figure 9.6: Wave evolution of a freed Gaussian wave function. In contrast with Fig. 9.4, this image shows the accurate time evolution of the freed wave function. Notice the general shapes of the two curves in the last several time steps. As time goes on, the shapes in the center remain but continuously get wider, while more oscillations appear at the wings.

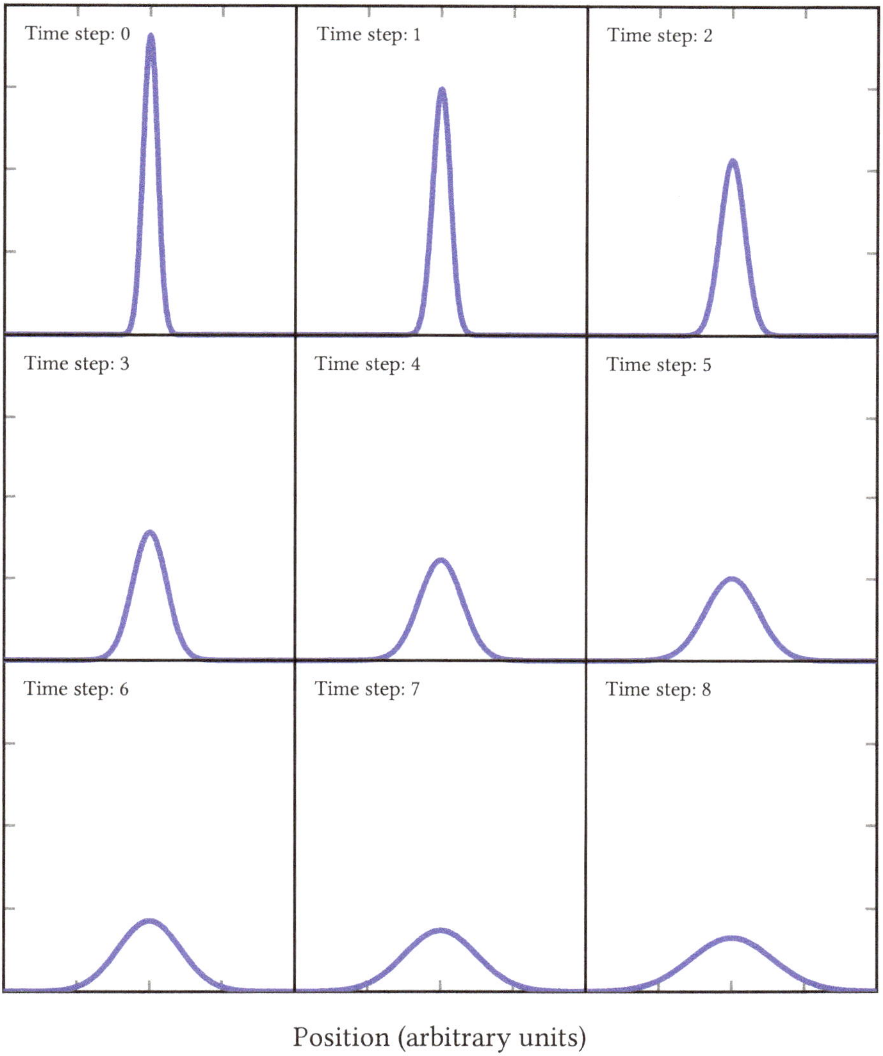

Figure 9.7: The matter-density evolution of a freed Gaussian wave function. These snapshots correspond exactly to the snapshots in the previous figure.

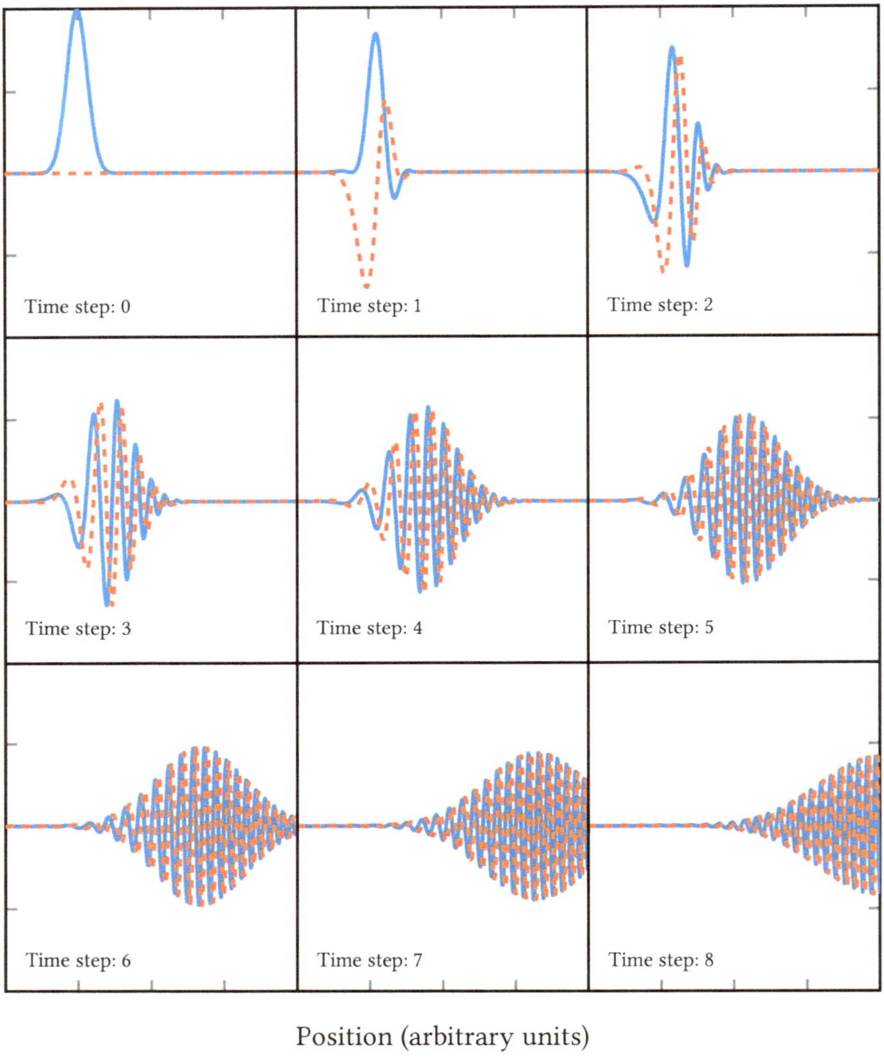

Figure 9.8: Wave function of a Gaussian wave function subject to a constant force.

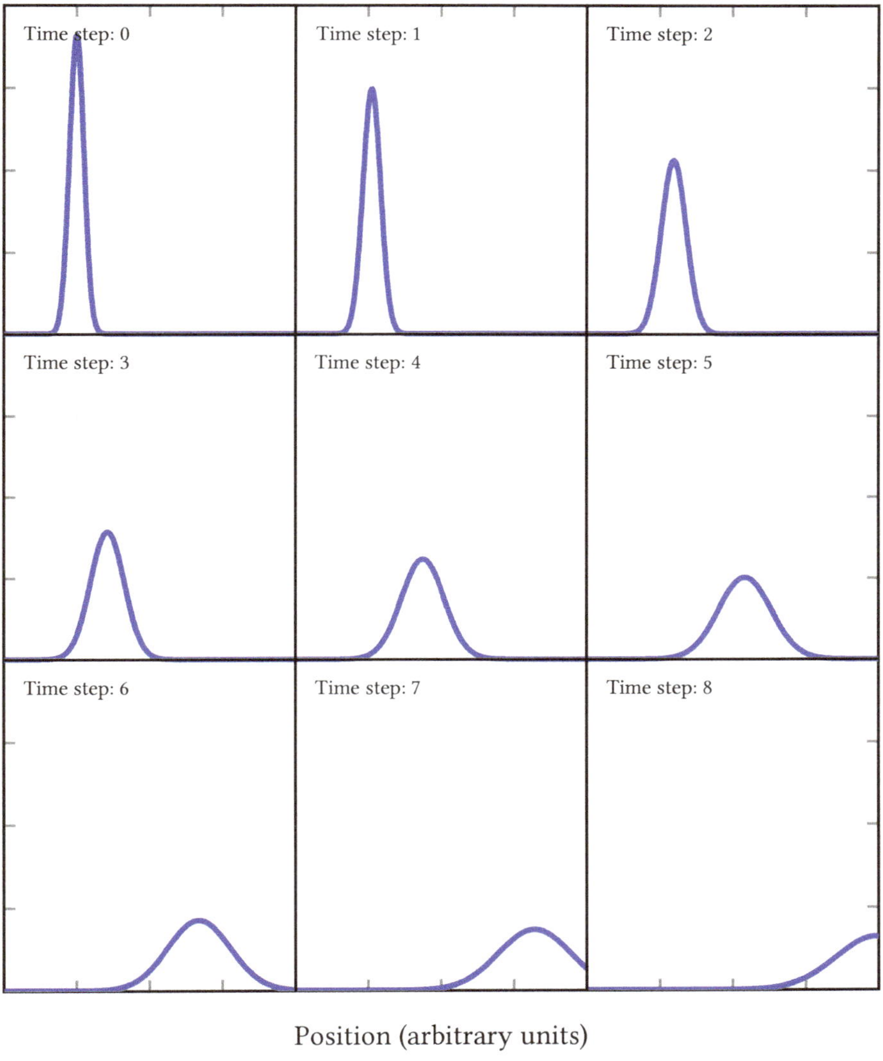

Figure 9.9: Matter/probability density of a Gaussian wave function subject to a constant force.

The calculations used to make the images in the next chapter are different than what you have seen in this chapter because there is no need to calculate amplitudes in computer-like time steps. Instead, the equations for the wave functions of hydrogen's stationary states are found by pen and paper methods. (On good authority, I am told that Schrödinger didn't have a personal computer in 1926).

Being able to write down an accurate wave function for a realistic scenario with a finite number of variables and constants (aka, a *closed-form equation*) is a rare situation. As soon as we add a second electron, we must resort to approximations, via pen and paper or via computers. There is no other element in the periodic table that has a closed-form equation of its electronic wave function. This is the reason that the *Gallery* chapter only shows hydrogen orbitals and the reason that the previous chapter used the example of ground-state of hydrogen to show the energy balance in the Schrödinger equation: I, like most physicists, don't have the skill set to model the wave function of, say, a helium atom.

10 Gallery of atomic orbitals

> *A formula is worth a thousand pictures.*
>
> — Edsger Dijkstra, computer scientist

Labels for atomic orbitals

In the first chapter, I showed you the matter density of the $3s$ orbital and one of the $3p$ orbitals without much explanation. If you haven't seen them before this book, this chapter is for you. Take note that, because it is difficult to accurately represent functions that are both three-dimensional and complex valued, I will primarily be showing you the real-valued matter density. After going through the matter density of the first four sub-shell types (s, p, d, and f), I will show the complex-valued wave functions.

Before reading much further, you must understand how orbitals are labeled by quantum numbers and how the number of orbitals proliferates with higher quantum numbers. If the review of quantum numbers that follows after this does not make perfect sense to you, then you need to read Appendix A again.

The **principal quantum number** n tells us both the energy of the orbital and the spread of the matter density in space relative to other orbitals. Let's euphemistically call this degree of spreading the orbital's size.

The **secondary quantum number** l groups the orbitals into similar shapes. In this chapter, I choose to group the orbitals by l so that you can more readily see the patterns in the orbitals.

The next quantum number m_l tells us the actual amount of orbital angular momentum (OAM) there is. It is commonly called the **orbital-magnetic quantum number**. This name makes some sense if you recall that with angular momentum and electric charge comes a magnetic field.

Let's consider m_l in terms of the three different p orbitals ($l = 1$ orbitals). An orbital with $m_l = 1$ will have $\hbar$ of angular momentum with its vector pointing in a particular direction. An orbital with $m_l = -1$ will have $\hbar$ of angular momentum that points in the opposite direction. Finally, $m_l = 0$ tells us that the orbital has no orbital angular momentum and no magnetic field arising from the shape of the orbital.

There is another quantum number m_s, which represents the **spin-magnetic quantum number**. It is commonly described alongside the three other quantum numbers. However, there is no need for it in this chapter.

In summary, we need three quantum numbers (n, l, and m_l) to label a particular hydrogen orbital. This label refers to a wave function and describes the spatial part of the state. If we had included the spin-magnetic quantum number, we would have fully labeled a particular state of a hydrogen atom.

Gallery of *s* orbitals ($l = 0$)

Images: clouds and slices

When trying to understand something complicated, rarely will one view of the subject be sufficient. To better understand the electron density of atomic orbitals, I give you three different ways of viewing them.

Looking at Fig. 10.1, you can see the first two ways. On the left are the clouds of electron density you've seen once before. To make one, I sliced up the charge density distribution and added the slices together. As I mentioned in the first chapter, you can imagine that you are looking through a translucent cloud of matter.

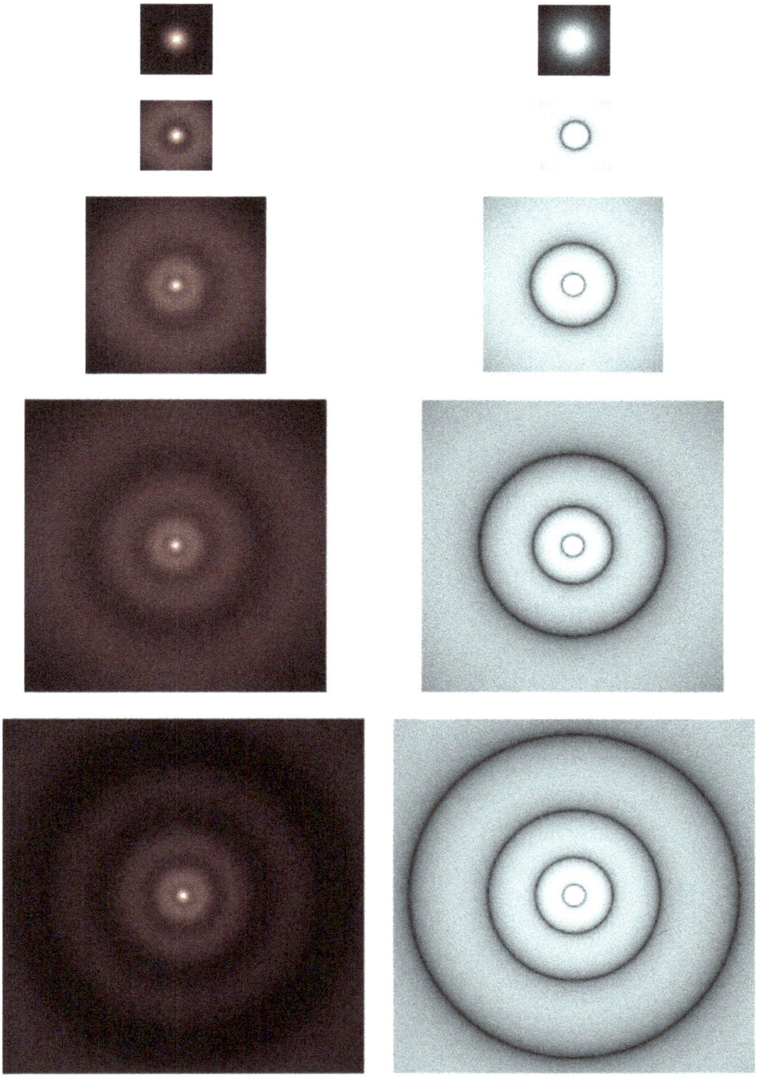

Figure 10.1: Cloud images and nodal slices of s orbitals. Images are of matter/charge/mass/spin/probability density. Going down the image, n increases from 1 to 5. The proton density is not included, but would be a minuscule dot at the center of each image. For all images of these types, the center is the origin of the coordinate system. The z-axis is vertical, the x-axis is horizontal, and y is in/out of the page.

On the right, I have taken a single slice through the center of the atom. I then took the logarithm of the density and used a different color map. Taking the logarithm doesn't make a different physical quantity. However, it does make the regions of near-zero density pop out as black lines and curves.

As you move down the picture, the principal quantum number goes from $n = 1$ to $n = 5$. Having read the appendix, you know that an *s* orbital is labeled with $l = 0$ and that this value tells us that m_l can only be zero for the *s* orbitals.

All squares are on the same length scale, so you will notice that as n increases, the density of the electron cloud is distributed farther from the center. Going down the figure, the square lengths are approximately 0.5, 0.5, 1.6, 2.6, and 3.1 nm.

Be warned that the color intensity in an image is only appropriate for that image. You can't take the lightest spot in one image, compare it to the same lightness in another, and then think it's the same density. That would be incorrect because the color map is scaled to emphasize the density in an individual image. For example, in the bottom left image of the figure, the electron density seems to have darker circles. This is because I cut this final image tighter than the others to fit it on the page, and as such, I cut off a significant fraction of the electron density from the calculation.

Nodal surfaces

To understand the logarithmic images on the right, imagine an axis running from the square's top center to the square's bottom center. This is the z-axis. Revolve the image about that axis in your mind (as you will see, *all* of the matter density images have this axial symmetry). The black circles will then delineate a spherical surface. Those surfaces are where the electron density (and the wave function) is exactly zero. We call any surface that represents a node a *nodal surface.* The ones you see here are called *spherical nodal surfaces,* or *radial nodes.* Take note that the nodal surfaces don't have a thickness even though the images make it appear that way; I remind you, in the light-colored images, I used a logarithm function to emphasize values close to zero (made them dark) to visualize the location of the node.

Understanding the rules for the patterns of the nodal surfaces and lines will help make the crazy patterns you will soon see much easier to understand. There are three rules that use quantum numbers to tell you which and how many nodes you will see: one for spherical nodes, one for nodes in the shape of planes and cones, and one for a line node.

The first of the three rules tells you the number of spherical nodes: $N_{\mathrm{Sph}} = n - l - 1$. You can easily confirm this works for the *s* orbitals. I'll introduce the two other rules when I introduce the *p* orbitals.

Half-volume images

The charge density changes smoothly from one point to the next. This means that points close to a nodal surface will have very small values. If I reject the smallest values from 3D space, I can create interesting images of solids showing volumes with the highest electron density.

Figure 10.2 shows the volume images of the same orbitals that appear in Fig. 10.1. I needed to cut the volumes in half so that you could see the inner *shells*. I chose to cap off the half shells with the electron density on the surface of the shells. To be clear, this new rainbow color map of density, which caps off the volumes, is a single slice and represents the same data as the light-colored nodal slice images; the difference is that I applied a logarithm to the nodal slice images. Don't read too much into the thickness of any of the volumes. For each image of this type in the chapter, I played with the rejection limit value until the volumes were visible and aesthetically pleasing.

Please don't be confused and think these images show an extra nodal surface outside the outermost shell. The electron density gradually decreases farther from the center, and my program rejects all points below a certain value, not just the ones near the nodal surfaces.

I must also point out that I am no longer putting the orbitals on the same length scale. I made them as large as possible so that you

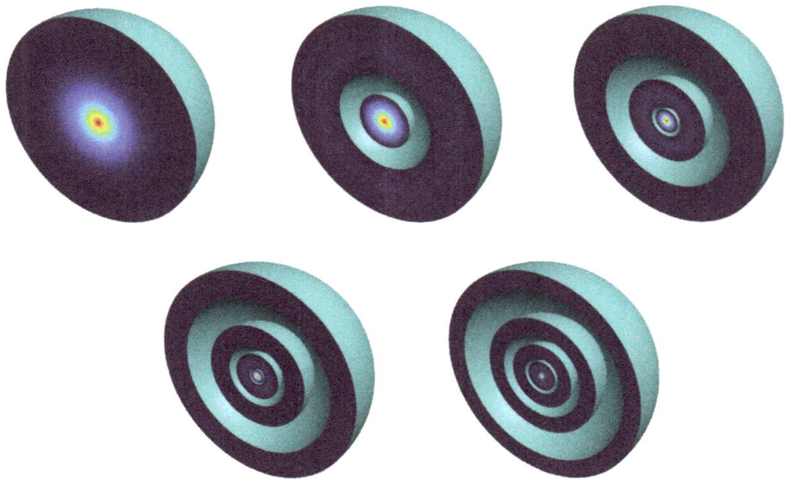

Figure 10.2: Solids showing the volumes of highest electron density in the $n = 1$ to $n = 5$ s orbitals.

can see the details. Across all the images of orbitals in this chapter, you can say that they are roughly the same size for a particular n, and you can use the lengths I mentioned earlier. Note that the previous sentence applies only to atomic hydrogen.

I say this because you can also consider these *s* orbitals as the shapes of the electrons most loosely bound (commonly called *the outermost electrons*) in helium and groups 1 and 2 of the periodic table. However, the sizes are not the same as in hydrogen. The elements of group 1 and 2 are lithium & beryllium ($n = 2$), sodium & magnesium ($n = 3$), potassium & calcium ($n = 4$), and rubidium & strontium ($n = 5$).

I won't list elements for each of the next three types of orbitals. Instead, you could google "*s*-block, *p*-block, *d*-block, and *f*-block elements", and you'll learn where each type of orbital comes to the fore in the periodic table. This then opens up a non-secret, which I can reveal to you: chemistry is the study of the outermost electron orbitals. Since elements going down a column are similar, they can often be switched in a chemical formula to predictable effect.

Gallery of *p* orbitals ($l = 1$)

The *p* orbitals have significantly more complexity than the *s*-orbitals, so I start with the half-volume pictures, which have the most information. We will then see how spherical nodes, plane nodes, and line nodes contribute to the pictures. Take a brief moment and attempt to deduce where those nodes are in the following figure.

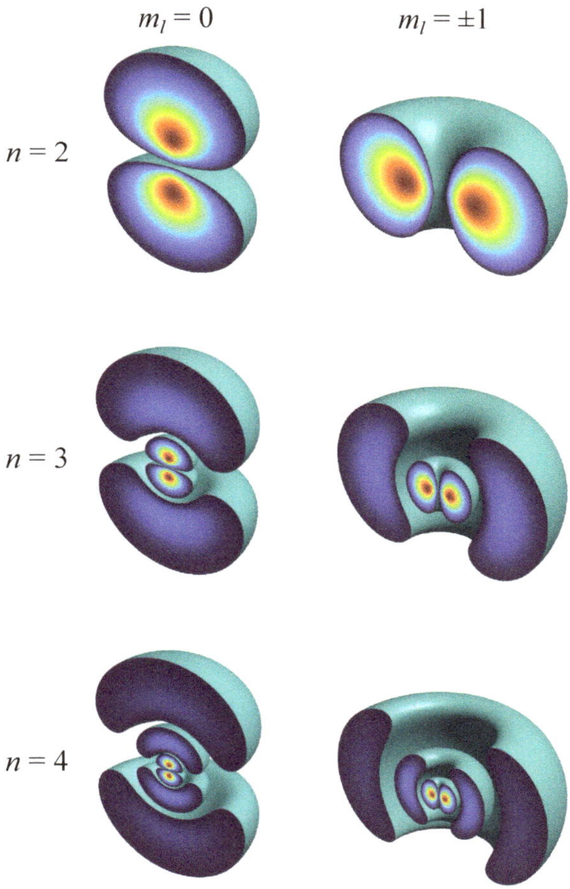

Figure 10.3: Volumes containing the highest values of matter density of some *p* orbitals.

Next, we look at the cloud and slice images. The first set contains the *p* orbitals with no orbital angular momentum ($m_l = 0$), which I show in Fig. 10.4. The new thing is what appears to be a horizontal black line in the images on the right. However, you know you can mentally rotate the image about a vertical axis to determine what that line represents. It is a nodal plane. It is where the electron density and the wave function are precisely zero.

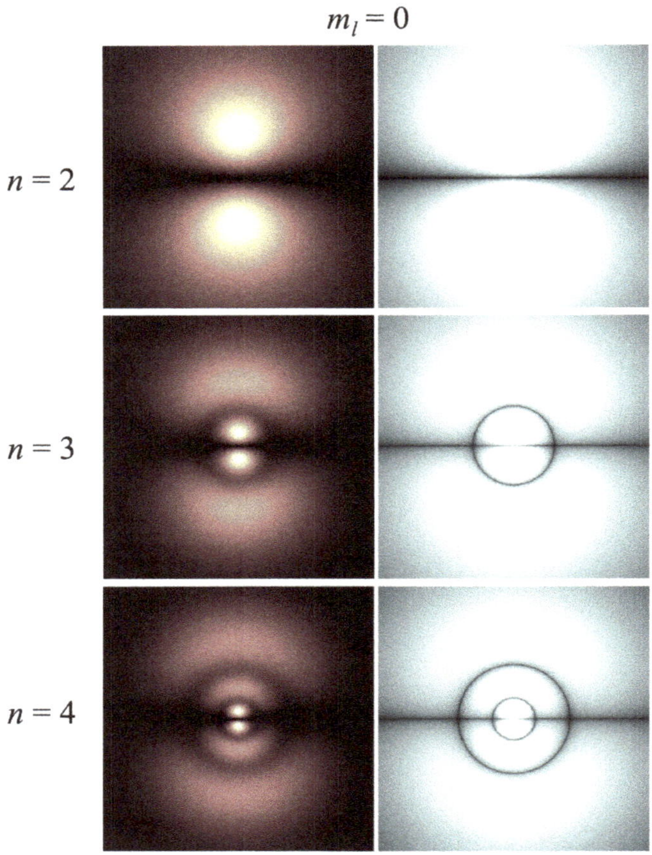

Figure 10.4: Matter density of *p* orbitals with no angular momentum.

The second nodal rule tells you the number of planes and cones $N_{\text{P \& C}}$. The rule is $N_{\text{P \& C}} = l - |m_l|$. There is only ever one nodal plane or no plane. Whenever there is one, it lies in the xy-plane. It occurs when $N_{\text{P \& C}}$ is odd. The cones come in pairs because the electron density has a mirror symmetry across the xy-plane. You will first see cones in the d orbital section.

Next, we turn to the p orbitals with $m_l = \pm 1$. The wave functions for $m_l = 1$ and $m_l = -1$ are *slightly* different. When the electron density is calculated, this difference disappears, and the electron density for them is exactly the same. I plot the cloud and slice images of these in Fig. 10.5.

In the newest image, l and $|m_l|$ are both 1, so the second rule says there will be no cones or horizontal planes. Instead, we now have gained a vertical line in the exact spot of our imagined rotation axis. Rotating the image in our mind shows that the line in 2D remains a line in 3D.

Looking to the left at the electron cloud images, it's not obvious that there is a vertical nodal line. This is why the nodal images are needed. Had I included an image looking down the axis of the nodal line, you would have seen electron clouds shaped like *bagels.* (Why do I say bagel and not doughnut? Later I will talk about slicing the orbital like I'm going to cut a bagel to toast it. Only a maniac would slice and toast a doughnut.)

The third and final rule dictates when a line node appears. The rule is: when $m_l \neq 0$, a line node will exist on the z-axis. A mnemonic for remembering this rule could be "whenever the orbital has angular momentum, there will be a nodal line on the axis where the angular momentum vector points along, which is the same as the axis of its magnetic dipole." (Please note I am not saying a line node causes OAM to exist. I am simply giving you a memory aid.)

I want to give you another aid to understand the pictures. Have a look again at the half-volume pictures in Fig. 10.3. In the first column, you should notice that the only shapes are vaguely teardrop shapes and mushroom caps. I refer to these collectively as *lobes.* The second column has the bagels I mentioned. They will still be called *bagels*

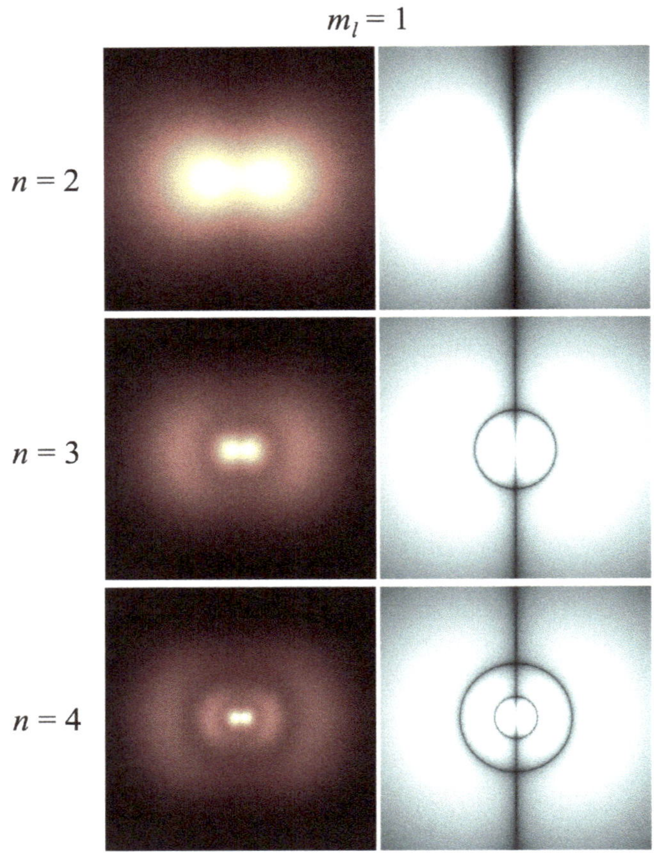

Figure 10.5: Matter density of *p* orbitals with $\hbar$ of angular momentum.

even when their cross sections are deformed from circles to various degrees. We still have two more types of orbitals to go, but this completes all the types of shapes you will see in the volume images: shells, lobes, and bagels. Knowing this classification (which I totally made up) should make it easier to understand the pictures. There are correspondences with the nodal rules which you could deduce. For example, when there is no nodal line, shells or lobes will be seen (bagels may or may not be seen). When there is a nodal line, it's all bagels.

Drawings in introductory chemistry classes

Many of you who've had a standard chemistry education and recall what your teacher or professor drew for p orbitals might be suspicious of the accuracy of my images. They aren't how you remember them. Typically, students are told that the charge densities for the $2p$ orbitals are representative of all higher p orbitals. I can agree with that for the purposes of this story. They are then told that the three p orbitals are called p_x, p_y, and p_z, which supposedly correspond to $m_l = +1, -1$, and 0, respectively. I can also agree that the p_z orbital corresponds to the $m_l = 0$ orbital: a two-lobed volume or a figure-eight-looking 2D drawing. (Some textbooks say the lobed shapes are wave functions; others say they are electron densities.)

Textbook authors go wrong when they say that the p_x and p_y orbitals are just like the lobed p_z orbital with the exception they have been rotated to lie on their sides aligned along the x- and y-axes. That is inconsistent with the bagel shapes I show you.

Chemists are actually using superpositions of the $m_l = +1$ and $m_l = -1$ states to make the p_x and p_y drawings. They do this because the superpositions are purely real and the new shapes tell a convenient story to the students of how wave functions come together to form bonds in molecules. Absent the convenient fiction, their textbooks would have to justify how bagel-shaped charge densities become a linear bond. (The justification is both easy and the same as what they already say: the atomic orbitals change shape when in molecules.)

The spatial symmetries of the chemist's orbitals do have some conceptual utility in molecules (e.g., energy-level splitting in ligand field theory). However, their approach is misleading because students are not told the truth about the shapes of the electron densities or the complex-valued nature of orbitals. The chemist's atomic orbitals also don't have an overall OAM, so they can't label p_x or p_y with the m_l quantum number, but many still pretend that they can. By the number of questions on physics and chemistry forums from confused students on this issue, it's clear that chemistry instructors should spend some time disclosing the fuzzy status of their heuristic bonding story.

Gallery of *d* orbitals ($l = 2$)

Starting with the node images (Fig. 10.6) of the *d* orbitals, you can see what appears as two diagonal lines intersecting in an X in the $m_l =$ 0 images. It's better to understand them as two lines going upward from the center and two lines going downward from the center. You then mentally rotate them about the imagined vertical axis. By doing this, you can see an upward- and a downward-opening cone. And the number of cones matches the $N_{P\,\&\,C}$ rule. The higher angular-momentum images follow the rules you already know.

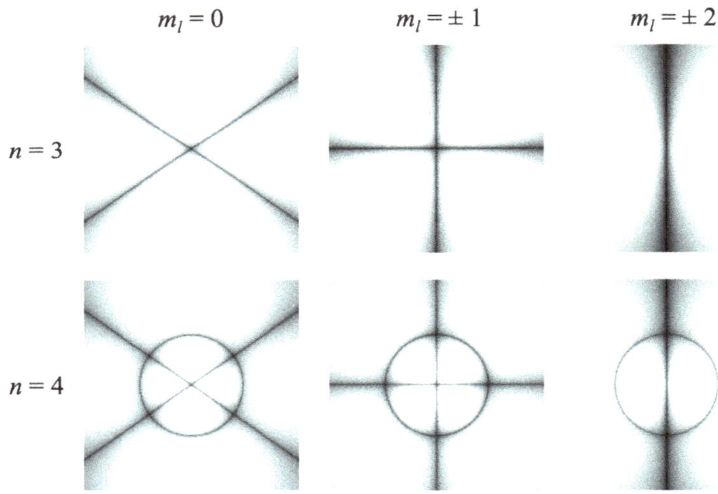

Figure 10.6: Nodal images of *d* orbitals. You should be trying to mentally rotate the nodes of the images to imagine the three-dimensional nodal surfaces before you peak at the volume images.

The electron cloud images for these orbitals are in Fig. 10.7. For the former chemistry students, looking at the volume images for the *d* orbitals (Fig. 10.8), you can see that they, too, don't match the hard-to-draw *d* orbitals you were taught, either. The reason is the same as for the *p* orbitals.

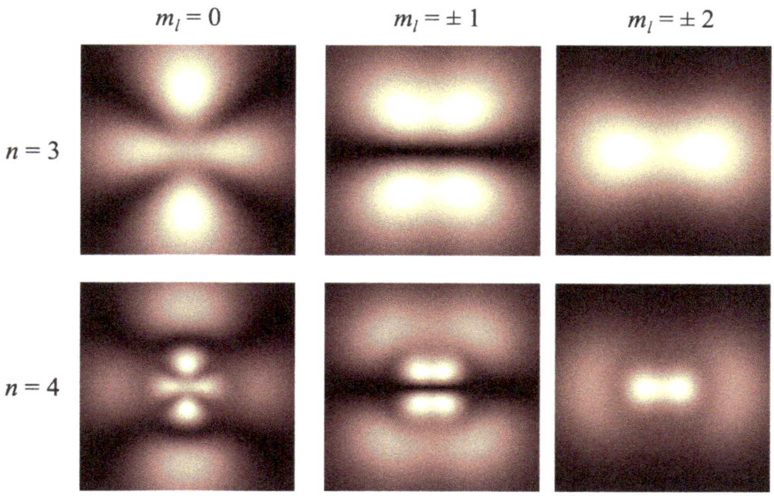

Figure 10.7: Cloud pictures of the *d* orbitals' density.

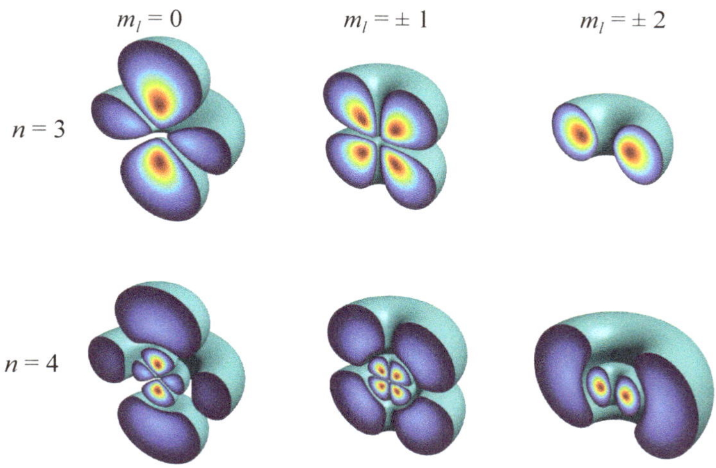

Figure 10.8: Volumes containing the highest values of matter density of some *d* orbitals.

Gallery of *f* orbitals (l = 3)

There is nothing new in the 4f orbitals (see Fig. 10.9). With the rules for nodes that have been established, you should be able to label the images with $m_l = 0, \pm 1, \pm 2, \pm 3$.

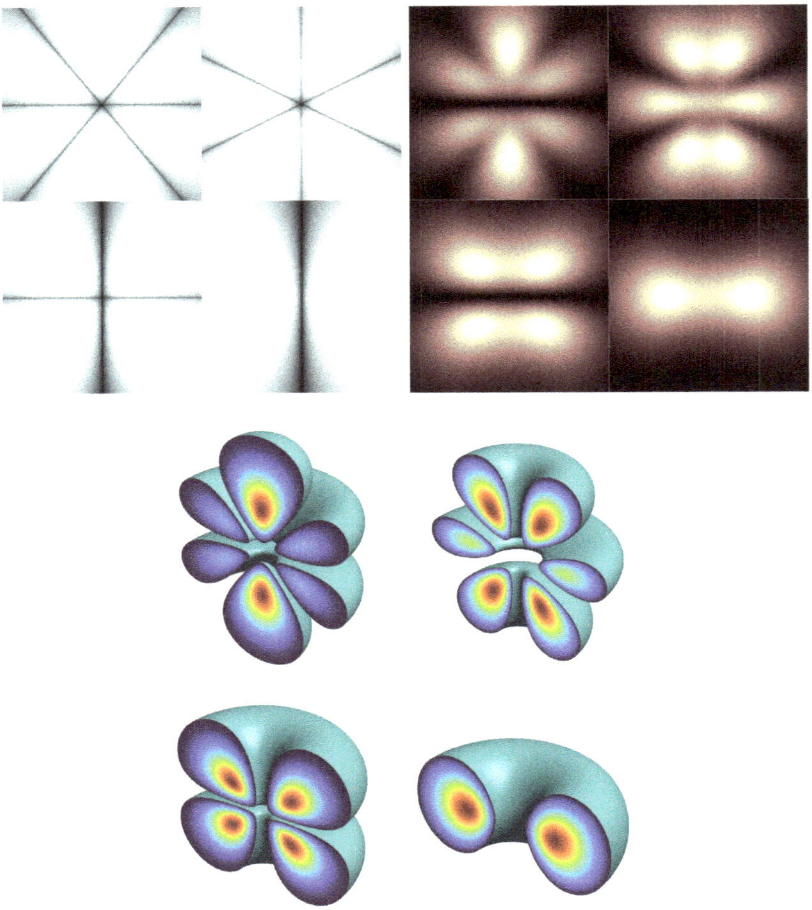

Figure 10.9: Various ways to view the matter density of the 4*f* orbitals.

Wave function gallery

Phase of the wave function

Figure 10.10: Phase of 2*p* wave functions. Slices are taken through the *xy*-plane of the $m_l = -1$ and $m_l = 1$ wave functions.

We now switch from pictures of the electron density $|\Psi|^2$ to pictures related to the wave function $|\Psi\rangle$. To start, have a look at Fig. 10.10. These are slices through the wave function of the $2p$, $m_l = \pm 1$ orbitals. The (full) volume image of these orbitals looks like a bagel. This slice is at the surface you would make to cut a bagel in half to toast it (the xy-plane). We are looking down the axis of the nodal line. I'm not showing you the amount of the wave function at each point yet (more appropriately called the **magnitude** of the wave function). Instead, I am plotting only the **phase** of the wave function, which I'll explain more fully in the next subsection after you have examined a few images.

In the *p* orbital section, I said there is a *slight* difference between the $m_l = -1$ and the $m_l = 1$ wave functions, and this is it. The ROYGBV order of colors is counterclockwise in the left image and clockwise in the right image (I am aware that this is not a true optical spectrum. Despite that, I will still use ROYGBV to refer to its order). I will refer to it again, so let's call the left image the *m_l = −1 color wheel*.

Because the color wheel has only one cycle, I know the orbital has one $\hbar$ of angular momentum pointing up or down along the z-axis. In Fig. 10.11, I plot the same slice for the $3d$, $m_l = 2$ and the $4f$, $m_l = 3$ orbitals. Of course, orbital-magnetic quantum numbers of 2 and 3 tells us there will be angular momenta of $2\hbar$ and $3\hbar$, but you can get the same information from the images by counting how many times you see a particular color. Because I need to refer to them again later, I will call them the *m_l = 2 and m_l = 3 color wheels.*

Figure 10.11: Phase of slices through the "bagels" of the 3*d* and 4*f* orbitals. The slice is in the xy-plane of wave functions with $m_l = 2$ and $m_l = 3$.

With these images, I've been very selective with the orbitals and the slice location. Had I chosen the $2p$, $m_l = 0$ orbital, the slice would be the horizontal plane node, and the wave function would be zero in the slices, which is not interesting at all because a value of zero has no defined phase. In that case, I would've had to make a completely white or black or whatever square.

Since the images have $m_l > 0$, you know that they have vertical line nodes. However, you shouldn't see them because a geometric line has no cross-section when you look down its length. Had I chosen orbitals with cone nodes, those nodes would only intersect the bagel slice at that same central pixel as the vertical node. If I had chosen a different orientation or location of the slice, those nodes would be possibly seen.

However, that all being said, I can show you images with spherical nodes with the same slice location as before. In Fig. 10.12, you can see the phase of the same slice through the $4p$, $m_l = 1$ orbital and the $4d$, $m_l = 2$ orbital.

Figure 10.12: Phase of wave functions with phase shifts due to sign changes across spherical nodes. Slice is in the xy-plane of the $4p$, $m_l = 1$ and $4d$, $m_l = 2$ orbitals.

Notice the color wheels have shifted in the different regions and the ROYGBV order is the same in an image. The rotated phase is because a portion of the equation I used changed from positive to negative. When the sign changes like that, the color wheel advances by 180 degrees of a single color wheel. That's easy to understand in the left image. In the right image, two full color wheel's worth of color go around a circle, so the two regions appear to differ by a 90-degree rotation.

Let me show you where the negative sign came from. The hydrogen wave functions are written in compact form like this:

$$\Psi_{n,l,m_l} = R_{n,l} \cdot Y_{l,m_l} \cdot \tau_n \tag{10.1}$$

The $R_{n,l}$ is called the *radial part*, and its variable (which I'm not showing in the symbols) is the distance from the nucleus. As you know from the previous sections, the forms of the wave functions change depending on the quantum numbers. The subscripts are indices that tell the physicist to go to a table of functions and look up the particular function corresponding to those two quantum numbers.

The sign changes and color shifts in these highly selective slices of phase come from the radial part. I plot several radial functions in Fig. 10.13. These functions start at the nucleus and go radially out in all directions. Notice how the functions oscillate and cross the zero level a few times. Those zero crossings make the spherical nodes. After a zero crossing, the value switches from positive to negative, or vice versa, and a shift of a color wheel occurs.

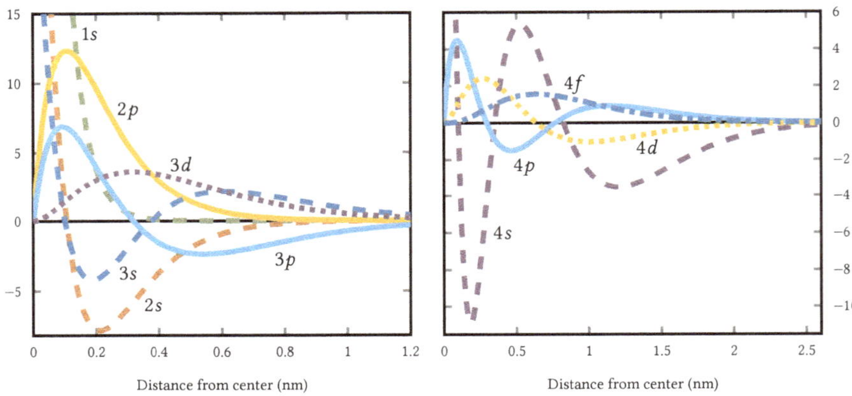

Figure 10.13: Radial parts of wave functions. I mentioned the principal quantum number n determines the "size" of an orbital. You can see that most easily by looking at the s orbitals. As n increases, the wave functions have their weight shifted farther from the nucleus.

I won't go into too much detail with the second portion of the wave function. It is Y_{l,m_l} and is called the *angular part.* Its variables are the polar angle around a vertical axis (like a compass direction) and the elevation angle above or below a horizontal plane (like elevating the angle of a cannon). The indices here also tell the physicist where to find the function in a table. All the nodes except the radial ones come from the wave function's angular part.

The τ_n can be called the *time part* or the *temporal phase function.* It takes time as a variable, and how fast it cycles depends on the energy level E_n, hence the subscript n. I'll say more about it below.

Magnitude and phase of a complex number

As I said a few times before, at every point in space, the wave function has a complex value (which is also known as the quantum amplitude at that point). Most often, one hears talk of complex numbers' real and imaginary parts. Another equivalent way to describe the same number involves a magnitude and a phase.

The idea is quite easy to understand because you only need to understand a triangle. Let's pick one complex value from some point in the wave function. Let's call this point z because that's the letter every mathematician, engineer, and physicist uses when they teach this. In Fig. 10.14, I show you this point in a plane made by the real number line and the imaginary number line. This is called the *complex plane*.

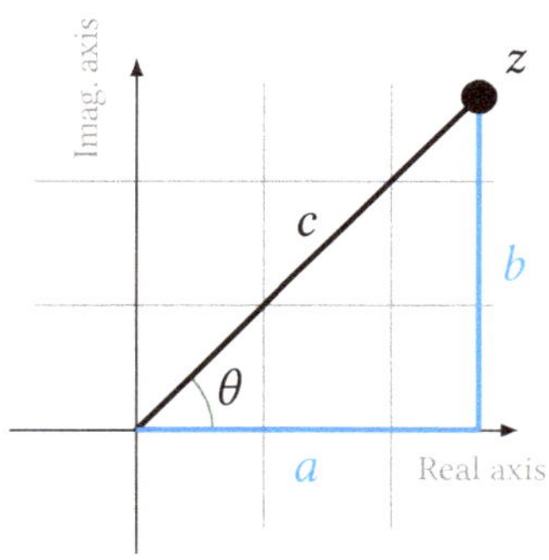

Figure 10.14: Two ways to represent a complex number *z*. The horizontal axis is the real number line, and the vertical axis is the imaginary number line. Whenever I said I was showing the squared magnitude of the wave function or probability/charge/mass density (which are all proportional to the function $|\Psi|^2$), I was finding c^2 at each point in space (to be clear, this diagram isn't set in physical space; it is an *Argand diagram/the complex plane*).

You can get to the point z in two ways. You can move horizontally a units and vertically b units. Or, you could point yourself straight at it and move the length of the hypotenuse c. The length of the hypotenuse is the magnitude of the complex number z. The length is found from the Pythagorean theorem using a and b. The **phase angle** θ

is found from trigonometry functions and a and b (do you remember SOHCAHTOA from high school trigonometry?). Our generic complex number z can be represented as a pair of coordinates in two ways:

$$z : (a,b)_{\perp} \quad \text{or} \quad (c,\theta)_{\angle} \tag{10.2}$$

The takeaway from this is that you can talk about either the real and imaginary parts or the magnitude and phase of a complex number. In this chapter, I find this new way is more enlightening, so when I made the images, I assigned the 0 to 360 degrees of the phase angle θ to the colors of the color wheel.

I must point out that, while they have the same information, the **phase** and the phase angle are not exactly the same thing. Using the magnitude c and the phase angle θ, a complex number z is calculated via:

$$z = c \cdot e^{i \cdot \theta} \tag{10.3}$$

The e is called the *exponential function.* It has many uses in math and science. Once the imaginary unit i is placed in its argument, the exponential function calculates a phase, so we say $\exp(i \cdot \theta)$ is the phase of the complex number z.

Math-oriented people often make no distinction between phase and phase angle because it is so easy to move between them. (I blurred the distinction in the sentence after Eq. 10.2 and may do it again.) Anyone who works with complex numbers makes the following substitutions in their head easily:

$$\begin{array}{rcr@{\qquad}rcr} \theta & \longleftrightarrow & e^{i \cdot \theta} & \theta & \longleftrightarrow & e^{i \cdot \theta} \\ 0^\circ & \longleftrightarrow & 1 & 45^\circ & \longleftrightarrow & (1+i)/\sqrt{2} \\ 90^\circ & \longleftrightarrow & i & 135^\circ & \longleftrightarrow & (-1+i)/\sqrt{2} \\ 180^\circ & \longleftrightarrow & -1 & 225^\circ & \longleftrightarrow & (-1-i)/\sqrt{2} \\ 270^\circ & \longleftrightarrow & -i & 315^\circ & \longleftrightarrow & (1-i)/\sqrt{2} \\ 360^\circ \rightarrow 0^\circ & \longleftrightarrow & 1 & & & \end{array} \tag{10.4}$$

It will be very informative for you to draw this out. First, draw a circle of radius 1 on the complex plane. Then, label the left column's phases on the circle using the real and imaginary parts (a, b) as coordinates on a grid. Draw a line from the origin to each coordinate, and label the phase angles. After this, do the same for the right column's phases.

While I am telling you about phase, I should also tell you that these images of wave functions are snapshots in time. Hydrogen orbitals are stationary states, so the electron densities and all the magnitudes (all the c's) of the wave function won't change as time goes on. However, the phase of the wave function cycles as time goes on. So, the colors of our color wheels will be rotating as if they were on a pinwheel if we let time progress. This comes from the wave function's time part:

$$\tau = e^{-i \cdot t \cdot E_n/\hbar} \tag{10.5}$$

The phase angle of the exponential is collectively all the symbols that multiply the i. Whenever a physicist sees an exponential with an i in it, they think something has a cycle in space or time. In this case, it is time t. So, the phase is advancing at some rate. That rate is $|E_n|/\hbar$. Working out the units of that fraction, it comes out to units of frequency, so $|E_n|/\hbar$ is the phase's cycles per second. The value of the exponential is the temporal phase (aka the "time part" of the wave function) at a particular point in time.

By the way, there is no convention for the symbol of the time part, so I chose τ. The τ (spelled "tau"; rhymes with "cow") is the Greek version of a t. You might guess it has units of time, but it doesn't. This is because functions like e^x, $\sin x$, and $\cos x$ are unitless quantities (also, whatever grouping of symbols that replace x must have all their units cancel each other).

The magnitude and phase of hydrogen wave orbitals

Bringing them together

Now, I will start by slicing up some hydrogen orbitals to show you how the magnitude and phase change throughout them. This part aims to give you an idea of how the magnitude and phase change throughout the three-dimensional volume of the hydrogen wave functions. I won't go through all the orbitals or give you a complete view of the wave function of any one orbital. The information is difficult to show in full, and that's why physicists don't often discuss it.

In Fig. 10.15, I show a horizontal slice (xy-slice) and a vertical slice (xz-slice) through the bagel of the $2p$, $m_l = -1$ orbital. Both images in the figure are snapshots of the same point in time. The horizontal slice is like the $m_l = -1$ color wheel I showed you, but it now has a volcano shape. The height above the plane tells you the relative magnitude of the wave function at that point.

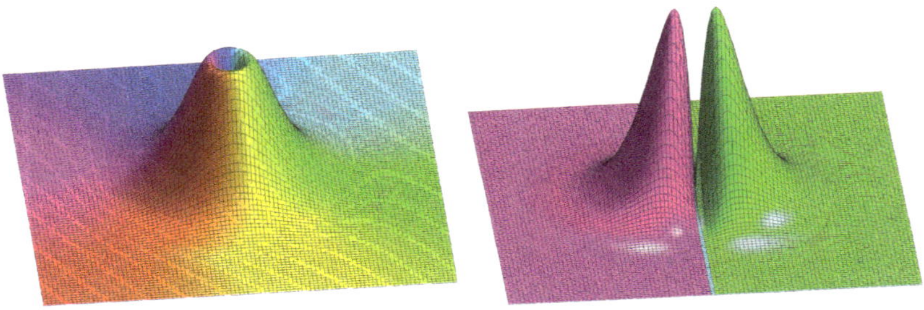

Figure 10.15: Magnitude and phase seen in slice of the wave function. The left is a horizontal slice, and the right is a vertical slice through the $2p$, $m_l = -1$ orbital.

To understand the vertical slice, let's return to the $m_l = -1$ color wheel in Fig. 10.10 on p. 189. We can imagine drawing a horizontal line through the center. It would be half magenta and then half green. That is where our vertical slice of Fig. 10.15 is, and you can see the colors match. The highest points of the double peaks in the vertical slice correspond to the dark red region in the half-volume electron-density images (see the bagel in Fig. 10.3). Additionally, compare the topography with the $2p$ radial function in Fig. 10.13.

Only two colors when there is no OAM

You saw the electron density images at the beginning of the chapter have an axial symmetry where rotation about this axis doesn't change the density. The same is true for the wave function's magnitude. Had I chosen any other vertical slice *through the center* of the $2p$, $m_l = -1$ orbital, the double-bump topography would be the same, but the two colors would differ.

All the colorful images of the wave functions or phases in this section have had an m_l greater than zero up to this point. In contrast, the $m_l = 0$ wave functions have only two phases corresponding to positive and negative real numbers. They don't have a rainbow of phases. The reason for both of these facts is a factor in the angular part of the hydrogen wave function of:

$$e^{i \cdot m_l \cdot \phi} \tag{10.6}$$

The Greek ϕ (spelled: "phi"; rhymes with "pie") is the polar angle around the z axis that I mentioned previously (p. 192). A physicist sees a cycle in space with this expression. It is a spatial phase function and contributes to the overall phase. When $m_l > 0$, this factor causes the phase to change with the angle about the vertical axis, and we get a color wheel.

When $m_l = 0$, it doesn't matter what the angle θ is because zero times an angle is zero, and e^0 equals one. And a factor of one does nothing to the rest of the angular function. Ignoring the time part, because we are considering a snapshot in time, what is left is a function of real numbers that are either positive, negative, or zero.

This means that you can rotate a slice of an $m_l = 0$ wave function around the vertical axis, and it will have the same magnitude and phase. Both the height in the images and the color won't change. This applies to the slices of the $5s$ orbital in Fig. 10.16 below and to the bottom image of the $4f$ orbitals in Fig. 10.17. The $4f$, $m_l = 0$ slice represents all vertical slices through the center of its atom; even though there is no line node, there is still an axial symmetry. The s orbitals are special because their angular parts have no angular dependence. Therefore, the slice of the $5s$ orbital actually represent any slice through the center of the atom.

If we stop ignoring the time part, then the initial two colors in the

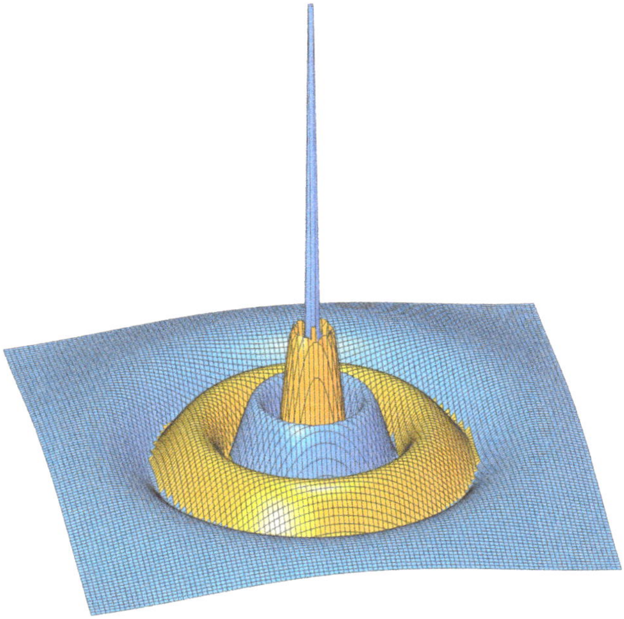

Figure 10.16: Magnitude and phase seen in every slice through the center of a 5*s* orbital. There are only two phases at any time in an $m_l = 0$ orbital.

$4d$, $m_l = 0$ orbital (which in the image are yellow and blue) would advance through all the color wheel colors. But, at any single point in time, there will only be two colors throughout the wave function.

Inputing $m_l = 3$ into the expression for the angular cycle (Eq. 10.6) will show us that we will have three times as many colors than the $m_l = 1$ case because as the *angle in physical space* goes from $\phi = 0$ to 360 degrees, the *exponential's phase angle*, namely $3 \cdot \phi$, does three complete revolutions.

4d orbitals

Let me walk you through that 4*d* orbital's color changes and topography. The first thing about the colors I should mention is that there is a thin, cyan line running from the bottom left to the top right in four of

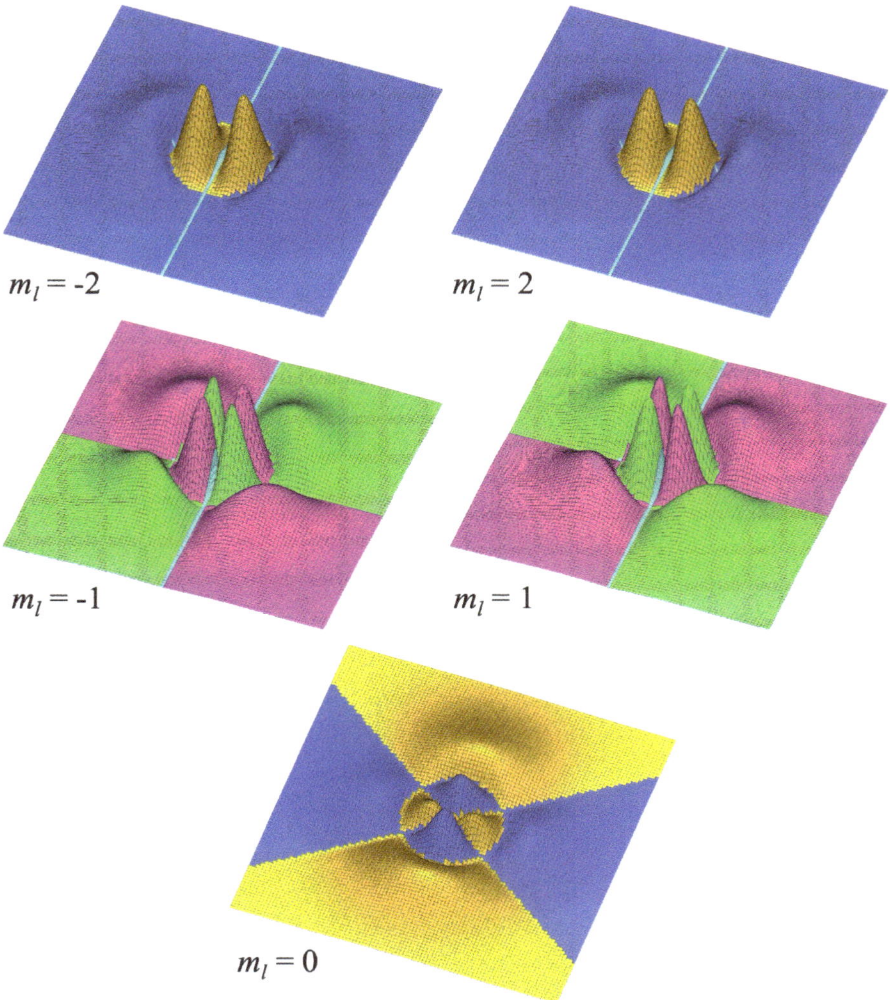

Figure 10.17: Magnitude and phase of 4*d* orbitals Slices are of the xz-plane (vertical slices). You should compare these to the nodal images of Fig. 10.6. In all the images in this section, I reduced the resolution of my images so that you can see the topography. This results in the axial nodal line having a visible thickness and being erroneously assigned a phase, which shows up as cyan.

the slices. This is the vertical nodal line, where the wave function is zero. Technically, zero is the only complex number that doesn't have a defined phase, so there shouldn't be any color. (To see the lack of a defined phase, go to the image of the complex plane, imagine the point z at the origin of the coordinate system, and see that you can't define a phase angle.) Why, then, is there a color? Computer codes tend to encounter faults when NaNs (NaN: not a number) are used in place of numbers, so the code I used assigns a phase angle to the number zero. Rigorously, those data points shouldn't be colored cyan. Perhaps they should be black or white. But, it's too much effort to remap the assignment of a phase to the number zero, and the cyan line is easy to see and orients the image in space for us. Why don't we see cyan for the other nodes you ask? It's because the other pixels did not perfectly align with the nodes to give a value of exactly zero.

The five slices represent five different states, yet there seems to be the same four colors repeating (i.e., yellow/blue and magenta/cyan). There's nothing significant in the seemingly coordinated colors. The current colors are an artifact of the fact that my code starts at the same relative phase in each one. In reality, if you pick random atoms, their phases won't be coordinated like in the images.

As for the different colors within a single slice, you know those have meaning. The combined effect of the radial and angular parts of the wave function gives the domains of colors. Crossing a node will shift the color by 180 degrees of a single-cycled color wheel.

Next, let's examine the topography by going out radially from the center of the images. First, you should examine again the $4d$ radial function in Fig. 10.13. You see a tall positive bump and a smaller negative dip. Mentally flipping the negative dip to a positive bump will give you the magnitude of the radial wave function. Therefore, thinking of the magnitude, going radially out from the center, there will be a tall bump, a node, and a small bump. Looking at *any* colorful slice in Fig. 10.17, you see exactly this if you start at the center and go in *any* radial direction except along a node. This then completes the level of understanding of the angular part of the hydrogen wave function that I will give you.

11 Spins, superpositions, and abstract spaces

> *[Q]uantum phenomena do not occur in a Hilbert space, they occur in a laboratory.*
>
> — Asher Peres in the preface to his textbook.

Setting the stage

First things

After being introduced to quantum mechanics, many are left wanting for a deeper understanding of the quixotic phenomena of spins and superpositions. As it has no counterpart in classical physics, it's quite reasonable to be confused by spin angular momentum.

A general understanding of superposition, on the other hand, remains elusive partly because people are led to believe follies like cats can be simultaneously dead and alive. Neither does it help that it is uncommon for professors to tell students an explicit interpretation of superposition. Their vocabulary is commonly restricted to talk of mathematical representations with little reference to physical interpretation. I think this is because anti-reality interpretations dominate our discourse. Nevertheless, the perpetually unintuitive and abstract spin spaces we use must take much of the blame, as you will see soon.

To give a rather complete account of spin, we will zig and zag across various topics in this chapter. It has a complex structure because I need to weave together various important concepts and also contrast spin-$\frac{1}{2}$ particles with other particles. Regarding the particles, I will first cover photons, then cover many aspects of spin-$\frac{1}{2}$ particles, and finally end the chapter with higher and lower spin particles. Throughout, I will introduce and use quantum amplitudes, superpositions, and physical and abstract spaces. I do this while making ample use of the Stern-Gerlach (SG) experiment and also while using misconceptions to clarify how things should be understood.

Before proceeding, I should mention that the electron is not the only elementary spin-$\frac{1}{2}$ particle. There are also the particles which are like medium and heavyweight versions of the electron: the muon and the tau particle. Their neutrino counterparts are also spin-$\frac{1}{2}$ particles: the electron, muon, and tau neutrinos. All six quarks (i.e., up, down, charm, strange, top, and bottom) are spin-$\frac{1}{2}$ particles. That being said, you should be aware when I wrote this chapter, I had electrons, atoms, and molecules in mind. For example, neutrinos have spin, but because they are elementary particles and have no electric charge, they don't have a magnetic dipole, so they won't respond to the magnetic field in a SG apparatus.

Physical and abstract spaces

We start with an example not concerning spin but about polarized light for a couple of reasons. First, light polarization is a classical concept, so its superposition is easy to understand. Secondly, it will allow me to contrast its physical space and abstract space pictures with the pictures of spin's physical and abstract spaces.

Let's imagine classical polarized light with a diagonal orientation of 45° from the vertical. In classical electrodynamics, we can make this by adding horizontal and vertical electric-field vectors, as I show in the Fig. 11.1. Its electric-field vectors superpose like this:

$$\vec{E}_D = \vec{E}_H + \vec{E}_V \tag{11.1}$$

The right-hand side is a superposition of what we call **basis vectors**. When we have basis vectors, it means we can scale the individual vec-

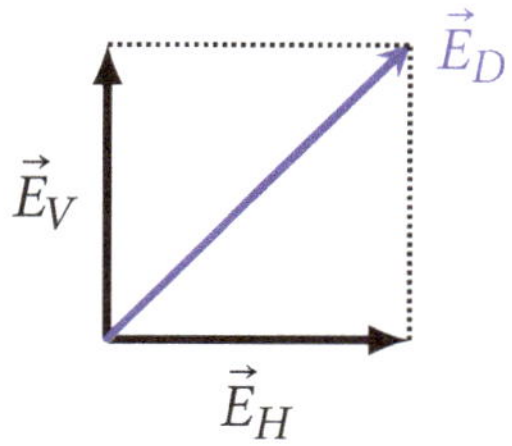

Figure 11.1: Diagonally polarized light can be viewed as being composed of horizontally and vertically polarized light.

tors by different magnitudes and superpose them to have a resultant vector point to any other location in the space under consideration. In this case, the space is the two-dimensional space of electric-field direction and magnitude. We could have chosen any two other (non-co-linear) arrows and scaled them so that, when laid from head to tail, they go from the tail of the $\vec{E}_D$ vector to its arrowhead.

Normally, we choose basis vectors to be **orthogonal**, like in the image above. In our 3D space, this means they are 90° apart. But the term has a more general meaning in non-physical spaces. If an abstract vector points purely in the "direction" of a particular basis vector, it can't have any component in an orthogonal basis vector. You implicitly know this in our physical space: if you are heading due east, you can't be getting closer to the north pole because you don't have any component along the north-south axis. The two directions due north and due east are orthogonal, or mutually exclusive.

Choosing to be more accurate and consider quantum mechanical light, we speak of photons. A photon with the same diagonal orientation of its electric field could be written somewhat incorrectly as:

$$|D\rangle \text{ "="} \ |H\rangle + |V\rangle \tag{11.2}$$

The quotation marks remind you that it is not completely true because I am ignoring the quantum-mechanical amplitudes. In analogy to the electric-field-vector equation above, the kets $|H\rangle$ and $|V\rangle$ are also orthogonal basis vectors. The difference is that kets live in an abstract space. You can see this represented in Fig. 11.2.

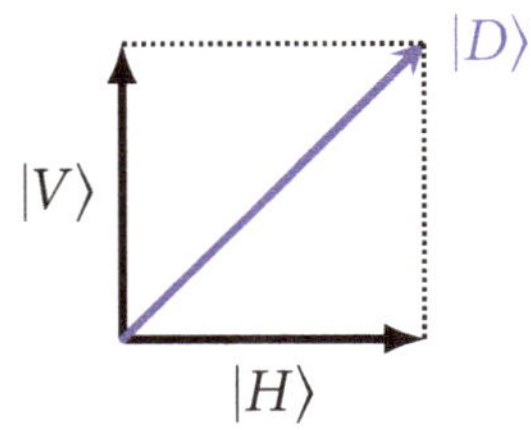

Figure 11.2: Abstract space for the photon-polarization kets. This picture is slightly inaccurate because the length of the arrow for $|D\rangle$ is too long for how things work in quantum mechanics (this will be explained in the next figure). However, the important point is the similarity between the basis vectors of the physical and abstract spaces. Both have two basis vectors that point in the same directions on the page.

This abstract picture is the same as the previous electric-field-vector picture but with the labels changed. The two previous equations also have the same form. These similarities are quite significant to what I want to show with spin.

Because the physical- and abstract-space images are boringly similar, the classical physics of polarizations ports over to quantum mechanics in a readily understandable manner and has fewer quantum-mechanical surprises. This is why physicists often use polarization experiments as examples for the public. Spin is not as well-behaved.

By the way, in quantum mechanics, the abstract space is called a **Hilbert space**. Please don't think of it as some mysterious physical space. When physicists talk about Hilbert space, they mean that particular math rules apply to the vectors. It's a mathematical space, and it can be broken into two subspaces. One is associated with spin and is often called **spin space**. The number of dimensions of spin space is equal to the number of basis kets. The other subspace is associated with the spatial part of the wave function. For each location in configuration space, which you first heard about from Heisenberg (p. 72), there is a basis ket that contributes to the **spatial part of Hilbert space**.

This space has an infinite "number" of dimensions because each variable in the wave function has a continuum of positions.

I want to reserve the word *abstract* for use in describing unintuitive Hilbert spaces, so now is time for an explanation about the two meanings of **physical space** that I use. In the first meaning, it is the three-dimensional space we exist in. And, in truth, regarding physical space, there can be only one. In the second meaning, it refers to spaces like that of Fig. 11.1. In this second case, the spaces are actually abstract because we can't visit them. However, these spaces have three dimensions, have real-valued quantities, have superpositions that add intuitively, and show vector quantities that exists at some point in the actual physical space that we exist in, so I will continue to call these abstract but intuitive spaces *physical*.

Time to finally deal with quantum amplitudes

There is a small problem with my previous figure. In quantum mechanics, we think of each ket symbol as representing a whole state, and *all* of these ket symbols should have a length of one when we think of them like vectors. The way I drew it in the figure doesn't show that. Notice the diagonal vector is longer than the two basis vectors. I need to make the arrows the same length.

I am ready now to drop the quotation marks from Eq. 11.2. This is where the **quantum amplitudes** (aka **amplitudes**, **quantum-mechanical amplitudes**, or **probability amplitudes**) come in.

From the Pythagorean theorem, we know some vector 45 degrees from the horizontal is $\sqrt{2}$ longer than the horizontal and vertical contributions. Applying this knowledge to our present situation, we can say the basis-vector contributions are $\sqrt{2}$ shorter than the $|D\rangle$ vector. This looks like:

$$|D\rangle = \frac{1}{\sqrt{2}} \cdot |H\rangle + \frac{1}{\sqrt{2}} \cdot |V\rangle \tag{11.3}$$

This superposition is represented in abstract space in Fig. 11.3.

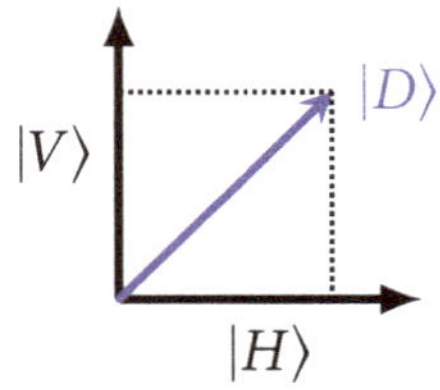

Figure 11.3: Abstract space for the photon-polarization kets. In this case, all three arrows are the same length. The tips of arrows all lie on a circle of radius one in this space. You can see $|H\rangle$ and $|V\rangle$ only contribute fractions of their lengths to $|D\rangle$. The fractional lengths are $1/\sqrt{2} \approx 0.71$, which is where the dotted lines meet the black vectors.

The way you interpret this superposition is similar to the way you interpret electric vectors superposing. However, you must consider the basis vectors in a superposition as contributing fractions of their state to the whole photon state. They aren't contributing fractions of their properties (e.g., electric fields) to the overall state. You are allowed to be confused or surprised by that statement because the physical and abstract diagrams you have seen so far sure make it seem obvious that the electric fields add, but that's because those abstract and physical spaces are boringly similar. Trust me. When we get to spin, you'll see physical quantities don't add in a superposition.

I would be negligent if I didn't say that the quantum amplitude also tells us the relative probability of finding a particular result in a measurement. Say we have a light beam of diagonally polarized photons. The quantum amplitude out front of each ket in the superposition can tell us the probabilities in measurements. To understand how, let's say we use a vertically-oriented polarizer in the light beam and place a detector on the downstream side. We would find 50% of the diagonal photons would come out vertically polarized and 50% would be absorbed by the polarizer. We get this percentage by squaring the magnitude of the amplitude:

$$\left|\frac{1}{\sqrt{2}}\right|^2 = \frac{1}{2} = 50\% \tag{11.4}$$

Inventing a mathematical convention

The problems with the amplitudes are that they, by tradition, occupy the same spot as the values returned by operators and they are both numbers. Students are endlessly confused by the two. A professor even wrote a quantum textbook that got these two consistently confused over eleven pages when introducing axiomatic quantum mechanics.

I don't like amplitudes in their standard location because it adds cognitive load for those who are new to quantum mechanics. For me, they also get in the way and make everything look like a notational jumble. By the end of the chapter, I will have built you up to understanding a pretty complicated notation involving spin. I think I should try to make reading these equations as easy as possible, so I will break with convention.

I could write Eq. 11.3 equivalently as follows:

$$|D\rangle \;=\; |H\rangle \cdot \frac{1}{\sqrt{2}} + |V\rangle \cdot \frac{1}{\sqrt{2}} \tag{11.5}$$

This moves the amplitude to a different position and is mathematically correct. But it still has ugly fractions that make your eyes look above, below, and along the reading line. Instead, I could write all amplitudes as square roots of the probabilities, or weights. Then we would have a nice readable equation:

$$|D\rangle \;=\; |H\rangle \cdot \sqrt{50\%} + |V\rangle \cdot \sqrt{50\%} \tag{11.6}$$

Part of me says to stop here, but I feel it will help you if I get typographically creative. What if we put the amplitudes below the kets so that kets are grouped on one line and the amplitudes on another?

$$|D\rangle \;=\; \underbrace{|H\rangle}_{\sqrt{50\%}} + \underbrace{|V\rangle}_{\sqrt{50\%}} \tag{11.7}$$

Note that the left sides of these equations have quantum amplitudes of 100%, but we physicists don't go writing down "1" beside all the lone kets.

Some experts may dislike this convention. But I insist this is okay for this book. You don't want to become a professor who writes a textbook and gets the amplitudes and eigenvalues confused. Once you've learned the physics, you can take the mathematical training wheels off in your future endeavors.

Besides, this is not far from how we write things in matrix mechanics. As a column vector, the diagonal polarization looks like:

$$|D\rangle = \begin{bmatrix} 1/\sqrt{2} \\ 1/\sqrt{2} \end{bmatrix} = \begin{bmatrix} \sqrt{50\%} \\ \sqrt{50\%} \end{bmatrix} \tag{11.8}$$

This notation hides the $|\mathrm{H}\rangle$ and $|\mathrm{V}\rangle$ basis vectors. But don't worry about learning column vectors now. It's just my sly way of previewing matrix mechanics before you get to that appendix.

Spin-$\frac{1}{2}$: the physical and abstract spin spaces

Now, we look at the physical and abstract spin spaces of spin-$\frac{1}{2}$ particles. But, first, here is a quick comment on the notation. In the *Math vs. Measurements* chapter, I used the $|+z\rangle$ ket to represent the spin-up state of a silver atom. In other places, you might see this state represented by various kets: $|\mathrm{up}\rangle$, $|+\frac{1}{2}\rangle$, $|+\rangle$, or $|\uparrow\rangle$. Because of how deep we will go, it's best to think about which direction in space the spin angular momentum points, so I will continue to use $|+z\rangle$. Although, I will apologize later for changing this type of notation, as it is only appropriate for spin-$\frac{1}{2}$ particles. The later notation is needed to understand different spin particles, but we use this easier notation to learn the basics.

In this first exposure to spin space, we won't look at spin superpositions yet. We will concentrate on the basis kets in spin space and how they relate to physical space. To start, let me emphasize for every spin ket, there is an associated angular-momentum vector, which has a direction. If we are working in the basis that uses $|{+z}\rangle$ and $|{-z}\rangle$, those angular-momentum directions are up and down in our physical space. (We don't need to concern ourselves with the exact magnitude of the vectors.) I will call these physical vectors $\vec{S}_{+z}$ and $\vec{S}_{-z}$. In contrast to these particular physical vectors that point up and down (180° apart), basis kets are always orthogonal in abstract space (90° apart).

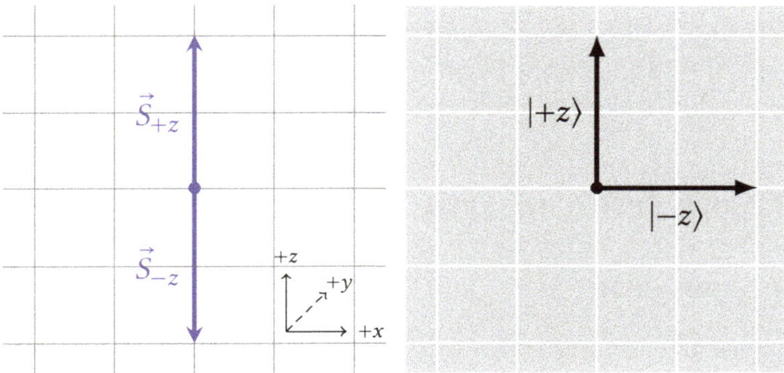

Figure 11.4: Physical and spin space for spin-up and spin-down particles. On the right is the two-dimensional spin space where ket addition happens. On the left side are the AM vectors in physical space. The large, blue vectors are not the space's basis vectors. Physical space's basis vectors are the x, y, and z Cartesian vectors. (By the way, if you've ever encountered the *Bloch sphere*, it would be on the left. Be aware that authors commonly label their AM vector with kets in their Bloch-sphere diagrams.)

An examination of Fig. 11.4 illustrates the difference between vectors in the physical and spin space. The two spaces are not boringly similar, like those for photon polarization. Because the kets and their corresponding AM vectors do not point in the same direction on the page, superpositions will be unintuitive. Additionally and importantly, because this spin space only has two basis vectors, it is two-dimensional. In contrast, physical space is three-dimensional. The differing dimensions add another layer of unintuitiveness.

Recall that I said before I think that the math on the page should be thought of as representing the quantum object. I still believe that. I think that the resultant ket, which is literally math on the page, represents the quantum object that exists in reality. You must understand, though, that the universe calculates with math that is unintuitive to your brain. The spin's property, angular momentum, which exists in physical space, is calculated by applying an operator and getting an eigenvalue. If the spin is up or down, that operator is $\hat{S}_z$ (you saw this abstract operator briefly on p. 101). However, I won't be using the spin operators in this chapter because the labels inside the kets are extremely informative as to what the AM direction would be. In contrast, in more general situations with non-spin kets, you won't be able to read off the eigenvalues from the kets' labels.

Learning from untrue claims: example #1

A superposition

I will critically examine the following incorrect claim to give us insight into superpositions: "There is a 50% chance the spin is in this state and 50% chance it is in that state." You will see that this interpretation leads to contradictions.

It's best if we have a specific state in mind. Suppose there is an electron, a spin-$\frac{1}{2}$ particle, with its spin vector aligned along the positive z-axis (i.e., spin up). There are multiple ways to mathematically represent a spin state. As counterintuitive as it seems, the $|+z\rangle$ state can be represented as a superposition of an equal weighting of spin-left $|-x\rangle$ and spin-right $|+x\rangle$ kets:

$$|+z\rangle = \underbrace{|+x\rangle}_{\sqrt{50\%}} + \underbrace{|-x\rangle}_{-\sqrt{50\%}} \tag{11.9}$$

Remember that Dirac's notation is ambiguous and hides the underlying math. You, the reader, cannot know this equation is true until you can plug in the math of matrix mechanics.

By the way, when discussing photon polarization, I mentioned that, in a superposition, the basis states contribute fractions of their state and not fractions of their physical properties. This equation is consistent with that explanation. You know that an AM vector pointing right and an equal-length AM vector pointing left, when added, cancel to give no angular momentum, which is obviously not equivalent to an AM vector pointing up. Therefore, it can't be the physical properties which add in a superposition.

The incorrect 50:50 claim

Alright, now, how about the claim in quotation marks? The claim could be interpreted in at least two ways.

First, someone may imagine that it oscillates back and forth between the two x-basis states in a 50:50 ratio. Maybe the person thinks it's flashing between them like a stroboscope. I think some ill-informed chemistry teachers contributed to the general idea of oscillating states by saying that some molecular wave functions oscillate between two resonance structures. That's at least what I was told in high school about benzene.

You may be aware that people alternatively think about superposition along these lines: a particle that is in a superposition of being in two boxes (or two wells of a magneto-optical trap) is really in one of the two boxes, and we just don't know which it is in. Or perhaps someone insists that an electron really does go through one slit when two are available, but we are ignorant of the truth. This was Schrödinger's view of superpositions around the time of his cat paper. The equations we write allegedly only tell us our degree of ignorance. Supposedly, the actual state is one of the basis states and is unknown until we measure it.

Both the stroboscopic and ignorance views can be refuted with the same argument. For concreteness, I choose to use the ignorance view as my target. After reading the refutation below, you can go over it again with the stroboscopic view in mind if you wish.

Those with the ignorance view would have to say that instead of an actual $|+z\rangle$ state, the superposition says that half of the $|+z\rangle$ particles that we are thinking of are actually in the $|+x\rangle$ state and the other half are in the $|-x\rangle$ state. This already denies that there is a spin-up state $|+z\rangle$, which is pretty absurd, but we can go further down their rabbit hole to get an outright contradiction from this view.

Another equivalent representation of the spin state is perfect for completely refuting the we-just-don't-know-its-actual-state interpretation of superpositions. A second equivalent way of representing our spin that points up along the z-axis is with a superposition of an equal weighting of the kets spin-out $|-y\rangle$ (spin pointing out from the page, from your point of view) and spin-in $|+y\rangle$ (spin pointing in towards the page):

$$|+z\rangle = \underbrace{|+y\rangle}_{\sqrt{50\%}} + \underbrace{|-y\rangle}_{i\cdot\sqrt{50\%}} \tag{11.10}$$

Don't be worried about the i, the imaginary unit. It turns into a 1 when you calculate the probabilities in measurements because its magnitude is 1 (i.e., $|i|^2 = 1^2 = 1$).

According to the we-just-don't-know-its-actual-state view, they now have to say the spin-up particle is really in one of either the $|+y\rangle$ or the $|-y\rangle$ state with 50% probability each. But that is at odds with the previous claim that the particle is in either the $|+x\rangle$ or the $|-x\rangle$ state with 50% probability each. They have a contradiction, and it is becoming more difficult to deny that the two kets on the right side contribute to a resultant ket.

Other spin-$\frac{1}{2}$ states

Basis kets, representations, and decompositions

Above, I showed you three different but completely equivalent ways to represent a spin-up particle using sets of basis kets. In this section, I will show you how we represent spins pointing in the other Cartesian directions. These representations will be helpful to us when we get to picturing vectors in spin space again as well as when we put spins through an idealized Stern-Gerlach experiment.

In the previous section, the basis kets I used to represent a spin-up state are:

$$\begin{gathered} |+x\rangle\,,\ |-x\rangle \\ \text{and} \\ |+y\rangle\,,\ |-y\rangle \\ \text{and} \\ |+z\rangle\,,\ |-z\rangle \end{gathered} \tag{11.11}$$

You didn't see the spin-down ket $|-z\rangle$, though. That's because its quantum amplitude, in the spin-up case, is zero. So, you can think of it as a lazy office worker. In a technical sense, it was present, but it didn't contribute anything.

It is informative to know that there are, in fact, an infinite "number" of other basis kets that we could use to represent a spin-up state. Pick any axis in three-dimensional space. The two directions along that axis define your two new basis kets. As one example, I could have chosen to represent the spin-up state as a superposition of spin states along the axis that bisects the z- and x-axes. The kets might be labeled as follows: $|+zx\rangle$ and $|-zx\rangle$. Physicists, however, don't discuss these other ones because they don't help much. Any single basis is usually sufficient for research purposes. The set of three is sufficient for most educational purposes because they seem the most natural to our human brains.

I will gather together the three natural representations of the spin-up state so you can refer to them as we look at other states:

$$\begin{aligned} |+z\rangle &= \underbrace{|+x\rangle}_{\sqrt{50\%}} + \underbrace{|-x\rangle}_{-\sqrt{50\%}} \\ &= \underbrace{|+y\rangle}_{\sqrt{50\%}} + \underbrace{|-y\rangle}_{i\cdot\sqrt{50\%}} \end{aligned} \tag{11.12}$$

To fully grasp this chapter, you must accept that those three mathematical representations written on the page all mean spin-up (caveat: as long as we aren't dealing with superpositions of spatially-separated wave packets). They all equally represent the spin-up state "out there" in the real world.

To help you, I will illustrate the idea with an analogy to something familiar. Suppose I want to discuss the *number* "five". I could represent it in various ways. I could use base-10 notation and write the *numeral* "5". Or I could use base-2 and write the numeral "101". I could even use base-1 and write "11111". There is no end to the ways we can represent numbers as numerals. Unless you are an ancient Babylonian time traveler, you probably find the base-10 system most convenient. A young child uses base-1 when counting on their fingers. A computer uses base-2. Still, all the numerals describe the same number. This situation is analogous to the quantum object we imagine to exist out in the real world and to the kets, in a particular basis, we write down to represent that object. We write down whatever basis is most convenient and conceptually easy for our limited brains in the present situation.

Before you see other spin-$\frac{1}{2}$ states, I want to point out some vocabulary points. You will see physicists use the words *representation* and *basis* in very similar ways. A **representation** of a state is when we write down a set of orthogonal basis kets and each ket is assigned a quantum amplitude (even when a quantum amplitude is zero and the ket is not written on the page). Therefore, Eq. 11.11 is a list of basis kets and is not a representation, but Eq. 11.12 has three representations. Considering their close meaning, I, like most physicists, will use the words *representation* and *basis* almost interchangeably when discussing spin. Beware that I also use *representation* in its everyday meaning.

Since we have some choice in which representation to use with spin, you may be wondering how it relates to the choice of decomposing a function into other functions, which I mentioned back in the *Waves and Superposition* chapter (p. 56). The choice associated with decomposing a function is about how we can choose to arbitrarily divvy up quantum amplitudes. We can do an analogous thing with spin states:

$$|+z\rangle = \underbrace{|+z\rangle}_{50\%} + \underbrace{|+z\rangle}_{50\%} \tag{11.13}$$

This is mathematically legitimate, but practically speaking, when dealing with spin, the right side is pointless most of the time. Nothing has been accomplished except for adding extra steps.

There is an exception to the previous sentence. I'm not showing them, but you know that there is a spatial ket attached to each spin ket by the tensor product symbol ⊗. (I will show an equation with ⊗ again later in Eq. 11.19.) When the wave packets associated with each decomposed spin ket on the right side of this equation are spatially separated, we can think of the identical spins as separate. There will be an incidence of this situation when I discuss 720° rotations of neutrons later in this chapter.

Finding the spin-down state via a 180° rotation

We will now examine the different ways to represent a spin-$\frac{1}{2}$ particle when its angular momentum is along the remaining positive and negative axes of the Cartesian directions. I want to emphasize that we are going to be considering new states. There are six ways vectors can point along the positive and negative Cartesian directions. That means we will be looking at how to represent six different states in total, and Eq. 11.12 above is one state represented in three ways. We have five more states to go.

Say I gave you a teacher's pointer stick pointing up. How would you get it to point down? You would rotate it by 180°. I'll do that too to the spin-up state to get the spin-down state. With spins, a rotation is normally accomplished physically by a homogeneous magnetic field, a radio-frequency electromagnetic pulse, or rotating by hand a macroscopic object. Mathematically, I need a rotation operator. A general rotation operator takes as an input the angle we want the state to be rotated through and the axis to rotate about. It does the necessary math to get the ket corresponding to that state. Applying a spin-$\frac{1}{2}$ rotation operator to the spin-up representations (not shown) and doing the matrix mechanics, I get the three "natural" spin-down representations:

$$\begin{aligned} |-z\rangle &= \underbrace{|+x\rangle}_{\sqrt{50\%}} + \underbrace{|-x\rangle}_{\sqrt{50\%}} \\ &= \underbrace{|+y\rangle}_{i\cdot\sqrt{50\%}} + \underbrace{|-y\rangle}_{\sqrt{50\%}} \end{aligned} \tag{11.14}$$

Take a minute to compare the spin-up and spin-down representations in Eq. 11.12 and 11.14 so that you may notice the minor differences between the quantum amplitudes, which lead to completely different states.

Finding the remaining states via 90° rotations

The four remaining states are spin-left $|-x\rangle$, spin-right $|+x\rangle$, spin-in $|+y\rangle$, and spin-out $|-y\rangle$. In the background, I will apply 90° rotation operators, which rotate about the appropriate axes, to get the four remaining spin states.

The spin-right state has these three natural representations:

$$\begin{aligned} |+x\rangle &= \underbrace{|+y\rangle}_{(1+i)\cdot\sqrt{25\%}} + \underbrace{|-y\rangle}_{(1+i)\cdot\sqrt{25\%}} \\ &= \underbrace{|+z\rangle}_{\sqrt{50\%}} + \underbrace{|-z\rangle}_{\sqrt{50\%}} \end{aligned} \tag{11.15}$$

The amplitudes with $(1+i)\cdot\sqrt{25\%}$ might look a little scary. They have the same weight as the amplitudes with the $\sqrt{50\%}$ when probabilities are calculated. That is to say, if you chose the magnitude of any amplitude on the right side of the equation above and squared it, you would find the answer to be 0.5 = 50%.

Moving on to the last three spin states in the Cartesian directions, you will see a similar pattern in the amplitudes. The spin-left state has the following three natural representations:

$$\begin{aligned} |-x\rangle &= \underbrace{|+y\rangle}_{(1-i)\cdot\sqrt{25\%}} + \underbrace{|-y\rangle}_{(i-1)\cdot\sqrt{25\%}} \\ &= \underbrace{|+z\rangle}_{\sqrt{50\%}} + \underbrace{|-z\rangle}_{-\sqrt{50\%}} \end{aligned} \tag{11.16}$$

The spin-in and spin-out states are:

$$\begin{aligned} |+y\rangle &= \underbrace{|+z\rangle}_{\sqrt{50\%}} + \underbrace{|-z\rangle}_{-i\cdot\sqrt{50\%}} \\ &= \underbrace{|+x\rangle}_{(1-i)\cdot\sqrt{25\%}} + \underbrace{|-x\rangle}_{-(1+i)\cdot\sqrt{25\%}} \end{aligned} \tag{11.17}$$

and

$$\begin{aligned} |-y\rangle &= \underbrace{|+z\rangle}_{\sqrt{50\%}} + \underbrace{|-z\rangle}_{i\cdot\sqrt{50\%}} \\ &= \underbrace{|+x\rangle}_{(1+i)\cdot\sqrt{25\%}} + \underbrace{|-x\rangle}_{(i-1)\cdot\sqrt{25\%}} \end{aligned} \tag{11.18}$$

The squared magnitudes of the amplitudes for these last three equations will also come out to be 0.5 = 50%.

I included six states with three representations each so that you may see how similar the math appears. More importantly, I want you to understand that there are multiple representations of a single state that are 100% mathematically equivalent. Though, everyone finds the "lone ket" representation (the left side of each of the equations above) to be the most intuitive.

In other sources, in place of amplitudes of the form $(1 + i) \cdot \sqrt{25\%}$, you will see neater, more compact amplitudes like those in the first line of the equation above. I have applied a single rotation operator to get the amplitudes in this chapter. Other authors will have applied an additional rotation after my single rotation. Their second rotation axis is along the new direction of angular momentum. The second rotation can be imagined as rotating the vector arrow about the direction it points. They found a rotation angle that makes the amplitudes compact and aesthetically pleasing. It is commonly said that they are exploiting the **phase ambiguity** of complex-valued amplitudes; therefore, they are allowed to do this. Neither method is wrong, per se. However, I think the phase-ambiguity explanation is somewhat misleading. Both methods will represent a spin state with the spin angular momentum in the same direction, yet different amplitudes show that the states are different. As far as I can tell, the only time that the difference in amplitudes will matter is when a particle superposes with another particle of the same type; the two scenarios will result in differing interferences of the amplitudes. I briefly talk about 720° rotations with neutrons further down, which is an example of this interference.

Quantum amplitudes and the SG experiment

First a review

You have seen many equations in this chapter. Now is a good time to show you how these spin representations relate to the physical world. I introduced you to the famous Stern-Gerlach (SG) experiment in the *Math vs. Measurement* chapter. In this case, we will go slightly deeper and get an idea of how the quantum amplitudes affect things.

To make the following discussion go smoothly, you need to have the essentials of the Stern-Gerlach apparatus firmly in your mind because I will verbally modify it later (see Fig. 11.5). There are important points to be aware of. There is a beam of silver atoms emerging from a furnace. Silver atoms are spin-$\frac{1}{2}$ particles and have no orbital angular momentum. All atoms emerging from the furnace have the same spin orientation in any one scenario. The beam always travels along the $+y$-direction in the coordinate system we imagine, which will never change.

In the figure, the magnetic-field axis is in the z-direction. I will be verbally rotating the electromagnet and its magnetic-field axis about

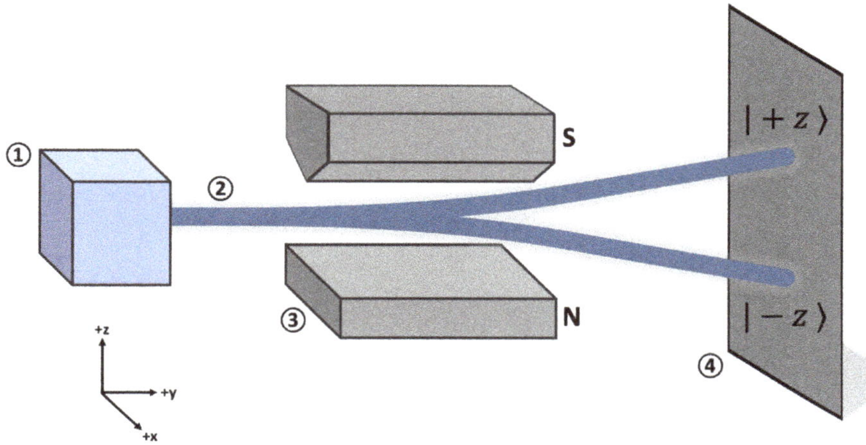

Figure 11.5: The standard Stern-Gerlach setup you saw in a previous chapter. (p. 104.) In this image, the magnetic-field axis (i.e., the north-south axis) is in the z-direction. Soon, I will talk about rotating the electromagnet around the axis of the atomic beam. ① Oven evaporating silver metal. ② Atomic beam. ③ Electromagnet. ④ Imaging detector.

the y-axis. This means the magnetic axis will point somewhere in the xz-plane. I will also restrict the atoms coming out of the oven to have a spin orientation somewhere in the xz-plane. If I included all spin orientations, the results of experiments wouldn't change, but my explanation would be longer.

You should pay attention to the fact that the poles of the electromagnet have different shapes, and they are also marked as north or south poles. The field is inhomogeneous in such a way that magnetic dipoles that point directly up or down experience a force moving them up or down (this is a fact from classical electrodynamics). The electromagnet's pole that a spin travels toward is determined by the magnetic field direction, the field inhomogeneity direction, and the direction of the magnetic dipole moment (which in turn depends on the direction of the angular momentum and the sign of the particle's charge [recall p. 24]).

In our particular setup, with the magnetic dipole moment coming from electrons (in silver atoms), angular momentum that is pointed toward the south pole of this magnet will move toward that pole. Similarly, angular momentum that is pointed toward the north pole of this magnet will move toward that pole. Hold on to that fact about the direction of movement throughout all the examples.

We know spin is distributed throughout the entire spatial wave function because of the tensor product symbol ⊗. Spin angular momentum is a property of matter, and you can't separate it from matter. I showed you the following wave function in my first explanation. This time, I have added generic quantum amplitudes *a* and *b*:

$$|\Psi_{\text{atom}}\rangle \;=\; \underbrace{|\Psi_{\text{wp1}}\rangle \otimes |+z\rangle}_{a} \;+\; \underbrace{|\Psi_{\text{wp2}}\rangle \otimes |-z\rangle}_{b} \tag{11.19}$$

This shows us that a wave packet (wp) is associated with each spin. Note that each wave packet ket represents a continuum of quantum amplitudes, so *a* and *b* can be seen as multiplying those and, if one desired, could be hidden inside those kets.

In the previous section, I emphasized the physical equivalence of the different representations. Once magnetic fields get involved, they aren't treated equally. The magnetic field will act on the representation that is in the direction of its magnetic axis (the north-south axis). We will take the last sentence as a fact because I can't give a satisfactory explanation for why it happens.

In our current scenario, the magnetic axis is in the z-direction, so it acts on the magnetic dipole moments of each z-basis ket, and the wave packets will move apart in space. In the end, the wave packets collide with the detector, and our best guess in this book is that the wave function somehow gets randomly localized inside one of the detector pixels.

Going forward, I will use a leaner schematic to represent the atomic beams than you see in the figure above. The lines will no longer show an accurate path. They will show that the beam split, but the paths aren't accurate. The line widths will be proportional to the fraction of the silver atom in the wave packet. This fraction is found by taking the magnitude of the amplitude and squaring it (i.e., $|a|^2$ and $|b|^2$).

'Lone' kets

In the first schematic image below (Fig. 11.6), I have sent a spin-up particle through the magnetic field. It has its magnetic axis parallel to the spin axis. This is a special situation where the atomic beam won't split but only get deflected because $|a|^2 = 100\%$.

Figure 11.6: A spin-$\frac{1}{2}$ particle transiting a SG apparatus with its spin aligned with the magnetic field. The beam travels from left to right, which is the $+y$-direction. Up on the page is the $+z$-direction.

Now, let's say I flip the electromagnet upside down. The magnetic field direction is now in the opposite direction. The spin-up particles are still attracted to the same triangular pole shape, so they go downward, as in Fig. 11.7.

This is fine with us because we are good scientists and are keeping track of the magnetic field. Even though the silver atoms go down, we know that the spin angular momentum started out pointing up and still points up.

Figure 11.7: A spin-$\frac{1}{2}$ particle transiting a SG apparatus with its spin aligned with the magnetic field. Everything is the same as in the previous figure except that the electromagnet has been flipped 180 degrees.

You should be able to see that if I had started this explanation with spin-down particles (i.e., $|b|^2 = 100\%$), the two images above would have been switched. Once you understand that, you must convince yourself that had I started with a purely spin-left or spin-right particle (i.e., $|+x\rangle$ or $|-x\rangle$) and additionally laid the SG apparatus on its side so that the magnetic axis was along the table, the two paths above are the possibilities, but the schematic paths are now top-down views. This is because a magnetic field along the x-axis will act on the magnetic dipoles of the x-basis kets.

An equally weighted superposition

Next, we return the Stern-Gerlach apparatus to its original orientation of a vertical magnetic axis along the z-direction. For the rest of this section, the magnet will remain in that orientation. So, when we look at a superposition, we want to look at the quantum amplitudes of the z-basis kets.

In the next example, we choose to input either a $|+x\rangle$ or $|-x\rangle$ state. Looking back at Eq. 11.15 and Eq. 11.16, we can see in both cases we will find an equal weighting of z-basis kets: $|a|^2 = 50\%$ and $|b|^2 = 50\%$. In the image (Fig. 11.8), via the line thickness, you can see that the matter density in each of the split paths is half of that in the original. After an experiment with numerous particles is complete, two regions of equal weighting will be measured on the detector.

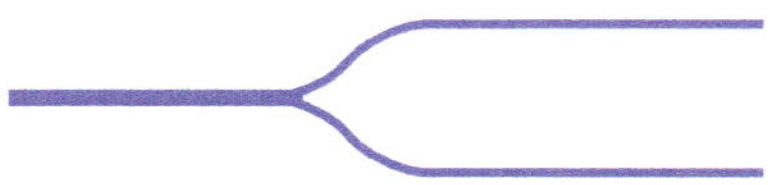

Figure 11.8: Schematic of a SG experiment for a spin-$\frac{1}{2}$ particle with the input spin polarization 90° from the magnetic-field axis. The beam travels from left to right, which is the $+y$-direction. Up on the page is the $+z$-direction.

It will be informative to contrast this with what one expects of a classical magnetic dipole aligned along the x-direction in this magnetic field along the z-direction. In classical physics, a dipole that is perpendicular to the inhomogeneous field will not experience a force; thus, there will be no deflection. Therefore, in classical physics, one expects a single spot on the detector in our present scenario.

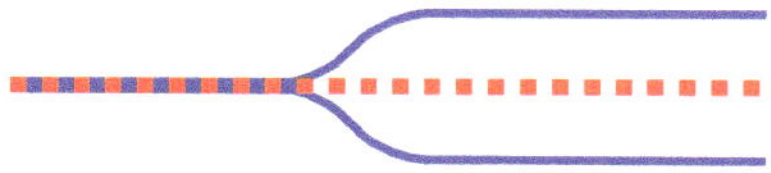

Figure 11.9: Same as previous figure, but with the dotted line, I show the result we expect if the magnetic moment behaved classically.

The striking difference between the classical and quantum expectations is what convinced Stern to ask Gerlach to do the experiment. Stern did not believe in the idea of quantized angular momentum and wanted to prove it wrong. After the results of this experiment came out in favor of the quantized mechanics, many skeptical physicists began to believe the world was not classical. I think this experiment tipped the scales because it was simple to understand and believable. Whereas, other aspects of *Old Quantum Theory* could be unbelievable or incomprehensible. I suppose, too, that the undeniable evidence of quantization contributed to the mindset necessary for Heisenberg to discover matrix mechanics a few years later.

Other superpositions

This time, we will input the $|+zx\rangle$ state I mentioned earlier. This input spin points in the direction that bisects the positive z- and x-axes. I want to draw the line thickness according to the information in the quantum amplitudes. To do this, I need to write $|+zx\rangle$ in the z-basis because the magnet splits the wave functions into that basis. How do I get the equation? After some thinking, I recognize the state $|+zx\rangle$ is the state $|+z\rangle$ rotated 45° towards $+x$-axis. So, I apply the appropriate rotation operator to $|+z\rangle$, to find:

$$|+zx\rangle = \underbrace{|+z\rangle}_{\approx\sqrt{85.4\%}} + \underbrace{|-z\rangle}_{\approx\sqrt{14.6\%}} \tag{11.20}$$

Squaring those amplitudes gives us the amount of the silver atom going down each branch (i.e., approximately 85.4% and 14.6%). The detector will randomly measure particles in its upper and lower pixels. On average, the distribution will be 85.4% in the upper and the remaining balance in the lower pixels (see Fig. 11.10).

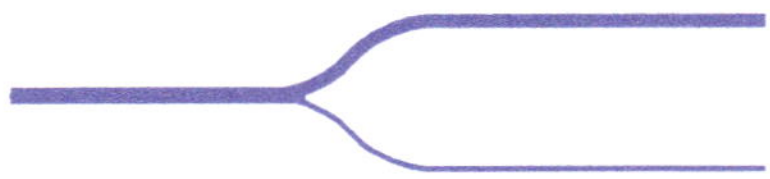

Figure 11.10: A SG experiment of a spin-$\frac{1}{2}$ particle with its input spin pointing 45° from vertical. The beam travels from left to right, which is the $+y$-direction. Up on the page is the $+z$-direction.

This scenario is at an angle intermediate to the orientations of the previous two scenarios, as are the weightings. We can imagine doing a series of experiments of numerous spin orientations that gradually go from spin-up to spin-right to spin-down. As the input states change, the weightings of the branches smoothly change from purely the upper channel (spin-up) to mostly the upper channel (a range of orientations) to equally both channels (spin-right) to mostly the bottom channel (a range of orientations) to purely the bottom channel (spin-down). I show you the weightings in the plot of Fig. 11.11. I will use this plot in the following section and show you more plots of this type, starting with spin-1 particles near the end of the chapter.

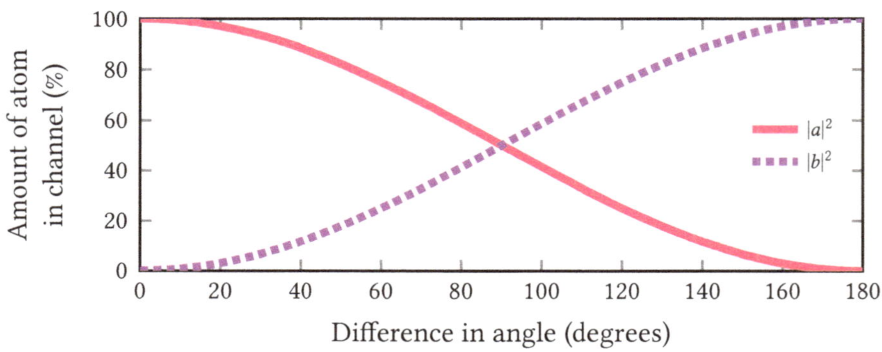

Figure 11.11: Amount of silver atom in the upper and lower channels of the SG apparatus at particular orientations of AM. The angle is measured from the positive z-direction. Therefore, 0° represents spin-up, 90° represents spin-right/left, and 180° represents spin-down.

Learning from untrue claims: example #2

The claim

I previously mentioned that some people think that superpositions only represent our ignorance of what is really happening. There is an idea worse than that. It is the idea the universe doesn't even know.

Brian Greene, in the PBS documentary *The Fabric of the Cosmos: Quantum Leap*, has partly contributed to this idea. Around the 27-minute mark, a colorful, glowing orb with an implanted spin-vector appears that is wildly jumping between many different orientations (perhaps this misleading visual should have been mentioned when I talked of stroboscopic jumping between states). At this time, Brian Greene says,

> Unlike a spinning top, an electron spin, as with other quantum qualities, is *generally completely fuzzy and uncertain*, until the moment you measure it. And when you do, you will find it is spinning clockwise or counterclockwise.

Of course, the spinning clockwise part is wrong; as you know, nothing is spinning. But that's not the point here. I emphasized with italics the part I want you to pay attention to. The message he is sending is not true at all if you are a realist.

From what he says in the first chapter of his book *The Fabric of the Cosmos*, I believe Greene describes himself as an anti-realist, so he is on safe ground in that interpretation. Because, in that interpretation, you are supposed to plug your ears and shout, "I'm not allowed to discuss what exists between measurements." I guess, with that philosophical position, one could say things are fuzzy and uncertain before a measurement. (Though anti-realism does seem to be at odds with Greene's visualizations of quantum mechanics and descriptions of the happenings in the quantum realm.)

The resolution

I say the electron exists at all times, so let me show you why I say an electron's spin is not fuzzy and uncertain. Consider a general spin-$\frac{1}{2}$ state with unspecified quantum amplitudes a and b (which, because of its simplicity, I write in the traditional convention):

$$a \cdot |{+z}\rangle \; + \; b \cdot |{-z}\rangle \tag{11.21}$$

One might think that this means that something is unknown or unknowable because we haven't shown the values of a and b. In actuality, the a and b are not precisely specified to keep the argument generally applicable. The statements in the following paragraphs work no matter which values are chosen for a and b.

Every spin-$\frac{1}{2}$ particle has an angular-momentum vector pointing in a definite direction. Always. There is always an orientation in space that you could orientate a Stern-Gerlach apparatus's magnetic axis so that it will measure a spin-$\frac{1}{2}$ particle in that apparatus's upper channel. Naturally, there is also a second orientation 180 degrees from that which would measure it in the lower channel.

How do I know this about spin-$\frac{1}{2}$ measurements? Say you have a spin-$\frac{1}{2}$ particle in a superposition like that in the equation above. A particular basis was chosen for that expression. There is always a different basis, which can be arrived at by rotation operators, where the spin-$\frac{1}{2}$ particle is represented purely by one ket:

$$a \cdot |{+z}\rangle \; + \; b \cdot |{-z}\rangle \; = \; |\chi\rangle \tag{11.22}$$

The Greek letter χ represents some direction in space that is the direction of the particle's spin angular momentum. We can get a feel for this using my previous analogy of numbers and numerals. Because 2 + 3 is a superposition of numerals in base-10 doesn't mean the number five is fuzzy and uncertain. The number five can also be written clearly as a single numeral in some basis (e.g., 5, 101, or 11111).

Some clarifying remarks

I should mention something that may have confused you. Earlier in the chapter, when I discussed the rotation operators, I described them as operators that forcibly change the state/object. This is called an *active rotation.* In contrast, in this case, I chose to think of the rotation operators as doing *passive rotations*, which means they change our view but leave the spin untouched. In other words, a spider on the wall can always skitter around the room until it finds a point of view where the spin is up or down.

We can co-opt the image in the previous figure (Fig. 11.11) to think in terms of passive rotations. Now, it's a proxy for finding the orientation where a quantum amplitude is 100% in a new representation. I also need to mention that this figure that I have co-opted was created from rotations in a single plane. The principle I describe truly works for any spin direction in three-dimensional space.

Say your atom has a pair of $|a|^2$ and $|b|^2$ values in some representation. Find the angle for those on the figure, and treat the angle as the orientation of your head. Then, you can find how much you must rotate your view to get to 0° or 180° where one of the amplitudes is 100%. You then have confirmed the state has a definite direction of spin AM. (By the way, I say more about the technicalities of passive and active rotations in the appendix on matrix mechanics.)

Back to the main point

The only time I've ever seen someone give a name to the definite-direction idea is when Leonard Susskind and Art Friedman call it the *spin-polarization principle* in their *Theoretical Minimum* book on quantum mechanics. Unfortunately, we can't make a similar statement that there is a guaranteed output of a Stern-Gerlach apparatus when using higher spins. There are certain states of higher spin particles that will, for all orientations of a Stern-Gerlach apparatus, have random outputs. (I will show you this when we get to higher spins.) If I could, I would suggest to Susskind and Friedman to add "1/2" to the name so that it is the **spin-$\frac{1}{2}$ polarization principle**, which is more restrictive and more accurate.

Finally, let me be very clear. In the absence of a collapse of a wave function, nothing is ever random about a spin-$\frac{1}{2}$ or a higher spin state (or any state for that matter). Time evolution under the Schrödinger equation is causal and absolutely predictable. So, when following the reality interpretation, you can safely ignore Brian Greene's message about uncertainty before a measurement is made.

Superpositions in spin space

Now that you have some understanding of representations and superpositions, we return to the abstract spin space. This time, I will show you how superpositions work in this space. Recall, though, that it's not going to be intuitive. Have a look at the spin-right state again:

$$|+x\rangle = \underbrace{|+z\rangle}_{\sqrt{50\%}} + \underbrace{|-z\rangle}_{\sqrt{50\%}} \tag{11.23}$$

Perhaps you are wondering if there are other spin-$\frac{1}{2}$ spaces because we have kets in different bases. Nope. They are all in the same spin space. The $|+x\rangle$ ket vector is made from two vectors with lengths of $\sqrt{50\%}$ determined from the quantum amplitudes. These lengths are approximately 0.71. We can draw this on the spin space.

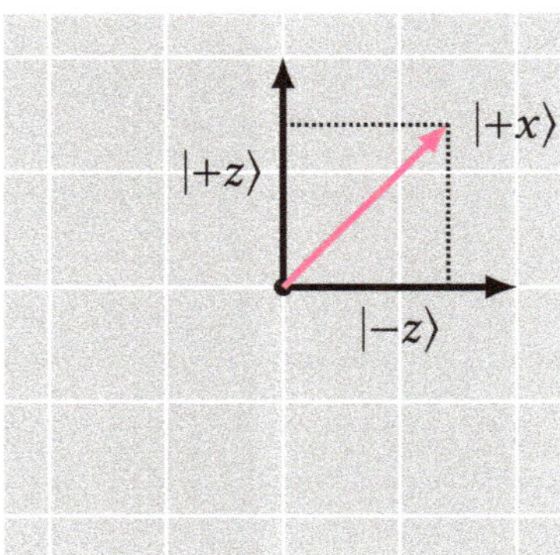

Figure 11.12: Spin space with spin-up, -down, and -right states.

We can add the other x-basis vector using the following equation:

$$|-x\rangle = \underbrace{|+z\rangle}_{\sqrt{50\%}} + \underbrace{|-z\rangle}_{-\sqrt{50\%}} \tag{11.24}$$

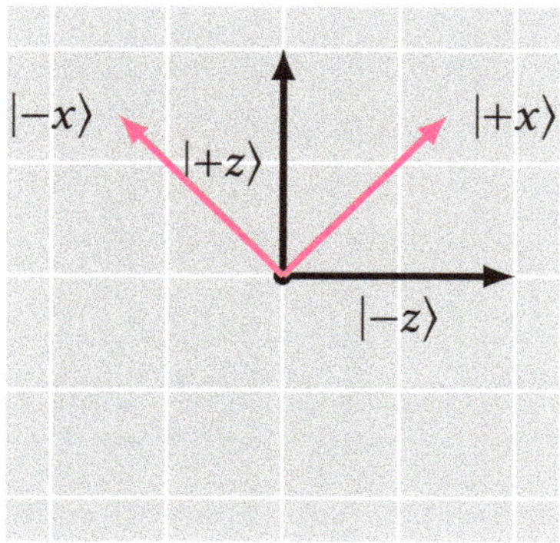

Figure 11.13: Spin space has all the other representations.

If we want to plot the y-basis on this same figure, we have a complex twist to what we have been doing. The quantum amplitudes now have the imaginary unit i them (see the top line in Eq. 11.17 and Eq. 11.18). But we can't draw vectors with imaginary lengths. Without the imaginary unit in those amplitudes, the y-basis kets would lie on the x-basis kets and would be indistinguishable from each other.

In an act of desperation, we could appeal to the color wheel and say the vectors have a color representing their phase. The imaginary axis is 90° from the real axis. Looking at the color wheel from a previous chapter, we see that orange and blue are 90° from magenta and cyan. We could then think of colored vectors. A problem arises because negative numbers now have colors. We would then have to stick everything in the upper right quadrant of spin space. These vectors of magenta, cyan, orange, and blue colors (i.e., $|\pm x\rangle$ and $|\pm y\rangle$) would all sit on top of each other. Perhaps you could imagine the kets as glowing vectors of particular colors and try to invent a color theory of mixing to make superpositions in spin space make sense.

The question we must ask is, "Is this going to help us?" Few would think this is a good use of time. Moreover, later, we will see states of spin-$\frac{3}{2}$, which have abstract spin spaces with four dimensions, which I suppose no one can imagine (and which I cannot draw for you).

Rather than trying to understand superpositions via visualizing spin spaces, physicists trust the algebra and crunch through the math to get the resultant ket and its properties. Having told you that, I say that you can visualize the superpositions in abstract space in limited circumstances (i.e., limited dimensionality and limited to real amplitudes). Regardless, in the end, do know that the abstract space is the reason you find superpositions unintuitive.

Phase ambiguity and a 360° rotation

There is a famous conundrum that involves rotating a spin-$\frac{1}{2}$ particle, such as an electron, by 360° and not getting back the state we started with. To accomplish this mathematically, we apply a rotation operator of 360° and get an eigenvalue of -1 in front of the ket. This tells us the phase of the state has shifted. We have to do another full rotation in space, for a total of 720°, before we get back to where we began, the original state, which is quite contrary to our traditional notions.

Cups and belts

There is a macroscopic analogy that illustrates a need for 720° rotation before returning to the original state. It is known as Dirac's belt trick or Feynman's cup trick. The cup analogy is easier to explain but requires good elbow and shoulder mobility. Your arm will be the phase of the spin-$\frac{1}{2}$ particle, and the cup will track the amount of rotation in physical space. A mirror will help you observe yourself.

Start with a cup that has something indicating a direction so that you know when it is pointing in the same direction again. A cup with a handle or a paper cup you have marked will do. (Honestly, you can forgo the cup and hold a pen). Have your palm face up and put the cup on it. The starting position of your arm will be out to the side, like you are a waiter carrying a tray. The cup's handle will point forward (or in your preferred direction). Throughout all this, your palm must always face upwards.

Begin rotating your arm to the front and inward toward your body. You need to contort your arm and body so that the cup goes under your armpit and out to the side. The cup will be lower than when you started, and you will be in an awkward position as if a self-defense expert put you in an elbow lock. When you reach the end of your range of motion, in the mirror, you should see that the cup's orientation is back to its original position. It did a 360° rotation, but your arm, representing the spin-$\frac{1}{2}$ particle, is not back to its original position.

To continue rotating, you will need to bring the cup above your shoulder or head level. It might take some trial and error. Don't hurt yourself. In then end, after 720° of rotation, the cup will be in its original orientation, and so will your arm.

Seeing it in spin space

Better than a cup and arm analogy is a visualization with images of physical and spin space as we rotate a spin from spin-up to spin-right to spin-down to spin-left and return to spin-up (Fig. 11.14). To return the ket to the original orientation in spin space, another 360° rotation would be needed in physical space.

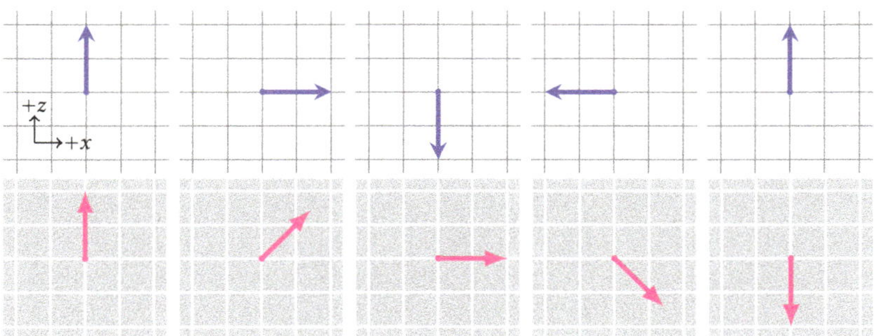

Figure 11.14: Phase shift from rotations Going from left to right, the AM is rotated 90° from its previous position in physical space (top row). In the abstract spin space (bottom row), we start at $|+z\rangle$ and end pointed in the opposite direction, which is $-1 \cdot |+z\rangle$.

This half-speed rotation in spin space is seen in all half-integer spin systems. They all pick up a phase of -1 after a $360°$ rotation. In contrast, integer spin systems will be back to the original state in physical and abstract space after a $360°$ rotation. The mathematical reason for the difference is that the rotation operators for the integer spins have an angle θ in the arguments of trig functions and the rotation operators for half-integer spins have a $\theta/2$.

The neutron-interference experiment

There is a notable neutron-interference experiment first conducted in 1975 that confirmed that a neutron needs to go through $720°$ of physical rotation to return to its original state. See Fig. 11.15 for a schematic of the neutron-beam path. Only a portion of what is called a *perfect-crystal (or perfect Si-crystal) neutron interferometer* is shown. It used a single crystal of silicon (Si) several centimeters across with several thin slabs jutting out of the base. Its crystal planes are nearly perfectly aligned. Understanding how the thin slabs split or combine the neutron beams requires understanding crystal diffraction, which is too specialized for this book. I will say though that neutrons and perfect crystals were needed because it is too difficult to precisely recombine atomic beams with magnets.

The experimenters used an electromagnet with a homogeneous magnetic field to rotate one of the neutron wave packet's spin to the desired amount. A homogeneous magnetic field will not cause a magnetic dipole to be attracted to one direction or the other. It will cause the magnetic dipole to rotate as long as it is in the field. The rate of rotation is proportional to the field strength. Knowing the speed of the neutrons, the electromagnet's field strength, the neutron's magnetic moment, and the path length through the field allowed the experimenters to calculate and thus choose the amount of rotation of the magnetic dipole in physical space. The actual experiment used all orientations of neutrons simultaneously. For mathematical simplicity, I show the input neutron beam as spin-up in the figure.

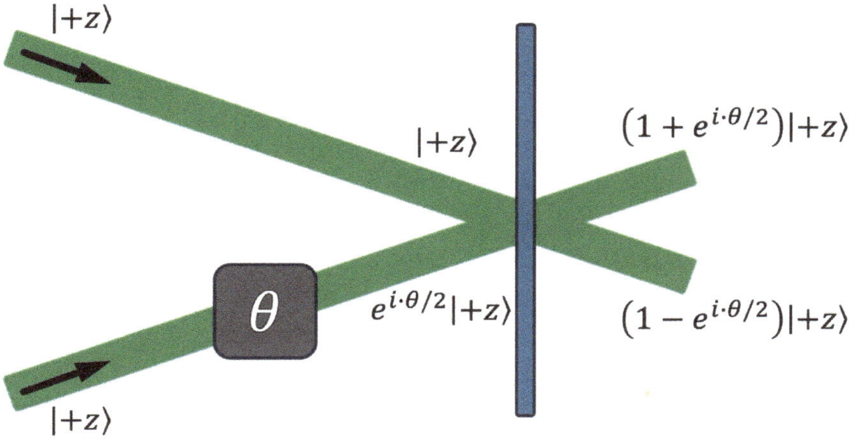

Figure 11.15: Idealization of the experiment and math that confirmed a neutron must go under 720 degrees of rotation to return to the original state. We are looking down from above, down the z-axis. Neutrons move left to right in the green, crossed beams. Much like the SG experiment, each beam has half of the neutron's wave function. Crucially, each half of the neutron enters from the left with the same polarization; I have chosen $|+z\rangle$ for its simplicity. The amplitudes are not normalized. The box labeled with θ is an adjustable-strength electromagnet with a homogeneous magnetic field along the z-axis that, via the magnetic dipole moment, rotates the angular momentum by θ degrees in physical space about the magnetic axis. In spin space, the phase angle of the ket advances only by $\theta/2$. The vertical bar represents a part of a perfect-crystal neutron interferometer, which changes the beams' polarization, mixes them, and splits them again. On the far right, only one output beam is needed to confirm the theory. The experimenters chose to show the results of both beams. The phase values in Eq. 10.4 on p. 194 may be useful to review. Beware that the phase angle on this page is $\theta/2$ and on that page it is θ, so you will have to divide the angles in the previous chapter by 2. Beware also that $\theta = 360^\circ$ appears to make the kets cancel out right before the crystal slab, but that ignores the continuum of amplitudes in the wave packets (recall the paragraph after Eq. 11.19).

For any spin, the direction of the dipole/angular momentum will rotate about the electromagnet's axis (not necessarily the spin's axis). This is a well-known phenomenon called *Larmor precession*. For example, an initially spin-right particle ($+x$ direction) immersed in a magnetic field along the z-axis will cyclically point in all directions in the xy-plane. It will cycle through spin-right, spin-out, spin-left, spin-in, and back to spin-right.

The classical explanation of this phenomenon involves magnetic torque due to the application of the field. In the classical case, there is no torque when the magnetic dipole and the field are perfectly aligned, and we think there is no change of state. In quantum mechanics, the rotation operator disagrees. In the case I am showing you, the magnetic field is applied in the same direction as the angular momentum, which is the z-direction. In addition, I have used the z-representation. The result is a phase factor in front of the ket: $e^{i \cdot \theta/2} |+z\rangle$. This special situation of co-linear spin and rotation axis has a phase ambiguity for all rotation angles. Whereas, non-co-linear situations rotate the spin polarization in other directions and each of those polarizations has a phase ambiguity at multiples of 360° of rotation (see Fig. 11.14).

Once finished the appendix on matrix mechanics, you could apply a rotation operator about the x-axis to a spin-right state written in the z-basis to see a less mathematically simple phase ambiguity. This is the second rotation I mentioned earlier (p. 217) that for certain angles gives aesthetic numbers.

In the neutron-interference experiment, the researchers chose various magnetic field strengths in order to induce various amounts of phase difference between the neutron's wave packets. Once they plotted the intensity of the measured beam versus the calculated rotation angle, a sinusoid with a period of 720° was observed. Therefore, the phase of a particle is physically significant, and a spin-$\frac{1}{2}$ particle needs to rotate by 720° to return to its original state.

Finally, as is common in this book, I must again admit that I did not mention something to you earlier. The inhomogeneous field of the Stern-Gerlach magnet will also cause Larmor precession. However, it has no effect on the results of the Stern-Gerlach experiments, so it is not commonly mentioned except in textbooks.

Why don't we see this phase ambiguity with our eyes?

In a previous chapter, I made the point that crystals, like diamonds, are well-understood, macroscopic quantum objects (p. 75 - p. 77). You may surmise that we should be able to see this phase ambiguity when we rotate macroscopic objects because objects are made of electrons which are spin-$\frac{1}{2}$ particles.

First, some solid-state physics caveats are due. It is not necessarily true that macroscopic objects will be spin-$\frac{1}{2}$. Electrons tend to pair up and cancel their spins. Isolated and at zero Kelvin, diamond will be in its ground state and all the spins will cancel out, making it a spin-0 object. However, a single or a few UV photons can easily change diamond to a higher spin state by promoting electrons to different states which break the pairing. That being said, I won't go into the solid-state physics of how that could happen. For this brief discussion, we will assume that the diamond we imagine we are rotating is a spin-$\frac{1}{2}$ quantum object. If you notice any other conceptual difficulties, such as ignoring the nuclei, you can assume the difficulties away so that you may get to the core of the lesson I am making.

When you rotate the diamond, you don't notice the phase ambiguity because the only way to see it is via interference. You might guess that the photons that reflect off diamond and enter our eyes could be used to somehow cause an interference that is somehow visualizable. Unfortunately, this is impossible. Even though photons and electrons (or any two other particles of different types) are distributed throughout the same physical space, their wave functions don't add (superpose) in the math on the page, so the two particles don't have quantum interference. In quantum field theory, one would say they are different fields.

We must superpose electrons with other electrons for the phase ambiguity to cause interference and a physical difference. But what happen you try to bring the electrons of the diamond into superposition with the electrons of your fingers? It's the same story again that I describe on p. 77. The electron clouds repel and don't superpose to a significant degree. Perhaps at the nanometer scale there are some interference effects where the surfaces of the diamond and your fingers meet, but your eyes aren't atomic-force microscopes, so you won't be able to notice any effects.

Higher and lower spin systems

That was a fairly comprehensive journey through what it means to be a spin-$\frac{1}{2}$ particle. To give you a truer appreciation of spin in general, you should see some of the topics above applied to other spin systems.

As you read through the material below, you will notice that the number of basis kets for each spin-s particle follows the rule $2 \cdot s + 1$, which parallels the $2 \cdot l + 1$ rule for the number of orbitals you saw in the appendix (p. 341). The quantum number l tells us the maximum and minimum values of the orbital-magnetic number m_l. The same is true for the two spin quantum numbers. The spin quantum number s tells us the maximum and minimum values of the spin-magnetic number m_s. The range of values of m_s are $-s, \dots, 0, \dots, s$ when s is an integer or $-s, \dots, -1/2, 1/2, \dots, s$ when s is a half integer. In both cases, there will be jumps of 1 between the m_s values.

Spin-0

The Higgs boson is the only elementary particle with spin-0. On the other hand, the list of composite particles with spin-0 is endless. An alpha particle is composed of two protons and two neutrons. In its ground state, all the spins cancel to make it spin-0. Adding two electrons to the alpha particle gives us a helium-4 atom, and it has a net spin of zero in its ground state. Pions are composed of quarks and antiquarks and are also spin-0 particles.

Since spin-0 particles have no spin-angular-momentum vector, they are the easiest to represent. Here is everything you need to know about their spin:

$$|0 \cdot z\rangle = |0 \cdot x\rangle = |0 \cdot y\rangle \tag{11.25}$$

It's all three natural representations. There is no plus or minus sign because a vector of zero length doesn't point in any direction. This state is not made by a superposition of other states, so there is no colored underbrace telling you the weighting. You know it is 100% for each representation.

Knowing the quantum amplitude will help you avoid getting confused. Some might think that $|0 \cdot z\rangle = 0$. That's incorrect. That says there is no state. The zero in the ket is a label that tells us what the spin's eigenvalue would be if we applied the spin operator. The ket is actually the following single-entry "column vector" of a quantum amplitude:

$$|0 \cdot z\rangle = \left[\sqrt{100\%}\right] = \left[100\%\right] = \left[1\right] \tag{11.26}$$

Okay, this ket notation with the zero in it is a little bit ugly, isn't it? This is where I apologize because I will change how I write spin kets. Up to this point, I have written spin-$\frac{1}{2}$ kets the way I did to make them easier to understand. But, with more complex spins, I need to make the change. From now on, for spins other than spin-$\frac{1}{2}$, I will move the basis sign outside the ket. And inside the ket, I will write the value of the spin-magnetic quantum number m_s. When multiplied by $\hbar$, the quantum number tells us how large the spin angular-momentum vector is. A spin-0 particle in this new notation will then be represented as:

$$|0\rangle_z = |0\rangle_x = |0\rangle_y \tag{11.27}$$

Both of Eq. 11.25 and Eq. 11.27 contain the same information. That information might seem to be nothing, but we can make a few conclusions. Perhaps you want to come up with a few before you read on.

First, since there is no spin angular momentum, you know that this spin state will not have an intrinsic magnetic moment. Second, there is only one ket in a basis, so the abstract space is one-dimensional. You might guess that there could be a changing phase associated with rotating this state. The rotation operator turns out to be multiplication by one, so there won't be any funny business about changing phases and interference. The last conclusion that I can come up with is that a spin-0 Stern-Gerlach experiment will not be something you excitedly write home about because there is no magnetic moment that can be used to deflect the spin-0 particle.

Figure 11.16: A spin-0 Stern-Gerlach experiment in any orientation. The atomic beam is a straight line because without spin angular momentum (and also no orbital angular momentum), the particle has no magnetic field and thus won't deflect.

Spin-1

The elementary particles with spin-1 are photons, gluons, Z bosons, and W bosons. There are no known elementary particles higher than spin-1. (Some pop-sci authors are fond of mentioning an elementary and spin-2 graviton particle as if it is known to exist. It is actually part of a hypothesis among many other hypotheses that attempt to make gravity quantum mechanical.) A deuterium (heavy hydrogen) nucleus is a composite particle with its proton and neutron spins aligned, resulting in it being a spin-1 nucleus.

The m_s = +1 state

If we consider a spin-1 particle with its angular-momentum vector aligned along the positive z-direction, I can show you how the z-basis and the x-basis are related:

$$|+1\rangle_z = \underbrace{|+1\rangle_x}_{\sqrt{25\%}} + \underbrace{|0\rangle_x}_{\sqrt{50\%}} + \underbrace{|-1\rangle_x}_{\sqrt{25\%}} \tag{11.28}$$

But I have created an ambiguity in our notation. If I didn't tell you this equation is about a spin-1 particle and you were in a hurry, you might think the symbol $|0\rangle_x$ belongs to a spin-0 particle. Alternatively, maybe you think $|+1\rangle_z$ belongs to a spin-2 particle. Neither of which makes sense. We need to add to our notation again.

In the final notation that I use, I must tell you that this is a spin-1 particle and its actual amount of angular momentum. This requires the two quantum numbers s and m_s. The values of s and m_s can only be integers or half of an integer. Our kets will have a form like:

$$|s, m_s\rangle_z \tag{11.29}$$

Using this notation on what you are familiar with, here are a few examples of spin-1/2 states:

$$\begin{aligned} |+z\rangle &= \left|\tfrac{1}{2}, +\tfrac{1}{2}\right\rangle_z & |+x\rangle &= \left|\tfrac{1}{2}, +\tfrac{1}{2}\right\rangle_x & |+y\rangle &= \left|\tfrac{1}{2}, +\tfrac{1}{2}\right\rangle_y \\ |-z\rangle &= \left|\tfrac{1}{2}, -\tfrac{1}{2}\right\rangle_z & |-x\rangle &= \left|\tfrac{1}{2}, -\tfrac{1}{2}\right\rangle_x & |-y\rangle &= \left|\tfrac{1}{2}, -\tfrac{1}{2}\right\rangle_y \end{aligned} \tag{11.30}$$

For completeness, I will also show you the spin-0 states from Eq. 11.27 in the final form of our spin notation:

$$|0\rangle_z = |0,0\rangle_z \qquad |0\rangle_x = |0,0\rangle_x \qquad |0\rangle_y = |0,0\rangle_y \tag{11.31}$$

We can then see that a spin-1 particle with $\hbar$ of angular momentum along the positive z-direction will look like this:

$$|1, +1\rangle_z \tag{11.32}$$

Now I can show you that spin-1 particle in that state in all three natural representations:

$$\begin{aligned} |1,+1\rangle_z &= \underbrace{|1,+1\rangle_x}_{\sqrt{25\%}} + \underbrace{|1,0\rangle_x}_{\sqrt{50\%}} + \underbrace{|1,-1\rangle_x}_{\sqrt{25\%}} \\ &= \underbrace{|1,+1\rangle_y}_{\sqrt{25\%}} + \underbrace{|1,0\rangle_y}_{-i\cdot\sqrt{50\%}} + \underbrace{|1,-1\rangle_y}_{-\sqrt{25\%}} \end{aligned} \tag{11.33}$$

To the uninitiated, a lot is going on there. Heck, even when I read it, I still have to slow down and read it carefully. So, I hope that by building up the notation like I did, you will be able to understand every single symbol. Look over that equation closely and see if you can do it right now before moving on. To save space, I won't write out all the possible spin-1 states corresponding to angular momentum along all the Cartesian directions.

A caveat is in order before we visualize the channels of a Stern-Gerlach experiment. Photons, gluons, and Z bosons have no charge, so they won't have magnetic moments. This means they won't deviate in a Stern-Gerlach experiment. Therefore, I say again, when thinking of SG experiments, you should only consider atoms and molecules. (Additionally, for "particle-physics reasons", the spin states for photons and gluons are not like the ones I show in this book: all the $m_s = 0$ kets, whether in a superposition or not, have zero amplitude.)

Inputting an atom in the state $|1,+1\rangle_z$ into a SG apparatus with the magnetic field along x will give us trident-esque split beams. The amplitudes in the first line of Eq. 11.33 tell us the weightings of the beams in Fig. 11.17.

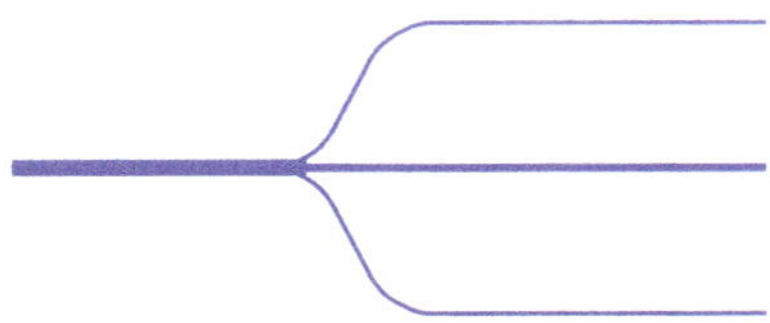

Figure 11.17: SG experiment with an $s = 1$, $m_s = +1$ particle transiting through a magnetic field with its axis 90° away from the spin AM vector. The top and bottom branches are twice as far apart as the spin-$\frac{1}{2}$ cases because, all other things being equal, $\hbar$ will produce twice the dipole moment as $\hbar/2$, so separation distance doubles.

As I did for spin-$\frac{1}{2}$, I show you how the beam weighting changes if we start with the $|1,+1\rangle_z$ atomic beam and gradually change the state to point in different directions in the xz-plane (Fig. 11.18). The $|a|^2$, $|b|^2$, and $|c|^2$ in the figure belong to the z-basis kets with $m_s = +1$, $m_s = 0$, and $m_s = -1$, respectively:

$$a \cdot |1,+1\rangle_z \; + \; b \cdot |1,0\rangle_z \; + \; c \cdot |1,-1\rangle_z \tag{11.34}$$

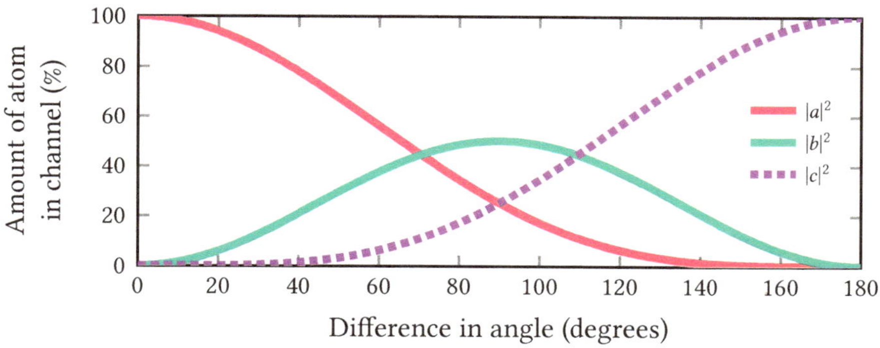

Figure 11.18: Spin-1 atom with $\hbar$ of angular momentum. The atom splits into three or one wave packets in the SG apparatus. The magnetic axis is always along the z-direction. Each angle represents the direction of angular momentum of the state that is input into the apparatus. The angle is measured from the positive z-direction.

The m_s = 0 state

As we make our way down the ladder of spins, an interesting thing arises when the angular-momentum vector is zero:

$$\begin{aligned} |1,0\rangle_z &= \underbrace{|1,+1\rangle_x}_{-\sqrt{50\%}} + \underbrace{|1,-1\rangle_x}_{\sqrt{50\%}} \\ &= \underbrace{|1,+1\rangle_y}_{-\sqrt{50\%}} + \underbrace{|1,-1\rangle_y}_{-\sqrt{50\%}} \end{aligned} \tag{11.35}$$

If I didn't know these equations, my naive intuition would have told me that the x- and y-representations would have had each one ket of $m_s = 0$. Instead, the very kets I was expecting to see are missing, and the kets I was expecting to be missing, I see!

In Fig. 11.19, I show this state going through a SG apparatus with the magnetic field pointing in the x-direction. In Fig. 11.20, I show the weights of the channels in a SG experiment as the input state's spin axis deviates from alignment with the magnetic axis.

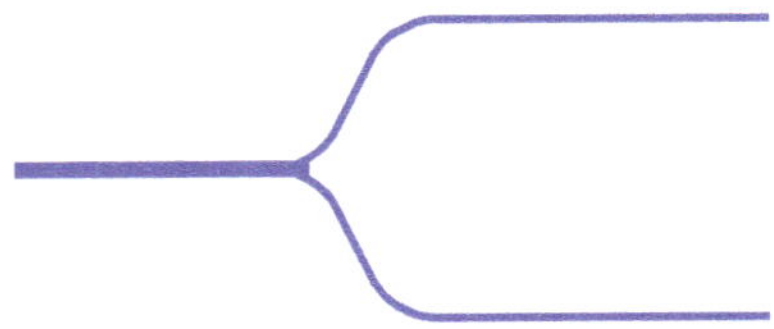

Figure 11.19: SG experiment with an $s = 1$, $m_s = 0$ particle transiting through the magnetic field 90° away from the spin AM vector.

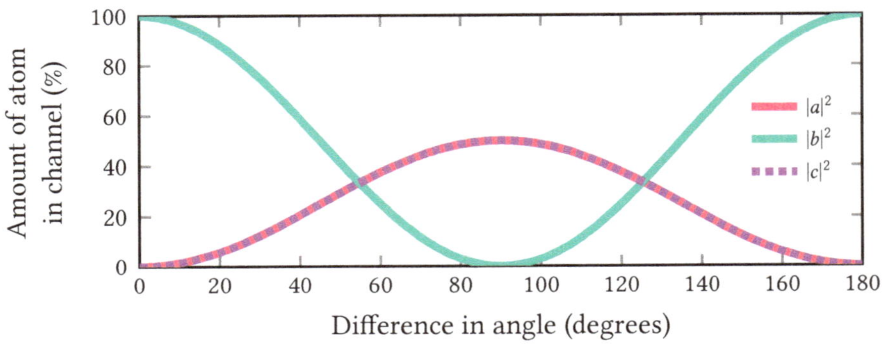

Figure 11.20: A spin-1 atom with zero overall angular momentum. The atom splits into one, two, or three wave packets in the SG apparatus.

A little bit of reflection will make the $m_s = 0$ state seem odd. Looking at the z-basis, we see a lone ket. It tells us that the state has neither angular momentum nor a magnetic dipole moment. If the SG magnet's axis is in the z-direction, there will be no deflection of the atomic beam. The wave function goes straight through. Yet, there is a continuum of orientations with respect to that axis that causes a splitting into two additional beams. Therefore, this state still has a defining axis despite having no spin angular momentum.

This is analogous to the orbitals in the previous chapter with $l \geq 1$ and $m_l = 0$ (the ones with lobes, not shells or all bagels); there is no AM in those states, but in the pictures, the spatial wave function has a definite orientation in space. While talking about analogous states, I should say the spin-0 state of the previous subsection is analogous to the $l = 0$ orbitals of the previous chapter. Those are the s orbitals, which have no defining axis.

According to our classical thinking, as the $m_s = 0$ states have no overall magnetic moment, the SG magnet has nothing to act on. On p. 220, I said we would accept as a fact that magnetic fields split spin superpositions into the basis kets corresponding to the magnet's field axis. This particularly confusing case is one reason why I wrote that.

The $m_s = -1$ state

To make the set complete, the three representations of spin along the negative z-direction for a spin-1 system are:

$$\begin{aligned} |1,-1\rangle_z &= \underbrace{|1,+1\rangle_x}_{\sqrt{25\%}} + \underbrace{|1,0\rangle_x}_{-\sqrt{50\%}} + \underbrace{|1,-1\rangle_x}_{\sqrt{25\%}} \\ &= \underbrace{|1,+1\rangle_y}_{-\sqrt{25\%}} + \underbrace{|1,0\rangle_y}_{i\cdot\sqrt{50\%}} + \underbrace{|1,-1\rangle_y}_{\sqrt{25\%}} \end{aligned} \tag{11.36}$$

These quantum amplitudes will give the same trident-esque splitting as in Fig. 11.17. This state will have a figure similar to Fig. 11.18, but the $|a|^2$ and $|c|^2$ curves will be switched.

The spin-1 abstract space

Now, how about the abstract spin space? Since there are three basis kets, the spin space will be three-dimensional. Does that give us hope that it will be intuitive? Sorry, no. The kets are all orthogonal to each other in that space still, but the AM vectors associated with $|1,+1\rangle_z$ and $|1,-1\rangle_z$ are 180° apart in physical space.

Moreover, the spin-1 abstract space is the first space to add the next layer of unintuitiveness. Just like it's impossible to rotate any physical vector (or rigid object) from having some length to zero length, you can't rotate an AM vector from $\hbar$ to zero angular momentum. In spin space, this means you can't rotate from, say, $|1,+1\rangle_z$ to $|1,0\rangle_z$. Sure, you can easily *imagine* a ket vector in this spin space rotating into another. But the carefully constructed rotation operators, which tell you what happens in spin space, can't do it. To access a state that can't be accessed via rotations, a different physical mechanism is needed, such as a quantum jump/collapse into the state.

There is no spin-1 polarization principle

This rotation restriction has an interesting consequence. If you have a spin-1 superposition in some basis with an amplitude for one or both of the $|m_s| = 1$ kets and an amplitude for the $m_s = 0$ ket, then a rotation can never purify the amplitudes to be purely one of the kets in that basis. This naturally segues into something I've said before. The spin-$\frac{1}{2}$ polarization principle doesn't generalize to higher spins.

Let's say we have an atom in some state $|\chi\rangle$. We will say in the z-basis that it has equal and non-zero amplitudes for the $m_s = 1$ and $m_s = 0$ kets:

$$|\chi\rangle = \underbrace{|1,+1\rangle_z}_{\sqrt{50\%}} + \underbrace{|1,0\rangle_z}_{\sqrt{50\%}} + \underbrace{|1,-1\rangle_z}_{\sqrt{0\%}} \tag{11.37}$$

We imagine again an atomic beam full of these atoms and do a series of experiments where we rotate the magnetic axis of a Stern-Gerlach apparatus. Figure 11.21 shows you the weights in the channels, and you see that no channel is ever 100%, which confirms the non-existence of a spin-1 polarization principle in the xy-plane. It's the same conclusion for any direction in three-dimensional space.

Spin-3/2

The list of elementary particles ends at spin-1. As for composite particles, there are plenty of nuclei with spin-$\frac{3}{2}$. As one example, on Earth, beryllium is essentially pure beryllium-9 (four protons and five neutrons), which has spin-$\frac{3}{2}$. The nucleus contributes the spin-$\frac{3}{2}$, and the four electrons have their spins cancel out. There is no OAM from the electrons because it's a $2s$ atom.

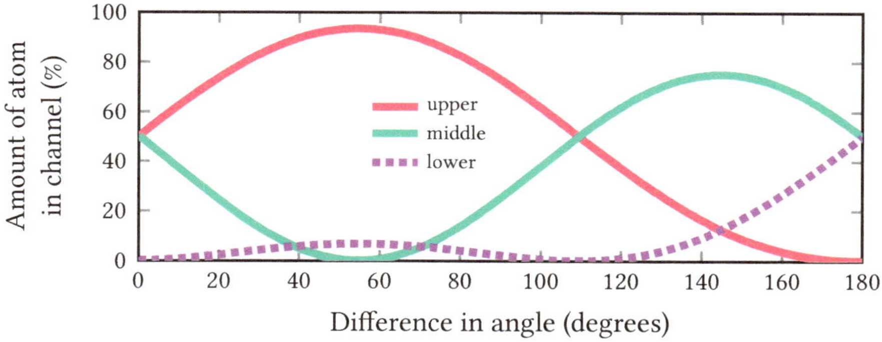

Figure 11.21: There is no spin-1 polarization principle. I will change the scenario slightly from previous examples. The input state $|\chi\rangle$ will now not change. Instead, the magnetic axis will be rotated in the xz-plane. The curves still represent the amount of the wave function in that channel of the SG apparatus at some particular angle. The basis will be whatever direction the magnetic field points in. You can see that for this particular spin-1 state, every orientation of a Stern-Gerlach magnet will result in a random measurement.

This time, instead of showing the equations for the representations first, I will lead with the different types of Stern-Gerlach beams possible. Remember that the states I am showing are specially chosen. I'm showing the states with the angular momentum 90° from the magnetic axis. Other superpositions will give more arbitrarily chosen beam weights.

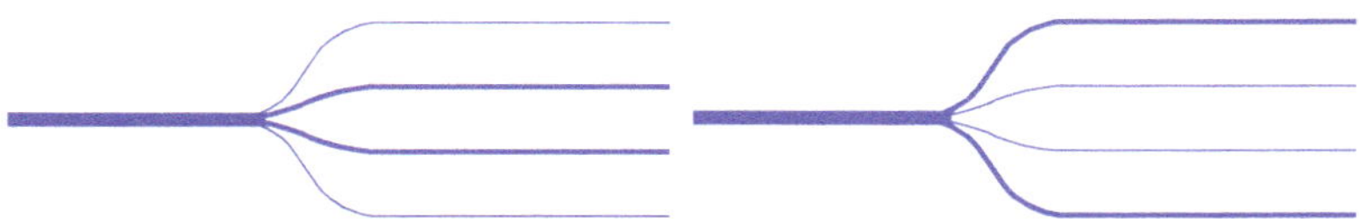

Figure 11.22: SG beams for spin-$\frac{3}{2}$ with angular momentum 90° from the magnetic axis. To save page space, the beams are closer than the previous images. We can say that we decreased the electromagnet's field so that the beams don't hit the vacuum chamber walls. The experiment on the left is $m_s = \pm\, 3/2$, and on the right, it is $m_s = \pm\, 1/2$.

Below, I show you the natural representations of a spin-$\frac{3}{2}$ particle. At this point, every bit of notation should be understandable to you. These long, dense equations should also illustrate why physicists adopt a simpler and more compact notation of column vectors to represent kets, which you will see, for spin-1 and spin-$\frac{1}{2}$, in the chapter on matrix mechanics:

$$\begin{aligned}\left|\tfrac{3}{2},+\tfrac{3}{2}\right\rangle_z &= \underbrace{\left|\tfrac{3}{2},+\tfrac{3}{2}\right\rangle_x}_{\sqrt{12.5\%}} + \underbrace{\left|\tfrac{3}{2},+\tfrac{1}{2}\right\rangle_x}_{\sqrt{37.5\%}} + \underbrace{\left|\tfrac{3}{2},-\tfrac{1}{2}\right\rangle_x}_{\sqrt{37.5\%}} + \underbrace{\left|\tfrac{3}{2},-\tfrac{3}{2}\right\rangle_x}_{\sqrt{12.5\%}} \\ &= \underbrace{\left|\tfrac{3}{2},+\tfrac{3}{2}\right\rangle_y}_{\sqrt{12.5\%}} + \underbrace{\left|\tfrac{3}{2},+\tfrac{1}{2}\right\rangle_y}_{-i\cdot\sqrt{37.5\%}} + \underbrace{\left|\tfrac{3}{2},-\tfrac{1}{2}\right\rangle_y}_{-\sqrt{37.5\%}} + \underbrace{\left|\tfrac{3}{2},-\tfrac{3}{2}\right\rangle_y}_{\sqrt{12.5\%}}\end{aligned} \tag{11.38}$$

$$\begin{aligned}\left|\tfrac{3}{2},+\tfrac{1}{2}\right\rangle_z &= \underbrace{\left|\tfrac{3}{2},+\tfrac{3}{2}\right\rangle_x}_{-\sqrt{37.5\%}} + \underbrace{\left|\tfrac{3}{2},+\tfrac{1}{2}\right\rangle_x}_{-\sqrt{12.5\%}} + \underbrace{\left|\tfrac{3}{2},-\tfrac{1}{2}\right\rangle_x}_{\sqrt{12.5\%}} + \underbrace{\left|\tfrac{3}{2},-\tfrac{3}{2}\right\rangle_x}_{\sqrt{37.5\%}} \\ &= \underbrace{\left|\tfrac{3}{2},+\tfrac{3}{2}\right\rangle_y}_{-i\cdot\sqrt{37.5\%}} + \underbrace{\left|\tfrac{3}{2},+\tfrac{1}{2}\right\rangle_y}_{-\sqrt{12.5\%}} + \underbrace{\left|\tfrac{3}{2},-\tfrac{1}{2}\right\rangle_y}_{-i\cdot\sqrt{12.5\%}} + \underbrace{\left|\tfrac{3}{2},-\tfrac{3}{2}\right\rangle_y}_{-\sqrt{37.5\%}}\end{aligned} \tag{11.39}$$

$$\begin{aligned}\left|\tfrac{3}{2},-\tfrac{1}{2}\right\rangle_z &= \underbrace{\left|\tfrac{3}{2},+\tfrac{3}{2}\right\rangle_x}_{\sqrt{37.5\%}} + \underbrace{\left|\tfrac{3}{2},+\tfrac{1}{2}\right\rangle_x}_{-\sqrt{12.5\%}} + \underbrace{\left|\tfrac{3}{2},-\tfrac{1}{2}\right\rangle_x}_{-\sqrt{12.5\%}} + \underbrace{\left|\tfrac{3}{2},-\tfrac{3}{2}\right\rangle_x}_{\sqrt{37.5\%}} \\ &= \underbrace{\left|\tfrac{3}{2},+\tfrac{3}{2}\right\rangle_y}_{-\sqrt{37.5\%}} + \underbrace{\left|\tfrac{3}{2},+\tfrac{1}{2}\right\rangle_y}_{-i\cdot\sqrt{12.5\%}} + \underbrace{\left|\tfrac{3}{2},-\tfrac{1}{2}\right\rangle_y}_{-\sqrt{12.5\%}} + \underbrace{\left|\tfrac{3}{2},-\tfrac{3}{2}\right\rangle_y}_{-i\cdot\sqrt{37.5\%}}\end{aligned} \tag{11.40}$$

$$\begin{aligned}\left|\tfrac{3}{2},-\tfrac{3}{2}\right\rangle_z &= \underbrace{\left|\tfrac{3}{2},+\tfrac{3}{2}\right\rangle_x}_{-\sqrt{12.5\%}} + \underbrace{\left|\tfrac{3}{2},+\tfrac{1}{2}\right\rangle_x}_{\sqrt{37.5\%}} + \underbrace{\left|\tfrac{3}{2},-\tfrac{1}{2}\right\rangle_x}_{-\sqrt{37.5\%}} + \underbrace{\left|\tfrac{3}{2},-\tfrac{3}{2}\right\rangle_x}_{\sqrt{12.5\%}} \\ &= \underbrace{\left|\tfrac{3}{2},+\tfrac{3}{2}\right\rangle_y}_{i\cdot\sqrt{12.5\%}} + \underbrace{\left|\tfrac{3}{2},+\tfrac{1}{2}\right\rangle_y}_{-\sqrt{37.5\%}} + \underbrace{\left|\tfrac{3}{2},-\tfrac{1}{2}\right\rangle_y}_{-i\cdot\sqrt{37.5\%}} + \underbrace{\left|\tfrac{3}{2},-\tfrac{3}{2}\right\rangle_y}_{\sqrt{12.5\%}}\end{aligned} \tag{11.41}$$

As for the spin space, it is now four-dimensional. All spins higher than spin-$\frac{1}{2}$ will have the issue of not being able to rotate into just any state. To give you another example to reinforce your understanding, we can rotate $\left|\frac{3}{2}, +\frac{3}{2}\right\rangle_z$ by 90 degrees into $\left|\frac{3}{2}, +\frac{3}{2}\right\rangle_x$ (note the difference in the subscripts). However, there is no way to rotate $\left|\frac{3}{2}, +\frac{3}{2}\right\rangle_z$ into $\left|\frac{3}{2}, +\frac{1}{2}\right\rangle_x$ because there is a difference in the magnitude of angular momentum. A projection into that state would be required.

I won't make more subsections for higher spins because it would be more akin to stamp collecting to list more. You should see some patterns in the examples that we have here.

Closing

Much of what you learned in this chapter applies more generally. It isn't restricted to spin. If you consider a system with OAM but no spin, most content in this chapter applies except that OAM uses the quantum numbers l and m_l, and l can only take on integer values. Be warned, though. Once you have spin and OAM, they follow some rules of addition that I won't explain. I should also mention that once we stop hiding things behind Dirac notation, OAM follows the math of wave mechanics and spin follows the math of matrix mechanics.

As for spin superpositions, understand that your hunter-gatherer's brain didn't evolve to comprehend ket addition in spin space. Mathematically, the behavior of spins is predictable, so rid your mind of any CGI visuals that lead you to think that spin chaotically tumbles around in a superposition.

12 Spring cleaning

> *The concept of 'measurement' becomes so fuzzy on reflection that it is quite surprising to have it appearing in physical theory at the most fundamental level ... does not any analysis of measurement require concepts more fundamental than measurement? And should not the fundamental theory be about these more fundamental concepts?*
>
> — John S. Bell, criticizing the place of measurement in quantum mechanics

Nowhere else to store them

Now is an opportune time to do some spring cleaning and clear away four topics that didn't fit with the flow of other chapters. This is not to say the topics are unimportant. In actuality, they are here because I didn't want you to miss out on them. Moreover, the second and fourth topics introduce you to the final three postulates.

There is no way to make smooth transitions between these disparate topics, so I make no attempt to do so. In this chapter, I first take issue with how classical physics is characterized when discussing quantum tunneling. Next, I discuss the negative kinetic energy that you saw when the kinetic energy operator was applied to hydrogen's ground state. Then, I give an "illuminating" example that demonstrates how we see superpositions around us all the time. Finally, I end with Heisenberg's issue with the number of variables in the wave function, which you saw in Chapter 3.

Quantum tunneling and the mischaracterization of classical physics

Quantum tunneling is usually held up as an example of an irrational thing that quantum mechanics can do that classical physics cannot. (For a brief refresher on tunneling, refer back to Fig. 5.2 on p. 64). Googling the words *quantum tunneling* will return such things as images of spherical particles that transit directly through a magical tunnel in a hill. The implication is that where classical particles need to travel over the top of the hill, quantum particles don't follow the same rules. The typical explanation a physicist might give is that classical particles with less kinetic energy than the potential energy of a barrier cannot cross said barrier.

A specific example of what they think may help. Let's say you have a zero-dimensional classical particle. You throw it upward and give it some speed. You have given it kinetic energy. From where you let go, the point object can travel upward a distance until all of its KE is converted to gravitational potential energy (GPE). Above that point is what some call the **classically forbidden region**. (It would be more accurate to say it is energetically forbidden.) This example gets turned into an object being forbidden from crossing over the top of a hill or a wall if it doesn't have sufficient kinetic energy. I want you to know that this explanation is based on a very narrow and incorrect conception of what occurs in classical physics. That story relies on thinking of the classical world as composed of objects that have no extent in space.

Let's now give the object a volume. To be specific, I will say it is a basketball. In this case, classical physics says that it is the center of mass that cannot travel into the classically forbidden region. But half of the mass of the ball is above the center of mass! That half travels into the "forbidden" region, which certainly makes it a lot less forbidden.

Some may instinctually protest, "But a classical object can't appear on the other side of a barrier when it doesn't have enough KE to cross the barrier, unlike a quantum object." Well, let's consider flexible objects, then. Think about Olympic-level high jumpers. To cross over the bar, they must contort their bodies into a U-shape. They do this because they have a problem. Their legs do not have the ability to generate enough KE for their center of mass to exceed the GPE at the bar's height. By changing their body shape, portions of their body cross over the bar while their center of mass always remains below the bar. At various points in time, each part of their body crosses into the so-called classically forbidden region. The end result is that they have crossed through the barrier with insufficient kinetic energy to get their whole body into the forbidden region all at once.

Wave functions do this distortion trick, too. When the wave function approaches a barrier of sufficiently high energy, it changes shape and a portion of the wave continues into the so-called (energetically) forbidden region. Inside the barrier, the wave function decays with distance in the barrier and forms a concave-upward shape. If it is a constant potential energy barrier, then I can be more precise and say the shape in the barrier is a decaying exponential function. Hopefully, this explanation shaves off some of the mystique from quantum tunneling.

The continuum of values of the KE operator

Some problems

A review

At the end of the first chapter dealing with the Schrödinger equation, I showed you the numerical values of the kinetic energy operator after I applied the operator to the ground state of hydrogen. As a reminder, applying a generic operator $\hat{O}$ to a single ket returns a single value or a continuous function, both of which I represent with the diamond symbol ◆:

$$\hat{O} \cdot |\Psi\rangle = \blacklozenge \cdot |\Psi\rangle \tag{12.1}$$

The numerical values returned by the energy operators are reproduced below. Focusing on the kinetic energy, you can see there are negative values. In a classical mechanics class, saying a negative kinetic energy exists would have you accused of sorcery and witchcraft. These troubling negative kinetic energies occur in more quantum-mechanical scenarios than just this one. I am unaware of a quantum textbook that attempts to deal with the existence of them. However, I have seen attempts by experts in online physics forums to confront the issue.

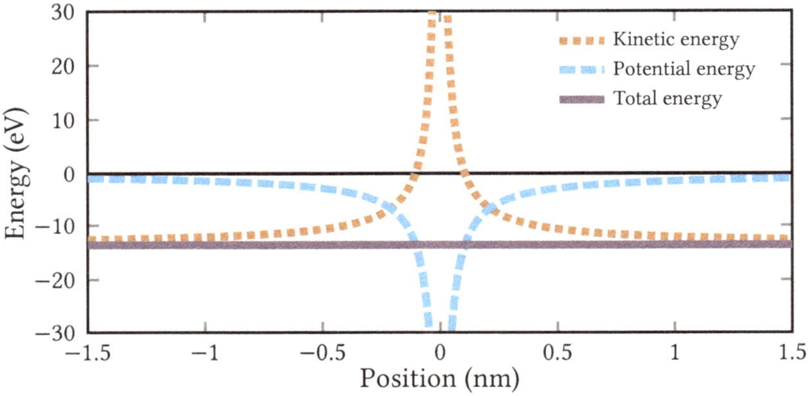

Figure 12.1: The numerical values returned by application of the energy operators when acting on the ground state of hydrogen.

Explaining away negative KE

The physicists in the online forums who recognize the problem have attempted various workarounds. Some say that when the KE is negative, this is the classically forbidden region that is commonly mentioned in the context of quantum tunneling. According to them, because it is forbidden, we could never measure a particle there. That is an odd claim to make, considering that we can fire particles into solids and we never have to take into account regions where particles can't be localized. It's also an incomplete claim because you can see in the figure that there is a region where the KE is negative and also greater than the PE, so this region would not be a classically forbidden region. More pertinent is the fact that they have added a rule that goes against axiomatic quantum mechanics.

Others say the measurement process has localized the particle in space. Sure, I can agree that might be what really happens. They say this localization will cause the wave function to have more positive curvature and, therefore, more KE. The total KE of the particle is now positive. However, this also goes against axiomatic quantum mechanics. I repeat: the postulates say that there is a finite probability of measuring one of the negative values returned by the KE operator.

It is apparent that any workaround tries to do away with negative KE and will reject part of axiomatic quantum mechanics.

Enormous positive KE

In addition to the negative values, there is another problem in the opposite direction. As the plot approaches the proton, the KE keeps getting larger.

A brief clarification is necessary before we continue this line of thought. In the mathematical approximation we use in undergraduate quantum mechanics, which I use in this book, the KE values of the hydrogen atom's ground state continue toward infinity as we consider positions closer to the center of the nucleus. In reality, there is the finite size of the nucleus that needs to be taken into account. I don't know the exact form the KE values will take near the nucleus when the *finite-nuclear-size effect* is included. Nevertheless, as long as we stay away from the region of significant proton density, the values will be accurate, and we will not need to do a different calculation. Picking the KE value at ten times the proton's "radius" to get an accurate value, I find it is approximately 170 000 eV. Therefore, according to the single-measurement postulate, for some measurements of the hydrogen's kinetic energy, the proton-electron system is measured to have enormous energies.

The axiomatic system is silent on how to measure the values returned by the kinetic energy operator. So I can't tell you how such an experiment could be carried out. Surely, by now, if it could be done, someone would have noticed such enormous energies coming out of atoms.

I can say, though, that the probability postulate says you will get a randomly measured value. Without further postulates describing measurements, the probability postulate suggests the result is not determined by the energy of whatever particle is used to probe this system. Since the total energy of the hydrogen atom is negative (i.e., −13.6 eV), a measurement that projects the system into a state of 170 000 eV must pull 170 013.6 eV of energy out of nothing. Therefore, the postulates of quantum mechanics severely violate the conservation of energy.

A realist's resolution

Clearly, I think that some of the postulates can't be used to consistently predict physical quantities. In the reality interpretation that I subscribe to, there is a straightforward and consistent way to find the physical values that a wave function possesses. In what follows, I show you how I see the values in the figure as a mathematical step towards telling us what the kinetic energy of the wave function is. This method generalizes to all physical quantities.

A classical example

Before I get into the quantum details, I will prime you first with a classical example. It will also be conceptually easier if we use potential energy for this. Suppose we want to find the gravitational potential energy of your body. How should we start to approach this problem?

Well, you are made of many tiny masses. We can find the GPE of a tiny mass via $m \cdot g \cdot h$. Here m is the mass, h is the height above some reference point (which we will take to be the floor), and g is the acceleration due to gravity, whose value doesn't matter for this.

What is your GPE? Your total mass runs from the floor to the top of your head. We could divide your body into infinitesimal heights and find the GPE of each slice. There are different masses at each height (I carry more mass around my waist than around my ankles). So, each slice contributes a GPE proportional to the amount of mass at that height.

To find your GPE, we need to average over all heights weighted by the amount of mass at each height. After some calculus, we find your GPE is determined by your total mass and a single height (which is your center of gravity). The point of this example is to illustrate that from many smaller parts, we can get a value that is representative of the whole.

Do it again, but quantum

In our minds, the difference between you and a quantum object is that you are divisible. A skin flake could fall off you, and the remaining amount of you has slightly less GPE. In contrast, we can't clip bits off of a wave function. We must consider it as a whole. This wholeness is the reason why those negative kinetic energies won't be a problem in the end.

To find the electrostatic potential energy of the electron in a hydrogen atom, we must follow a procedure similar to the classical example. From the PE function, we have a PE value for each point in space (Eq. 8.9 on p. 140). We will scale each value by a weighting factor. This weighting factor is how much electric charge (i.e., the fraction of the electron's total charge) is present at each point. You might be tempted to say it is the matter density $|\Psi|^2$ at each point. You are close. The amount is matter density times volume. (You can understand this with an analogy: The mass of water that one has is equal to the water's density times the volume.) Once we have the weighted PE for each point in space, we simply sum them up.

If you remember the chapter on misconceptions about electrons well, you will notice I've been a little loose with terminology. When I say a point in space, I actually mean a volume element, or an infinitesimal volume. I will use the symbol $\mathrm{d}V$ for this. (Was this a rhetorical technique to remind you of the difference, or did I slip into a bad habit? Perhaps a little of both.)

Now I am using V to represent volume. If I were to repeat the use of the letter V for the function of the $\hat{PE}$ operator as is convention, this would get confusing. So, let's use ♥ instead. While we are being carefree with symbols, I hope you don't mind if I use a big wonky S (like this: ∫) to represent the instruction to do the sum over all volume elements.

To recap, we are going to find the electric potential energy of the electron's wave function as a whole. As a realist, I say this single value is the potential energy of the quantum object. We find it by summing (∫) over all the individual potential energies (♥), which are weighted by the amount of wave function there ($|\Psi|^2 \cdot \mathrm{d}V$).

As you know, written equations can be turned into English, and an appropriately worded English sentence can be turned into an equation. Obviously, I was setting us up to write out an equation. It is an integral like this:

$$PE = \int \heartsuit \cdot |\Psi|^2 \cdot \mathrm{d}V \tag{12.2}$$

Technically, as written, this is an indefinite integral. I should specify the volume to integrate over, but for physicists, the integration is usually obvious and is often all of space. To be clear, PE is a single value found from a calculation of the various values in ♥ and $|\Psi|^2$. For the ground state of hydrogen, the answer is −27.2 eV.

The kinetic energy (or any quantity) is found in a similar way. Its integral is:

$$KE = \int \clubsuit \cdot |\Psi|^2 \cdot \mathrm{d}V \tag{12.3}$$

For the ground state of hydrogen, this value comes out to +13.6 eV.

In quantum mechanics, each of the two equations above is called an **expectation value**. We adopted the name from statistics because quantum mechanics deals with probabilities. It's a fancy way of saying **weighted average**.

Compact notation

Before we go further, let's ask: who would want to move around large equations like those above? Certainly not Dirac. In Dirac notation, the expectation-value equations become the following:

$$PE = \langle\Psi|\, \hat{PE}\, |\Psi\rangle \tag{12.4}$$

$$KE = \langle\Psi|\, \hat{KE}\, |\Psi\rangle \tag{12.5}$$

The right sides of the equations show a type of Dirac **bra-ket**. Don't worry about how to break down this notation. You should treat the right side of this type of equation as one large symbol that is a stand in for the integrals above. Speaking of large equations, for a physicist in a hurry, these require too many strokes of a pen. We often shorten these to the following:

$$PE = \langle\hat{PE}\rangle \tag{12.6}$$

$$KE = \langle\hat{KE}\rangle \tag{12.7}$$

Allow me to point out that this exemplifies why there is a great barrier to understanding physics. We have a very compact notation that needs to be conceptually unpacked. In this case, you would have needed to pay attention to which wave function the physicist was talking about because it's not explicitly written in these last two equations. You also need to recognize that KE is the single quantity you want and $\hat{KE}$ is the operator. After that, you must remember what math the Dirac notation actually represents. After that, you must remember how to do complicated integrals from calculus class. Dirac notation is much like a cryptic cipher to carry out some instructions that only those who are initiated into the mysteries understand.

The reality interpretation of KE

You could rightfully ask, "If it is at rest, how can a hydrogen atom have 13.6 electronvolts of kinetic energy, a quantity associated with motion?" (Perhaps you have been wondering this for a few chapters now.) I say you must accept the fact that until you read this book, you were only aware of KE associated with the motion of the center of mass of an object. Wave functions with their centers of mass at rest also have KE. Both of these show up as curvature in the wave function, as you saw in the final two examples of Chapter 7, where Gaussian-shaped starting wave functions were shown to evolve with time (starting on p. 170).

I must admit that I don't think that the KE of a state at rest can be measured directly. Though, I believe it can be deduced to exist based on our model of hydrogen. With an example, I can show you (with much math hidden from you) that the KE-curvature connection is needed to account for energy exchanges in hydrogen.

When a hydrogen atom goes from the first excited stationary state to the ground state and emits a photon, the photon carries away 10.2 eV of energy. Via two calculations of expectation values of PE, the potential energy drops by 20.4 eV, when transitioning from the higher to the lower energy state. This energy must have gone somewhere. From the 20.4 eV of energy, 10.2 eV goes to the photon. The only other place the remaining 10.2 eV can go is into the KE. Two calculations of KE using the expectation-value equation show that the KE does increase by 10.2 eV. Thus, unlike in axiomatic quantum mechanics, the principle of the conservation of energy is preserved in the reality interpretation, and an atom at rest has a kinetic energy associated with the curvature over its whole wave function.

That's how the kinetic energy of a bound state at rest can be accounted for. I can show you an even more intuitive picture of KE by comparing two states with their centers of mass at rest with the additional criteria that they are not bound by any potential energy function.

In Fig. 12.2 and Fig. 12.3, I plot the wave functions and matter densities for two particles evolving with time. The first column of each figure corresponds to the first particle. It has some set amount of kinetic energy which does not change with time because there is no potential energy function to change it. The three rows correspond to three points equally spaced in time. A similar explanation applies to

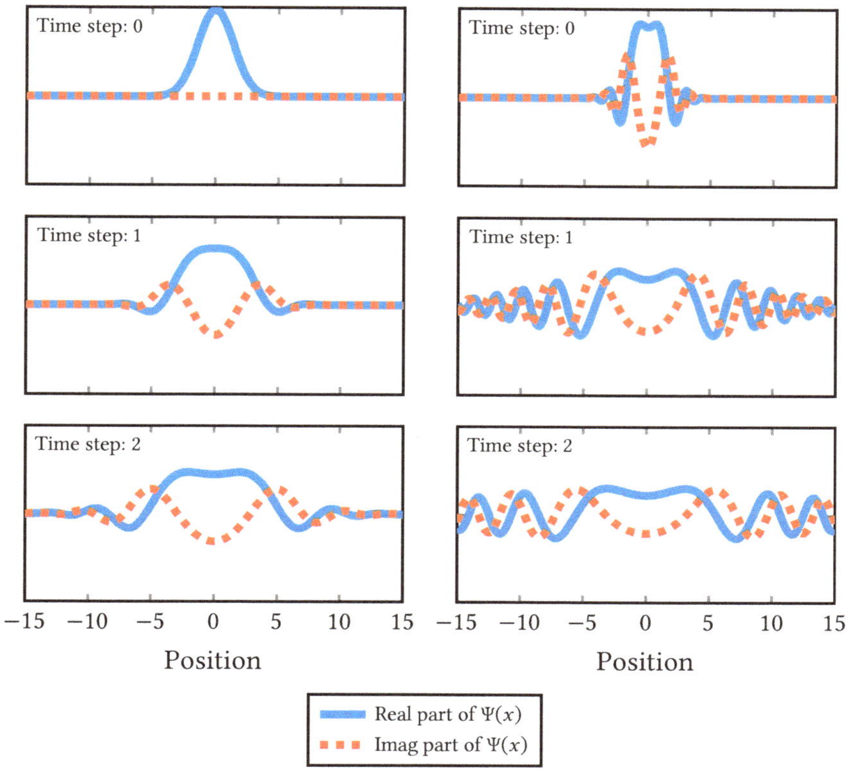

Figure 12.2: Wave functions of two particles both at rest with different KE but identical starting matter densities. In both cases, the potential energy function is zero. The particle in the second column has nine times the KE of the particle in the first column. Each row represents a single snapshot in time.

the second column and the second particle, except its kinetic energy is nine times greater than the first particle's.

To make the comparison that follows fair, I have given them the exact same matter distribution at the first point in time, which you can see in the top row of Fig. 12.3. As we allow the particles to evolve in time, we can see that the one with higher kinetic energy disperses more quickly. I have quantified the dispersions by indicating the matter densities' standard deviations with vertical lines.

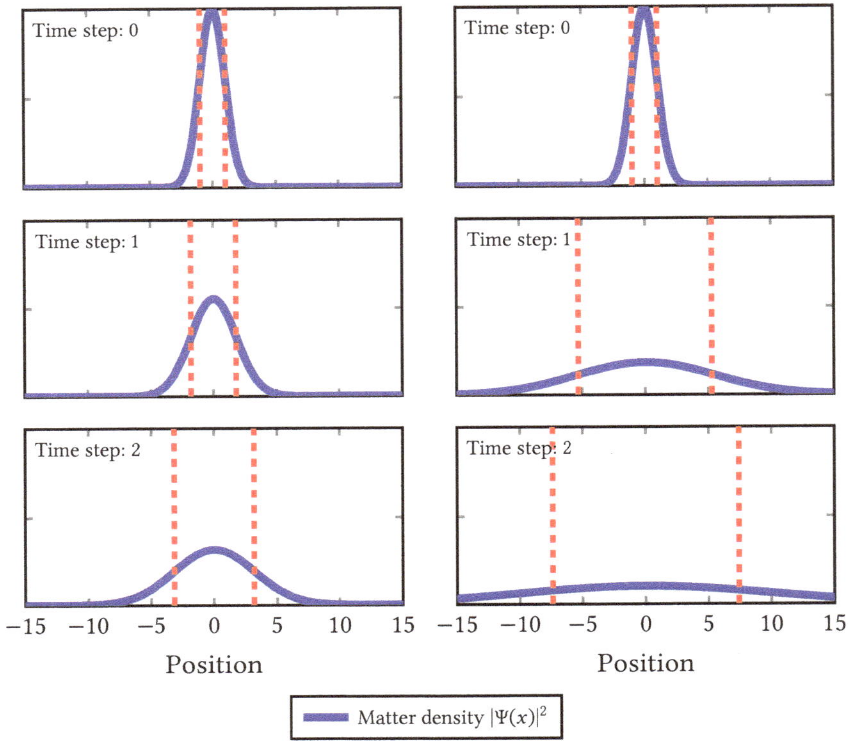

Figure 12.3: Matter densities found from the wave functions of the previous figure. Vertical lines represent one standard deviation from the center of mass.

A standard deviation is a type of measure of the distribution, or dispersion, of a set of numbers. (It is not the only measure of dispersion.) There is an equation that defines what a standard deviation is. I show you the quantum version of that equation in the next chapter. Instead of showing you that now, I will give you two facts. For matter densities, it is a length measured from the center of mass. For Gaussian distributions, which are the distributions in the matter density figure, a region bracketed by one standard deviation on either side of the mean contains approximately 68% of the matter density.

I can now tell you an interesting detail about these figures. Recall that the ratio of kinetic energies is nine. As time goes on, the second particle's standard deviation gets closer and closer to three times the first particle's standard deviation. If I had chosen the kinetic energy to be 16 times greater than the first particle's kinetic energy, then the standard deviation would approach being four times greater as time goes on. Why is this detail significant? Given a length and a time, we have an average speed v, an average speed of matter flowing outward. Interestingly, the ratio of the average speeds of the standard deviation is what is expected from a ratio of the classical KE equation: $KE = m \cdot v^2/2$. Therefore, in this case (I don't know how general this result is), when there is no potential energy function, the kinetic energy tied up in the curvature of a particle at rest will cause the overall matter density to disperse at an average speed appropriate for the relative amount of kinetic energy.

The reality interpretation of other quantities

In the reality interpretation I follow, all physical quantities can be found via expectation values: find the numerical values, give weight to each point appropriately, and sum up all the products of those two. You then have calculated your physical quantity. However, when the operator returns a single value, an eigenvalue, there is no point in doing the full calculation. When the eigenvalue of an operator exists, it gives the same answer as the expectation value of that operator when acting on the same state because it's a fact that the average value of a single value is the single value.

Recall from the last chapter, I said that a spin-$\frac{1}{2}$ particle always has an angular momentum pointing in a definite direction. I called it the *spin-$\frac{1}{2}$ polarization principle.* In my reality interpretation, calculating expectation values for the three spin operators or the OAM operators along the Cartesian directions resolves the problem of where any QAM is pointing.

I'll outline how expectation values show in which direction the

QAM points by using the following state:

$$|+zx\rangle \;=\; \underbrace{|+z\rangle}_{\approx\sqrt{85.4\%}} + \underbrace{|-z\rangle}_{\approx\sqrt{14.6\%}} \tag{12.8}$$

You've seen this state before (Eq. 11.20 on p. 223), and you know that the label "$+zx$" in the ket tells us that the spin's angular momentum vector bisects the positive z- and x-directions. However, I want you to forget that knowledge. In place of the ket $|+zx\rangle$, I want you to imagine a ket labeled by the symbol χ:

$$|\chi\rangle \;=\; \underbrace{|+z\rangle}_{\approx\sqrt{85.4\%}} + \underbrace{|-z\rangle}_{\approx\sqrt{14.6\%}} \tag{12.9}$$

The only information you have is now contained on the right side. Since, with our limited brains, we can't intuit the direction of the spin via the quantum amplitudes, we need to do some calculations.

During the calculation of the expectation values that follow, I did not use an integral equation like in the ones I showed you above. The integral equations apply to wave mechanics. Spin requires matrix mechanics. As you can see in the appendix on matrix mechanics, spin expectation values are calculated with a row vector, a square matrix, and a column vector. If you work through that appendix and are adept at trigonometry, then you will have the skill to find the following expectation values of spin along the Cartesian direction of the state $|\chi\rangle$:

$$\begin{aligned}
\langle\chi|\hat{S}_x|\chi\rangle &= \frac{\hbar}{2}\cdot\frac{1}{\sqrt{2}} \\
\langle\chi|\hat{S}_y|\chi\rangle &= 0 \\
\langle\chi|\hat{S}_y|\chi\rangle &= \frac{\hbar}{2}\cdot\frac{1}{\sqrt{2}}
\end{aligned} \tag{12.10}$$

Choosing to use the more compact notation that doesn't show χ, this angular momentum points in the direction of the vector with components:

$$\left(\langle\hat{S}_x\rangle\,,\,\langle\hat{S}_y\rangle\,,\,\langle\hat{S}_z\rangle\right) \;=\; \frac{\hbar}{2}\left(\frac{1}{\sqrt{2}},\;0,\;\frac{1}{\sqrt{2}}\right) \tag{12.11}$$

The $\hbar/2$ is the magnitude/length of the vector, and the triple of numbers tells us the angular momentum points in the direction that bisects the x- and z-directions, as expected.

I mentioned that those who are adept at trigonometry will have the ability to find these expectation values after working through the appendix. However, in Eq. 12.9, I only show approximate quantum amplitudes. With those, you can only find approximations to the expectation values in Eq. 12.10. To find the exact values, you need to know the exact quantum amplitudes in the column vector. For those motivated to try, I put the column vector here:

$$|\chi\rangle \;=\; \begin{bmatrix} \dfrac{\sqrt{2+\sqrt{2}}}{2} \\[2ex] \dfrac{\sqrt{2-\sqrt{2}}}{2} \end{bmatrix}_z \;\approx\; \begin{bmatrix} \sqrt{85.4\%} \\ \sqrt{14.6\%} \end{bmatrix}_z \tag{12.12}$$

Those adept at trigonometry can also find these exact quantum amplitudes using a rotation matrix (Eq. C.15 on p. C.15) after working through the appendix.

The expectation-value postulate

In contrast to the realist's physical picture, axiomatic quantum mechanics postulates a different use of expectation-value equations that has to do with numerous measurements. The **expectation-value postulate** requires some preliminaries before stating it. We have to imagine a large number of particles that are all prepared to be in equivalent states. In a real experiment, this is sometimes a problem. Other times, it is not. For example, a gas of helium atoms at room temperature would have the vast majority of atoms in the lowest electronic ground state, so it is easy to do repeatable experiments on these atoms.

When we imagine doing a measurement, we do so on one particle and record the value. We then conceptually discard that particle and move on to the next one.

The expectation-value postulate says that after a large number of measurements of an observable corresponding to $\hat{O}$, the average value we expect to find is equal to $\langle \hat{O} \rangle$.

To be clear, this postulate (in conjunction with other postulates) says the hydrogen atom's negative kinetic energies and enormously positive kinetic energies would be measured, show up on our apparatus's display, and the average would come out to a reasonable value.

On the other hand, the reality interpretation says that the expectation value is the value that a quantum object had the whole time before the collapse of the wave function projected it into a new state. Regarding our hydrogen atom example, the negative values and the enormous values exist only as intermediate values in the math, not in reality.

Transient superpositions

Our math is meant to describe the physical world

The majority of quantum mechanics textbooks are full of equations, discussions of the equations, and relations between equations. The books don't touch on physics and reality as often as I think they should. Nor are they written in plain language that allows a reader to think physically.

There is an odd belief among many physicists that we only observe, find, or measure photon-emitting objects that have been projected into energy eigenstates (equivalent to stationary states). This view seems obvious if one thinks superpositions are ghostly and impossible to observe. It's likely the projection postulate and a physicists' early education that emphasizes atomic spectra produced from quantum jumps have also biased their thinking. By the end of this section, you will see why this way of thinking is not correct.

The incorrect view would not occur if we spoke more directly about the physics of the math of quantum mechanics. I'll use text from Griffiths' book to set up an illuminating example. He writes:

> Finally, a set of functions is **complete** if any *other* function (in Hilbert space) can be expressed as a linear combination of them:
>
> $$f(x) = \sum_{n=1}^{\infty} c_n f_n(x)$$
>
> If the functions $\{f_n(x)\}$ are orthonormal, the coefficients are given by...

With regard to math, this isn't quite right. He made it limited to be simpler. There should also be an integral term on the right side to be generally applicable. With regard to physics, it misses out on clearly telling the reader what the consequences are for wave functions. In my opinion, I think he should have explicitly and emphatically explained that when we have a basis that is **complete**, we can either make *any physical shape of a spatial wave function* or make *any physical spin state* for the particular quantum system under examination by choosing each quantum amplitude c_n appropriately.

Often, instead of using the word *complete*, it is equivalently said the basis functions/vectors/states *span* the space. The space that is spanned can be abstract or the physical one we are in. Three vectors in each of the x-direction, y-direction, and z-direction are sufficient to span our three-dimensional physical space. Normally, we consider them to have a length of one unit and consider them to point in the positive direction. Using three vectors that span our space, we can compose all the vectors that represent the various points in our space by scaling their lengths appropriately.

Given a spin-s particle, the $2s + 1$ basis states I showed in the previous chapter make a spin space complete. With those basis states and by choosing various quantum amplitudes, all physically realizable states of the spin-s particle can be represented.

Soon, I will use the spatial wave function of hydrogen to construct an illuminating example. Recall the energy-level and multiplicity diagrams of the hydrogen atom from Appendix A (Fig. A.1 on p. 338 and

Fig. A.2 on p. 341). Each stationary state of hydrogen is labeled by n, l, and m_l. The states would be the set of kets of the form $|n, l, m_l\rangle$. In practice, it would be difficult to do so, but the continuum of unbound states could be imagined to have appropriate labels too.

Because it's long and difficult to prove, students should be simply told the infinite set of stationary states labeled by those quantum numbers and the infinite set of unbound states together form a complete basis for a lone hydrogen atom's electronic wave function. These functions span an infinite-dimensional function space. (The stationary states would be the set $f_n(x)$ beside the summation sign above. The unbound states would be included in the unseen integral in Griffith's equation above.) Any particular situation in which we find the hydrogen's electronic wave function will be described with that complete set of basis functions and some particular quantum amplitudes. We will soon realize why the physical explanation is significant for our worldview.

Photons transiting through matter

In what follows is an abridged description of what occurs during a particular thought experiment using photons. As an inexact outline, it has some shortcomings that will be obvious to a physicist. A fuller explanation would include mention of reflected/scattered waves and the idea of the collapse of the wave function at some point in the narrative. However, the main point is valid without all the details.

Let's say we have a gas of hydrogen atoms occupying some sizeable length appropriate for the experiment. I will approximate the atoms as being separate from each other and assume the atoms are in the ground state. Neither of these requirements are physically necessary. They are only used for educational purposes. For simplicity, we will also say a vacuum surrounds the gas.

A green laser produces a pulse of photons that is incident on the atomic hydrogen gas. Looking at the Balmer series in the first chapter (Fig. 1.2 on p. 15), you can confirm that there is no green emission line. The lack of an emission line tells me that an atom will not go from the ground state to a higher stationary state via the absorption of one of the green photons.

We measure the energy of the photons after they emerge from the gas and confirm we gathered green photons. It's well known that photons are delayed by the gas because of the gas's index of refraction. They didn't get to the detector as if they traveled at the speed of light *in vacuum* from the laser to the detector. We could have even timed the delay between the emission and detection of the pulse to confirm what is well known. From this, we can conclude the photons interacted with the gas.

Specifically, the photons' electromagnetic fields interact with the electric charges in the gas, inducing oscillations in the charge density. The photons give up their energy, linear momentum, and angular momentum to the gas to induce the charge oscillations. In an extraordinary feat of coherence, the wiggling charge in one part of the gas induces changing electromagnetic fields to keep the energy, linear momentum, and angular momentum localized so that the wiggling propagates to the next portion of the gas an infinitesimal distance away. Albeit, the propagation is at a speed slower than the speed of light in vacuum.

The photons have temporarily lost their existence as elementary particles and have been absorbed by the gas of hydrogen atoms. This must be so because the gas and its electromagnetic field are now excited above the ground state, and the excess energy, linear momentum, and angular momentum required to achieve this excited state came from external photons.

Eventually, the propagating disturbance reaches the end of the gas. The electromagnetic fields induced by the final portion of the gas are not absorbed by any matter. Therefore, the gas emits the electromagnetic waves into the vacuum. The photons are reborn, so to speak, and propagate toward the detector.

Because the charges wiggled, we know the hydrogen atoms were in non-stationary states. Using my exposition about non-stationary states in the first section of Appendix A, via deduction, we conclude we observed a mass of matter in a superposition of stationary states.

We, however, don't know the exact superposition the gas had because we don't know the exact oscillations of the electrons.

If one understands the states that occur in quartz, glass, or any transparent gas, liquid, or solid, this explanation would be easily generalized to those cases. The photon-induced, non-stationary superpositions are general phenomena. Everyone sees a superposition whenever they look through glass, water, etc. Importantly, none of the transient superpositions require a quantum jump into a stationary state (aka an energy eigenstate). Thus the odd view among some physicists is misplaced.

Configuration space vs. 3D space

In the *The cat, the keg, and the cut* chapter, you saw a quote by Heisenberg regarding the reality of the wave function (His "in no way" statement on p. 72). He claims that wave functions can't be elements of reality because once you have more than one particle, the number of variables exceeds the number of spatial dimensions of our universe. His description of the math is true: the wave function for a multi-particle system has a number of spatial variables equal to the number of particles times the number of spatial dimensions.

I say Heisenberg's claim of an unreal quantum realm that follows from concentrating on the dimensions of wave functions is not well thought out. To show this, I will give a well-known example of two electrons in a universe that is a one-dimensional line segment. Many physics students learn this example. To ensure it is simple and no computer simulation of Schrödinger's equation is needed, this example always lacks an external potential and pretends the electrons don't interact through the electric force. Taken together, this means $\hat{PE} = 0$.

Because there are two particles and one spatial dimension, the spatial wave function will have two spatial variables (i.e., x_1 and x_2). In configuration space, time is not counted as a variable or dimension. We will be looking at a snapshot in time of the wave function. Heisenberg would say that the wave function is a two-dimensional configuration space. I think a better description is a two-*variable* configuration space.

Before I handle the disagreement with Heisenberg, we must go through the physics of two very interesting postulates. Both change how we write down the wave function of our example. The first one, the composite-system postulate, you'll see creates the multi-variable configuration space. The second one, the symmetrization postulate, further constrains how wave functions may be written. Once we get past those two postulates, I can show you how to get from the two-variable configuration space wave function to the one-dimensional electron density.

Constraints on wave functions: an interlude

The second-from-last postulate

We will start by looking at the wave functions of electrons as if they were isolated from each other. For simplicity, I will show snapshots at some particular time when the wave functions are real (Fig. 12.4). The first plot shows a wave function $|A\rangle$, which vaguely resembles an A shape. This is the ground-state wave function if the particle were alone in this one-dimensional, line-segment universe. The second plot shows a wave function $|B\rangle$, which vaguely resembles a B if you imagine the dip as a hill. This is the first excited, stationary state above the ground state if the particle were alone in this one-dimensional universe. To understand what follows, you need to remember that the letters correspond to the wave function shapes.

When discussing multi-particle systems, kets like $|A\rangle$ and $|B\rangle$ are often called *orbitals* or *single-particle wave functions*. I will instead call them *input kets*.

As a point of contrast for what is coming up, the left image of the next figure (Fig. 12.5) shows what one might naively think the total wave function would be when the two input kets exist on the same line segment. Specifically, I've plotted:

$$|A\rangle + |B\rangle \tag{12.13}$$

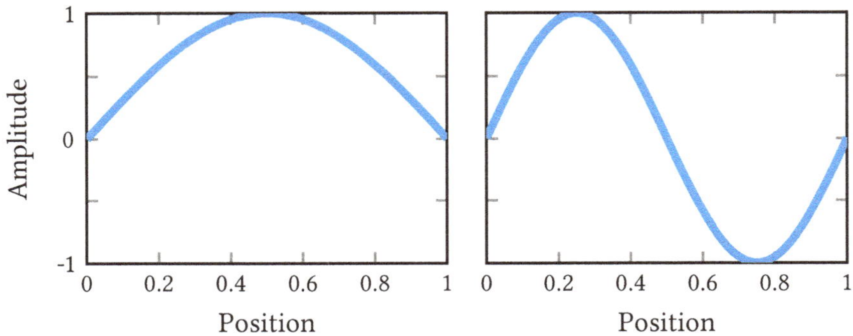

Figure 12.4: Our two input kets. The $|A\rangle$ ket is on the left, and the $|B\rangle$ ket is on the right. I used $\sin x/2$ and $\sin x$ to plot these. Note that all wave functions and densities in this final section of the chapter have their maximum values equal to one; this is not physically appropriate. If I were to be accurate, the functions would be normalized so that the area under their density curves equals one, which would change the y-axis values.

The right image of Fig. 12.5 shows the magnitude squared of this superposition, which is the charge/mass/spin/probability density. These two images do not correctly capture what happens in a composite system because the superposition is not the correct first step when the wave functions overlap so much.

The **composite-system postulate** says to write multi-particle systems as tensor products (aka Kronecker products) of the form:

$$|AB\rangle = |A\rangle \otimes |B\rangle \tag{12.14}$$

When there are kets multiplied by tensor product and no superposition, it is called a **product state.** You've seen the tensor-product symbol $\otimes$ before when a spatial wave function and spin state were on either side of the product (p. 46 and p. 103). I said then that the spin state is distributed over each point of the wave function. The distribution happens in this case too. But, now, because we're comparing two one-dimensional continua of finite length, the "multiplication" tells us we are making a two-variable "square-shaped area" called

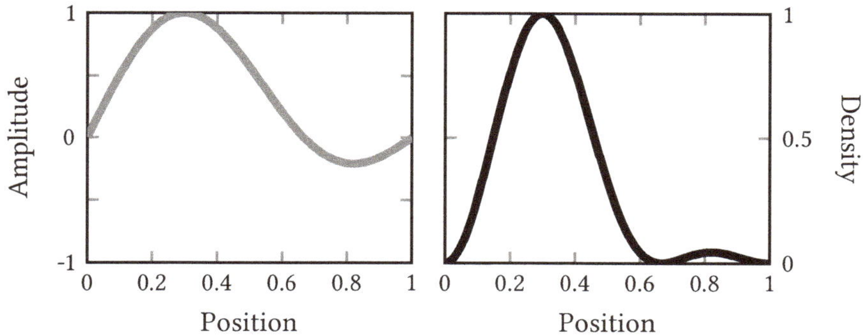

Figure 12.5: Naive guesses of the superposition and the electron density for our two-particle system. The superposition and density have been scaled so that the peak values are equal to one.

configuration space (have a peek at the left image of Fig. 12.6). However, written as is, this equation is not complete. We need to modify it using the next postulate.

The final postulate

Imagine you had the supernatural ability to label the electrons. You could say electron 1 has an A-shaped wave function and electron 2 has a B-shaped wave function. My notation convention for this labeling is as follows: reading from left to right, if a ket appears first in a tensor product, it represents electron 1; that is to say, it is notionally labeled with a 1. The second ket is then electron 2, which means it is notionally labeled with a 2. Going forward, it is important to keep in mind that the numerical labels each input ket has are implied by the order in which I write the letters for the wave shapes A and B.

Writing a product state as $|AB\rangle$, as in the equation above, means that I am claiming that electron 1 has a wave function with an A-shape and electron 2 has a B-shape. Had I written $|BA\rangle$ instead, the kets on the right side of Eq. 12.14 would be switched. And I would be saying that electron 1 has a wave function with a B-shape and electron 2 has an A-shape.

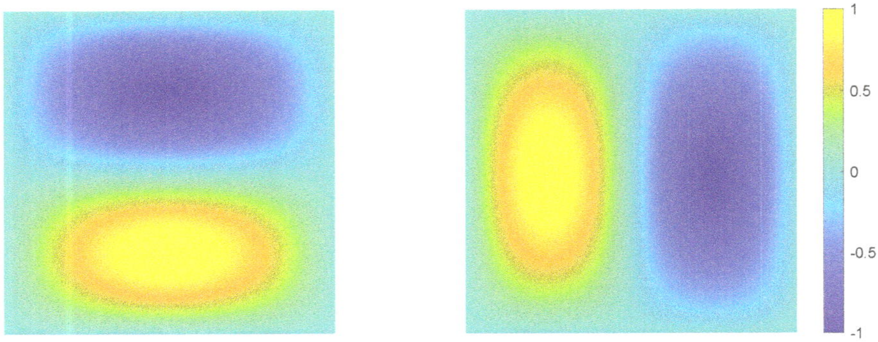

Figure 12.6: Configuration space images of the two product states. Each square is how to picture the tensor product of two input kets. The product state |AB⟩ is on the left, and |BA⟩ is on the right. The horizontal axis is x_1 and the vertical axis is x_2 in both images. The coordinates $x_1 = 0$ and $x_2 = 0$ are in the bottom left corners. Now, I would like to emphasize that the tensor product isn't so mysterious when we only consider spatial wave functions (that is, we ignore spin). If we instead called the axes x and y to make it more intuitive and let their values run from 0 to 2π, then these two squares are [left square] $\sin(x/2) \cdot \sin y$ and [right square] $\sin x \cdot \sin(y/2)$, where the dot means normal multiplication. This is so straightforward you could make low-res versions of these in Excel or with graph paper, a calculator, and colored pencils.

In configuration space, these two product states look slightly different. In Fig. 12.6, I show you how these two product states appear. I will call the two axes x_1 and x_2 and keep them in the same orientation in the next figure. The subscripts correspond to electron 1 and 2. In each image, you can deduce which input ket (i.e., |A⟩ or |B⟩) is associated with which axis, which in turn tells you the label (1 or 2).

There is a law of the universe known as the **symmetrization postulate** that says that you can't differentiate particles of the same type with notional labels like this. The law puts a constraint on the allowed wave function when dealing with multiple particles of the same type. By *same type*, I mean they have the same charge, mass, and spin quantum number (plus any other quantum numbers if you are doing nuclear or particle physics). I guess you could say there is no additional quantum number for enumerating each electron in the universe.

This law, or constraint, is why we can't stop with the simplest form of the wave function that the composite-system postulate allows. It forbids me to look at a particular configuration space (the left image in Fig. 12.6) and say x_1 is associated with the A-shaped wave function and x_2 is associated with the B-shaped wave function. But we can, in fact, do exactly that with that image. Therefore, the images in the figure are not allowed as configuration spaces.

For a system of two particles of the same type, not just electrons, the constraint that needs to be satisfied is one of the two of the following:

$$|\Psi_{AB}\rangle = \pm|\Psi_{BA}\rangle \tag{12.15}$$

This equation is the two-particle version of the symmetrization postulate. This equation says that when we switch the labels (i.e., switch the positions of the A's and B's) inside ket state, we get back the original wave function multiplied by either $+1$ or -1. Note that $|\Psi_{AB}\rangle$ is not a product state; it instead represents a total wave function that happens to also be labeled with letters. (I show an example below that makes this all clear.)

The plus sign is for when the input kets $|A\rangle$ and $|B\rangle$ are integer-spin particles (**bosons**) of the same kind, like two He-4 nuclei. We say the total wave function is symmetric with respect to the notional label exchange. The minus sign is for when the input kets are half-integer-spin particles (**fermions**) of the same type, like two electrons. We say the total wave function is antisymmetric with respect to the notional label exchange.

A consequence of this constraining equation is that the system is represented by a more comprehensive wave function than either product state I mentioned (i.e., $|AB\rangle$ or $|BA\rangle$). When physicists consider particles of the same kind, wave functions that satisfy the constraint above must be used in the Schrödinger equation.

There is an algorithm for generating wave functions that follow this constraint from some given input kets. You don't need to know it in this book because I will give the final wave functions. For the curious, in the case of fermions, you can look up *Slater determinants*.

In the case of bosons, it's the same thing with all the negative signs turned positive.

Let's turn back to our two electrons. It turns out that the symmetrization constraint applies to the total wave function (spatial plus spin) of the multi-particle system. I can hide the spin state if I simply declare it beforehand. Let's say both spins are spin-up. In that case, using a Slater determinant, the total wave function with hidden spins is found to be:

$$\begin{aligned} |\Psi_{AB}\rangle \text{ “=” } & |A\rangle \otimes |B\rangle - |B\rangle \otimes |A\rangle \\ = \; & |AB\rangle - |BA\rangle \end{aligned} \tag{12.16}$$

For the most part, I will use the notation of the second line, since it is cleaner and more compact. Using my equal sign in quotation marks signals that I'm leaving out the quantum amplitudes, which by this point you could guess both involve a $\sqrt{50\%}$.

We can confirm that switching labels of the electrons will give us back the original wave function multiplied by -1, which is required for fermions of the same type. In the convention of this book, the notional switching of labels on electrons is accomplished by switching the positions on the written page:

$$\begin{aligned} |\Psi_{BA}\rangle \text{ “=” } & |BA\rangle - |AB\rangle \\ = \; & -1 \cdot \big(|AB\rangle - |BA\rangle\big) \\ = \; & -1 \cdot |\Psi_{AB}\rangle \end{aligned} \tag{12.17}$$

In the final line, you can see that we get back the wave function before the switch and it is multiplied by -1.

Something familiar from the unfamiliar notation

If I stop hiding the spin state, the total wave function with both spins up appears this way:

$$|\Psi_{AB}\rangle \text{ “=” } |A\uparrow, B\uparrow\rangle - |B\uparrow, A\uparrow\rangle \tag{12.18}$$

Since I will stay in the spin's z-basis, I use a simpler up/down arrow notation here than I had in previous chapters. The letter and arrow on one side of a comma are glued together. When switching things inside a product state, the letter and arrow get moved together. You can confirm again this follows the fermions' minus-sign constraint.

I want to be absolutely sure I haven't lost you with this new notation. When you look at a product state of the form $|A\!\uparrow, B\!\downarrow\rangle$, you need to know that there are three tensor products in there. In a sense, one is weaker than the other two. You should see it as:

$$|A\!\uparrow, B\!\downarrow\rangle \;=\; \big(|A\rangle \otimes |\uparrow\rangle\big) \;\otimes\; \big(|B\rangle \otimes |\downarrow\rangle\big) \tag{12.19}$$

Spatial and spin kets inside parentheses are collectively an input ket. When we switch the notional labels, input kets don't change, but they can be moved. When we switch labels, we switch positions across the middle tensor-product symbol. Therefore, you should be able to see that when we switch labels (i.e., switch the written positions of the input kets), we get:

$$|B\!\downarrow, A\!\uparrow\rangle \;=\; \big(|B\rangle \otimes |\downarrow\rangle\big) \;\otimes\; \big(|A\rangle \otimes |\uparrow\rangle\big) \tag{12.20}$$

At this point, I must give you perpetual permission to admit to being confused by the math we use. Physicists really bungle the explanation of tensor products and do not mention to students what I am about to tell you. The two types of $\otimes$ act slightly differently. When a $\otimes$ joins a spatial ket and a spin ket — two things with different forms — it doesn't matter which order you write them. When a $\otimes$ joins two input kets — two things of the same form — the order matters. (To be honest, there is another common notation that uses numbers for the labels. In that case, the order doesn't matter.)

With that out of the way, let me ask you what happens when, instead of having different spatial wave functions, we make both electrons (fermions) have the same A shape? While we do that, we make sure to keep both spins up. It is easy to do:

$$|\Psi_{A\uparrow A\uparrow}\rangle \text{ ``='' } |A\!\uparrow, A\!\uparrow\rangle - |A\!\uparrow, A\!\uparrow\rangle \;=\; 0 \tag{12.21}$$

The total wave function is zero and thus can't exist. In other words, the two electrons can't occupy the same exact state. This is an example of the familiar **Pauli exclusion principle**. This principle is now a consequence of the more general symmetrization postulate, even though, historically, Pauli's principle came first.

If instead we say the input kets are bosons, then we need to switch the minus sign to an addition sign. The wave function won't equal zero, and this tells us that bosons are allowed to occupy the exact same state.

Helium

Now, let's use our knowledge from chemistry class and confirm helium's ground state wave function can exist. We know the spins will be in opposite directions and both electrons will have the same wave function shape (same orbital shape). I could use "1*s*" to represent the orbital shape, but I'll stick with "A". This will give us:

$$|\Psi_{\text{Helium GS}}\rangle \text{ ``=''} \; |A\uparrow, A\downarrow\rangle - |A\downarrow, A\uparrow\rangle \tag{12.22}$$

You ought to confirm that switching the labels/positions in the equation returns the original wave function times -1, and is, therefore, an allowed wave function.

There's much potential commentary to be had about the physical consequences of the symmetrization postulate. I barely mentioned bosons. Nor have you seen further implications of the Pauli exclusion principle, which is necessary to understand multi-electron atoms, molecules, and solids. You can google those or wait until I write a few books on those topics.

Back to configuration space

With that groundwork laid, we return to our A- and B-shaped input kets and our quest to find the electron density. I can't illustrate the spin state in the images of configuration space, so I will say that they are both spin up and leave them out of the notation that follows. The total wave function for our two-particle system in one dimension is:

$$|\Psi_{AB}\rangle \text{ “=” } |AB\rangle - |BA\rangle \tag{12.23}$$

In Fig. 12.7, I show this wave function as a two-variable configuration space. I additionally show an image of $|\Psi_{AB}(x_1, x_2)|^2$ on the right.

Note that, before this chapter, I mostly discussed wave functions that represent one particle. Additionally, I was referring to $|\Psi|^2$ as the matter density. I didn't include the position variable in the parentheses previously because it made the notation simpler for you. The example of two electrons we are now considering exists in a universe with one spatial dimension, so the electron density must also have one spatial dimension. Therefore, $|\Psi_{AB}(x_1, x_2)|^2$ with its two variables is not the electron density. I will call it the *squared magnitude of configuration space.*

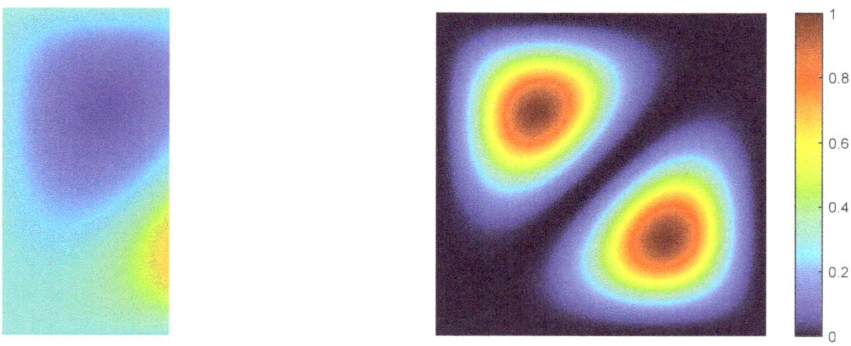

Figure 12.7: The wave function/configuration space of our two particles that exist in one space dimension (left) and the squared magnitude of configuration space (right). Color bars are scaled to the highest values in each function.

The electron density is found from $|\Psi_{AB}(x_1, x_2)|^2$ by compressing its data down to one dimension. In most unspecialized quantum mechanics textbooks, they are unaware of this step. However, the electron density is not an esoteric or controversial concept and is well known to people who calculate the properties of atoms, molecules, and materials. It is the foundational part of a technique called *density functional theory (DFT)*, which has been around since the 1960s, with precursors dating back to the first days of quantum theory.

To find the electron density from the right image above (i.e., from $|\Psi_{AB}(x_1, x_2)|^2$), you start by selecting either one of the variables (i.e., x_1 or x_2). They are both equivalent. For concreteness, let's pick x_1, the horizontal direction. At a single point along that configuration-space dimension, sum up all the points along the line extending in the other dimension (technically, it's an integral and not a sum). This will give a value proportional to the electron density at that point in space. As an example, let's pick a point approximately one-quarter of the way from the bottom left corner along the horizontal axis. Starting at the bottom and going up, we would add up virtually nothing for some time because we are in the dark blue (see the color bar), then add up the values as we cross over and down the multicolored peak. This gives a single value.

Repeat this summing of values along vertical lines for each point along the horizontal axis. This procedure compresses a two-variable space to the one-dimensional space of the universe we are working in. Multiply this final function by the number of electrons in the system to get the electron density. Since only one dimension remains, we rename x_1 to x. For those interested in the integral, the electron-density $n(x)$ is calculated this way:

$$n(x) = n(x_1) = 2 \cdot \int |\Psi_{AB}(x_1, x_2)|^2 \cdot \mathrm{d}x_2 \qquad (12.24)$$

I plot the electron density in Fig. 12.8. Notice how this is a different result from the naive guess at the electron density in Fig. 12.5.

In the most general situations, we have three spatial dimensions and multiple electrons. In that case, you integrate across all but one of $x1$, $x2$, $x3$, $x4$, ... to get the x dimension. It doesn't matter which one you leave out of the integration because whichever choice you make, we get the same answer in the end. You do the same for $y1$, $y2$, $y3$, $y4$,... and $z1$, $z2$, $z3$, $z4$,... to get the y dimension and z dimension. In

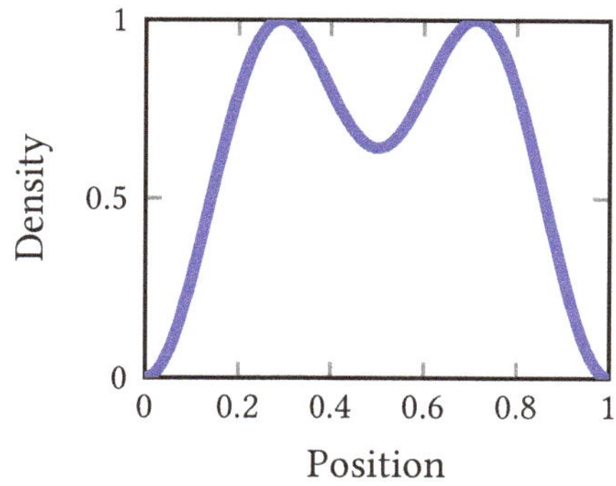

Figure 12.8: The electron density that comes from our two input kets. The actual density distribution should have its area equal to two because there are two electrons. Note this is a snapshot in time. Because the input kets have different energies, this is a non-stationary state. As time evolves, the density will slosh around cyclically.

the end, you will have a three-dimensional electron density at every point in space. Therefore, contrary to Heisenberg's supposed empty quantum realm, these three-dimensional electron densities can exist.

To recap, I have illustrated how, from the multi-variable wave function, we can get objects that exist in three dimensions and thus can be an element of reality. This is sufficient to rebut Heisenberg's anti-reality claim because it shows something three-dimensional can exist in the quantum realm. However, I should end this chapter with some remarks about three-dimensional wave functions because, in the first chapter, I said that I consider wave functions to be elements of reality.

Here be dragons

In this short comment, I switch to speculation, so you should be skeptical of the hypothesis I present. However, as I said in the first chapter, compared to the views held by some physicists (e.g., reality doesn't exist, infinite densities in black holes, multiple universes, and 11 dimensions in string theory), what I am about to say is rather mild.

First, I must switch to three-dimensional coordinates, so instead of x, I will be using a position of three coordinates $r = (x, y, z)$ inside the functions. The electron density $n(r)$ is then $n(r) = |\Psi(r)|^2$. Since we now have this three-dimensional function, our discussion of functions of excess dimensions, like $|\Psi(x_1, x_2)|^2$, is over and won't appear below. Also, for this discussion, it seems more natural not to switch to Dirac notation, so I won't use $|\Psi\rangle$.

I have said that, in my opinion, a wave function probably exists as an element of reality; it could be an object or a property of an object. If we insist that it is the object, how hard is it to go from $n(r) = |\Psi(r)|^2$ to $\Psi(r)$? (To be clear, I am looking for the total wave function, not two single-particle wave functions.) Well, I need a wave function that has a magnitude and phase at every point in space. In symbols, I am looking for $\Psi(r) = |\Psi(r)| \cdot \mathrm{Ph}(r)$, where $\mathrm{Ph}(r)$ is my way of writing a phase function.

The first step then is to take the square root of the electron density function. We instantly have half of what we need. We have the magnitude of $\Psi(r)$. The remaining step is to find the phase function. I don't see this step being easy for realistic systems. One could deduce some information based on the nodes in $|\Psi(r)|^2$ if they are present. But, in general, a search algorithm would be needed. The search would be constrained by the expectation values of $\hat{E}$, $\hat{KE}$, $\hat{PE}$, spin, and OAM.

Though it may not be practical in all cases, I think this can be done in principle. So, I think it is more likely than not that we could find three-dimensional wave functions if we put effort into finding them in some simple cases (e.g., helium). Then, the reality interpretation would have a stronger case.

There is another way to see that it is likely that three-dimensional wave functions exist. To find the electron densities, computational physicists and chemists use a simple superposition of *three-dimensional* orbitals (or other functions) as an approximation to an actual wave function, which is then used to find the electron density. They don't think the approximation represents the exact wave function. But, based on the fact that experiments tell us we are accurately modeling many materials with three-dimensional orbitals, the success is highly suggestive that a three-dimensional wave function could exist.

13 The uncertainty principle is not what you think it is

> *You keep using that word. I do not think it means what you think it means.*
>
> — Inigo Montoya in the Princess Bride

A road map is needed

I wouldn't include a chapter on the uncertainty principle if it wasn't so overhyped and so many wrong things were said about it. It's not nearly as important as people make it out to be. In fact, when someone uses it to justify something in quantum mechanics, their use of it is a red flag warning us they may not know what they are talking about. In actuality, the behavior of a wave function is completely determined by the Schrödinger equation (when uninterrupted by the hypothesized collapse of the wave function). There's no room for another rule to govern the behavior of a quantum object's state.

The number of topics I must touch on in this chapter necessitates a map of where I will take us. After this short introductory section you are reading, I briefly challenge the wording of *Heisenberg uncertainty principle*. That section hints at many of the topics you will see in this chapter. Next, I outline what Heisenberg originally said about his measurement-disturbance relation so that you can see it is not the same as what you will learn. Then we move onto the focus of the chapter, which is called the **position-momentum uncertainty relation**

(Eq. 13.1) or, even more accurately, the **x-p_x uncertainty relation** (the x is a stand-in for whichever axis in space you choose). To learn about that relation, I explain the symbols, show you that there is a mathematical wave function in momentum space, teach you about Fourier transforms, and explain what I mean by the term *flow function*. With that, you will have been introduced to the essence of the x-p_x uncertainty relation. The second half of the chapter covers a few serious mistakes physicists make when describing the x-p_x uncertainty relation.

Questioning every word of *Heisenberg uncertainty principle*

You likely saw the following inequality numerous times before starting this book. It is commonly and erroneously called the *Heisenberg uncertainty principle*:

$$\sigma_x \cdot \sigma_{p_x} \geq \frac{\hbar}{2} \tag{13.1}$$

Heisenberg did not come up with this inequality. He instead articulated the idea of the observer effect and gave, among other thought experiments, a rough idea of the amount a particle would be disturbed by a theoretical gamma-ray microscope. Proper credit for this formula goes to two physicists, E.H. Kennard and H. Weyl, who discovered it. So, the name inside the title *Heisenberg uncertainty principle* is inappropriate. Therefore, because credit isn't due to Heisenberg for the inequality, I drop his name.

Many physicists think the connotation of *uncertainty* conveys the incorrect message. In regular conversation, the use of the word sometimes has the connotation of a lack of knowledge. Critical thinking physicists don't want to give you the idea that the inequality is related to the existence of quantities which we don't have information about. The new preferred word for some physicists is *indeterminacy* because the results of measurements are not, in general, determined beforehand. Unfortunately, they too are misguided. In this case, they erroneously think that the uncertainty principle is inherently about measurements.

In truth, the inequality has meaning before we even make a measurement. Sean Carrol says the same: "But the essence of the uncertainty principle isn't about measurements at all." Looking in a quantum textbook, one will find textbooks do not utilize anything related to measurements when deriving the inequality. Moreover, the symbols we derive on the left-hand side of the inequality are two mathematical measures of the state of the particle at one point in time. One is a measure of the spread of the matter density $|\Psi(x)|^2$. The other is a measure of the spread of an analogous quantity $|\Psi(p_x)|^2$ in momentum space. The measure I am referring to is the standard deviation you encountered briefly in the previous chapter (p. 260).

The common term for $|\Psi(p_x)|^2$ is **probability density in momentum space**. But I'm not a fan of that name. I want to use fewer words and give you a physical interpretation, so I will be bold and call it a **flow function**. I will give a motivation for this name later. Be warned: no one else calls it this.

Knowing the inequality is not at its core about randomness or ignorance, it's easy to see why I am not completely comfortable with *indeterminancy* or *uncertainty* as tags for the inequality. As for the final word, some contend that our inequality is not a principle because the inequality is derivable from the founding postulates of quantum mechanics. Besides that, there are other more strict inequalities, which are not famous because they were not the first to be discovered (Schrödinger was the first to find a more strict version). So, the famous inequality above is neither unique nor the most powerful form available. For all these reasons, I too think *principle* is inappropriate. The inequality is better characterized as a constraint that is only sometimes useful.

Since all three words are inappropriate, I tried to think of a better term for the topic of this chapter, but they either didn't roll off the tongue or weren't general enough: standard-deviation inequality, matter-density-and-flow inequality, or comparison-of-underlying-distributions inequality. Without a replacement and due to our familiarity with it, I will continue to use the term *uncertainty principle.* In addition, it seems more natural to refer to the formula above as a *relation* or an *inequality* rather than a *principle,* so I will sometimes do that when referring to it.

What Heisenberg originally said

To ensure the facts are clear, I will repeat what I have said a few times throughout this book: Heisenberg deserves credit for examples of the observer effect and not for the uncertainty principle. The observer effect is about measurements. Specifically, it is about the violence a measurement does to a state.

Because you will see it brought up often and conflated with the x-p_x uncertainty relation, you should know what Heisenberg's article of 1927 actually said. The expression that Heisenberg came up with, in the most common modern notation, is:

$$\Delta x \cdot \Delta p \;\approx\; h \tag{13.2}$$

This is his measurement-disturbance relation. This has some similarity to the inequality in Eq. 13.1. But this is math, and things are often not the same when symbols are different.

The first thing to notice is that because of the ≈ symbol, this is neither an equality nor an inequality. It can be called an *approximation* or an *estimate.* In defense of 1927's Heisenberg, he did not claim to be doing things precisely in his original paper. It is abundantly clear that it is a semi-quantitative argument. For example, the most precise definition he gave of Δx is "is approximately the average error" in which we know the position x.

It is later writers who kept and keep misrepresenting what Heisenberg originally wrote. It probably didn't help that post-1927 Heisenberg was one of those later writers who confusingly said that the first inequality of this chapter and his loose approximation both involve the same physics. They do not. Again, I say, the inequality at the beginning of the chapter is true before measurements get involved and applies to a quantum system in one particular state. In contrast, Heisenberg's paper required disturbances that changed the state of the particle. There is a state before and a state after. On top of that, some of his thought experiments additionally required multiple equivalently prepared particles.

Interestingly, his paper ends with an addendum where he thanks Bohr for pointing out that a gamma-ray microscope will cause further uncertainties because he didn't fully treat gamma rays as waves and thus didn't consider the acceptance angle of the lens. So, his estimate in Eq. 13.2 is suspect. (This was one of several oversights Bohr pointed out to him.) However, his general observation that photons used for position or momentum measurements will disturb the states of atoms to some degree is still valid.

Understanding the x-p_x inequality

The symbols

To understand the x-p_x uncertainty relation as a whole, we must first examine the parts that make it up. Since we are talking about the symbols, I write again the inequality here:

$$\sigma_x \cdot \sigma_{p_x} \geq \frac{\hbar}{2} \tag{13.3}$$

By now, you ought to recognize $\hbar$ if woken from a sleep and know that it is a number (technically, it's a quantity because it is a number plus a unit). The greater-than-or-equal-to sign $\geq$ tells you that the right-hand side of the inequality, namely $\hbar/2$, is the lowest bound possible for whatever is on the left-hand side.

If you are in this quantum game long enough, you'll see some physicists are rather loose with math and do things like treat the inequality as a strict equality to prove or justify a result they want. When they do this trick – and I do mean it is deceptive – physicists say they have given a hand-waving explanation. Hand-waving is what illusionists do to misdirect you. A hand-waving explanation is a euphemism for "they haven't proven it", so be on the lookout for the inappropriate change to an equal sign or the implication of equality in other places.

The σs (pronounced and spelled: "sigma") are the conventional symbols for **standard deviations** in statistics and also serve the same function in quantum mechanics. Standard deviations are one mathematical measure of many that describe how spread out a set of numbers is. The σ_x is a measure of how spread out the matter density is. More specifically, it is a length measured from the mean position of the matter density. In Fig. 13.1, I show the standard deviations of two matter density distributions.

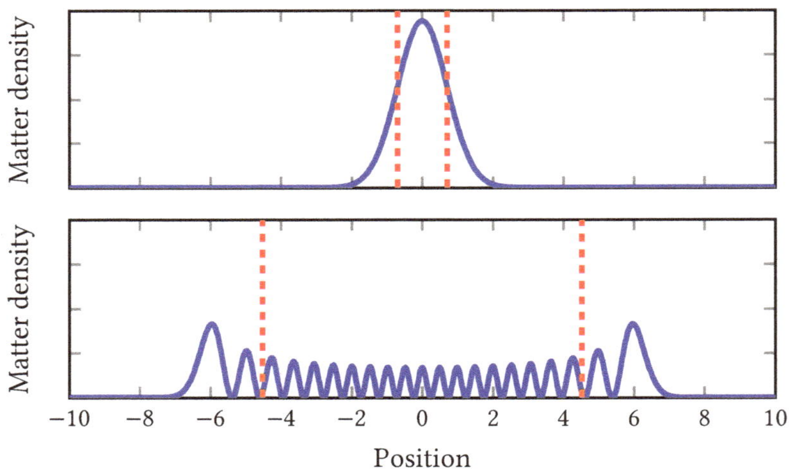

Figure 13.1: Matter density of a particle in a parabolic potential. Potential function not shown. I chose these as examples because of their dramatically different distributions. The top image is the ground state (the shape is a Gaussian), and the bottom image is the 20th stationary state above the ground state. The red, vertical, dashed lines are one standard deviation σ_x from the center of mass of the matter density.

In quantum mechanics, the standard deviation of position σ_x of some state $|\Psi\rangle$ is calculated using the expectation value of the squared-position operator $\langle \hat{x}^2 \rangle$ and the square of the expectation value of the position operator $\langle \hat{x} \rangle^2$:

$$\sigma_x = \sqrt{\langle \hat{x}^2 \rangle - \langle \hat{x} \rangle^2} \tag{13.4}$$

Even though I show you the equation, it's not important for you to understand it in depth. As a substitute, I will tell you two important details.

First, because of how the integrals of expectation values work, the standard deviation is a measure of $|\Psi(x)|^2$ and not the wave function (recall the explanation starting on p. 255). Second, one can conclude that σ_x isn't inherently about measurements because of the $\hat{x}^2$ operator. Classical physics particles don't possess a quantity known as x^2, so the observable postulate tells us this isn't an operator corresponding to an observable.

Next, we want to understand σ_{p_x}, which requires an understanding of the **momentum-space wave function** $|\Psi(p_x)\rangle$. Before I explain what that is, I need to mention a vocabulary shift that is about to occur. When physicists talk of momentum-space wave functions, we adjust our language and use the term **position-space wave function** in place of the previous term *spatial part of the wave function* $|\Psi(x)\rangle$. The word *spatial* refers to space, and we need to differentiate which space we are discussing.

Now, I must finally reveal what I've been hiding from you for the whole book (well, except for many spots in the glossary). Since they contain the same information, the position-space and momentum-space wave functions are two representations of the same state of a quantum object. For educational purposes, I have been writing as if only the position-space representation exists and have glossed over some things. The most pertinent is that with the new representation, there is a whole new set of operators. I'll show you these in a table in the next chapter (p. 317). For now, know that the abstract operator symbols with the hats stay the same; it's what is on the other side of the equal sign that is different. To avoid any confusion, I note that momentum space is not a physical space that you or I can visit; it is a mathematical construct.

Momentum-space wave functions

Two spaces and their two wave functions

To understand the relationship between the two wave functions, four foundational ideas are needed. Two are relatively straightforward to explain.

Let's consider three cosine waves that exist in position space at some snapshot in time:

$$A_1 \cdot \cos \frac{x \cdot 2\pi}{\lambda_1} \quad \text{and} \quad A_2 \cdot \cos \frac{x \cdot 2\pi}{\lambda_2} \quad \text{and} \quad A_3 \cdot \cos \frac{x \cdot 2\pi}{\lambda_3} \tag{13.5}$$

These three functions will be the basis functions in a superposition that will be the focus of our discussion. In all of this, I will ignore the units. Let's say that the peak amplitudes A_1, A_2, and A_3 of these waves are 14, 8, and 3. (The A_n can be negative. I just happened to not pick a negative one.) The wavelengths λ_1, λ_2, and λ_3 will be approximately 6.28, 1.57, and 0.628. In Fig. 13.2, I show you these basis functions and their superposition.

But we want to talk about momentum. Calling back to a previous chapter (p. 112), Louis de Broglie tells us that the wavelength of a matter wave λ is related to magnitude of momentum p via the equation:

$$|p| = \frac{h}{\lambda} \tag{13.6}$$

Using this and the fact that $\hbar = h/(2\pi)$, the cosines can be rewritten to be:

$$A_1 \cdot \cos \frac{p_1 \cdot x}{\hbar} \quad \text{and} \quad A_2 \cdot \cos \frac{p_2 \cdot x}{\hbar} \quad \text{and} \quad A_3 \cdot \cos \frac{p_3 \cdot x}{\hbar} \tag{13.7}$$

By dropping the absolute value sign around the p_n's, I've made the waves have momentum in the positive x-direction. If I wanted to describe the superposition in the figure, I could summarize it by specifying the pairs (p_1, A_1), (p_2, A_2), and (p_3, A_3). These pairs look like the coordinates of a graph, and I will graph them in a momentum space (see Fig. 13.3).

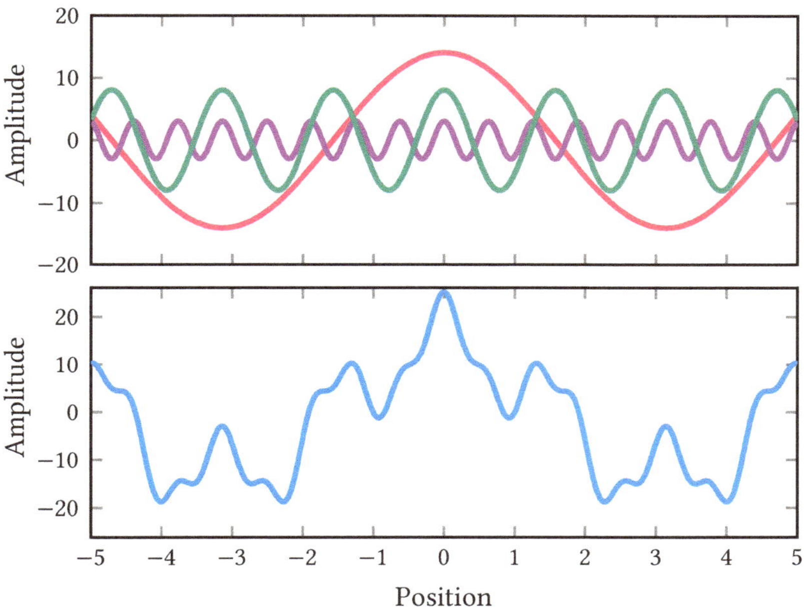

Figure 13.2: Cosine waves in position space The top image has the three cosines plotted independently. The bottom image is their superposition.

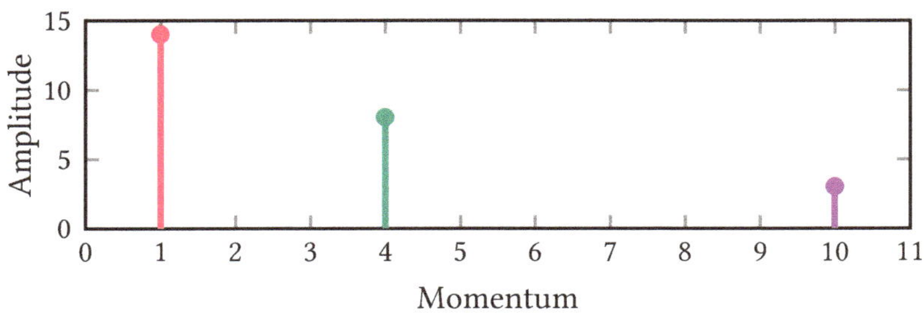

Figure 13.3: Momentum-space representation of the three cosine superposition. I set h equal to one for this plot. It should be apparent by their round numbers that I picked the values of the momenta and not the values of the wavelengths.

The construction of a momentum-space wave function from pairs of coefficients is the first and most important foundational idea. To be clear, this foundational idea says that the *peak* amplitude A_n of a single member of the cosine superposition in position space is the *particular* amplitude at a particular point p_n in momentum space: $\Psi(p_n) = A_n$. The two figures above contain equivalent information if we have a mutual understanding of how to make one from the other.

Although the example shows the first fundamental idea, physically acceptable wave functions do not have discrete values in momentum space like in the figure above. The discreteness in momentum space is related to the fact that the position-space wave function (bottom image of Fig. 13.2) repeats the same pattern forever beyond the plotted region. (The peak at position 0 repeats at positive and negative integer multiples of 6.14.)

Physical waves do not repeat the same pattern forever. They gradually fade in from one direction and fade out in the other, just like many of the images I have shown you in this book. Using a physically accurate position-space wave function would ensure that the momentum-space wave function is continuous. Let's call the need for a physical, and thus continuous, wave function in each space the second foundational idea.

When both wave functions are continuous, we need to work with some special integrals called **Fourier transforms** to get between the two spaces. Before getting to those integrals, we have two more foundational ideas. Their explanations involve some math that I gloss over and math that I move quickly over. There is no need to get caught up in the mathematics. Reading the conclusions is sufficient. If you know about traveling waves and the addition of even and odd functions, the explanations should be mostly tractable.

Two more foundational ideas

Notice that I have only plotted the momentum space's positive axis. That is because the negative axis is a mirror of the positive axis when the cosine basis is as I have written it. I'll show you an example that gives you the idea. You will need to know the trigonometric (aka trig) identity: $\cos(b) = \cos(-b)$. Let's say we have, among other basis functions, a wave that has the two following basis functions in the superposition:

$$A_1 \cdot \cos(4 \cdot x) \; + \; A_2 \cdot \cos(-4 \cdot x) \tag{13.8}$$

I have chosen a single magnitude and opposite directions for the momenta. The peak amplitudes are left as some unspecified values that are, in general, *different*. I have temporarily set $\hbar = 1$ to make the equations cleaner. With the trig trick I gave you, we can say $\cos(4 \cdot x) = \cos(-4 \cdot x)$. Substituting this in and grouping the amplitudes, we find:

$$\begin{aligned}(A_1 + A_2) \cdot \cos(4 \cdot x) &= B \cdot \cos(4 \cdot x) \\ &= \frac{B}{2} \cdot \cos(4 \cdot x) \; + \; \frac{B}{2} \cdot \cos(-4 \cdot x)\end{aligned} \tag{13.9}$$

In the final line, I have used the trig identity again to get back both basis functions with the peak amplitude divided evenly between them.

In momentum space, at the momentum points of 4 and -4, the amplitudes will be equal. Of course, we could have chosen any other pairs of momenta, so the argument generalizes. Therefore, momentum-space wave functions will be symmetric if the cosine basis states are as I have written them. However, this symmetry is a problem because, to handle all the possible physical states, asymmetries must be allowed to occur in wave functions.

If you go back to when I first introduced the cosine basis, you can see I said these were snapshots in time. We can break the curse of the trig identity by recognizing that time should be taken into account and the cosines are supposed to be traveling waves. Allowing the time to progress, a general basis function will be:

$$\begin{aligned}A_n \cdot \cos\left(\frac{1}{\hbar} \cdot \{p_n \cdot x \; - \; KE_n \cdot t\}\right) \\ = A_n \cdot \cos\left(\frac{1}{\hbar} \cdot \left\{p_n \cdot x \; - \; \frac{p_n^2}{2 \cdot m} \cdot t\right\}\right)\end{aligned} \tag{13.10}$$

With basis functions of this form, you could try to repeat the procedure I showed you. You will find that the trig identity can't be used to make a pair with the same amplitude unless A_1 happens to equal A_2 already. The third foundational idea is the need for both positive and negative axes in momentum space regardless of the symmetry of the momentum-space wave function.

To keep things simple for the next few pages, I will remove the term with time in it and go back to the original form. In all cases where time isn't mentioned, you should think we are looking at wave functions frozen at a particular point in time.

The fourth, final, and most interesting foundational idea involves phase shifts ϕ_n. When we include the phase shift in the cosine functions, a general basis function looks like this:

$$A_n \cdot \cos\left(\frac{p_n \cdot x}{\hbar} + \phi_n\right) \tag{13.11}$$

Now, have a look again at the bottom plot in Fig. 13.2, and note that it is a superposition made of cosines without any phase freedom. If we restrict ourselves to using only cosine basis functions to make the superposition, the general shape in the figure is unique to the three basis functions I showed you. What if I wanted the same shape but with its highest point located at position -2? It's reasonable to expect to be able to describe any physically reasonable shape possible. There is no other combination of cosine basis functions that can make it. If I insist on making that shape with its peak at position -2, I must allow the basis functions to shift to make a new superposition. I must then dial in the phase shift for each of the three basis functions appropriately. No problem there. I could get the desired shape with a peak to the left of the zero position.

To make what follows conceptually easier for you, we will pretend that the momentum-space wave function is continuous. Now, we must ask: how we are going to encode this new information from all the phase shifts into a momentum-space wave function? We have a two-dimensional plot, and we are already using A_n and p_n to plot the points of the curve, but we need to encode A_n, p_n, and ϕ_n at every point in momentum space. Until a moment ago, we had the negative axis free, and that could have given us hope, but it's used up now too.

By using a few other trig-identity tricks, it can be shown that our cosine basis with phases can be turned into a cosine and a sine basis, both without phases in the arguments of cosine and sine. Why can this be done? Looking one more time at our original superposition (Fig. 13.2), we notice that the function is an even function because there is a mirror symmetry across the vertical axis at position 0. When we shift the peak, it breaks the symmetry and gives a function that is neither even nor odd. The rules of even and odd functions say that an even function can be considered to be composed of a series of even functions. Often, one chooses cosines. Similarly, an odd function can be considered to be composed of odd functions, which are often sine functions. With a neither-even-nor-odd function, one needs both series. This is enough basis functions to cover a continuum twice.

Following the strategy from the first fundamental idea, we will have two pairs of coordinates (p_n, C_n) and (p_n, D_n). Therefore, this shows that we are forced to plot two new real-valued functions in momentum space. One will be an even function, and one will be an odd function.

If you are unhappy about having those two functions in momentum space, we don't need to stop there, as there are more math tricks that can be done. Glossing over the details of how to do it, we can do what everyone else does and replace the two functions with a single complex-valued function. Therefore, demanding that we be able to make a real-valued position-space wave function that is neither even nor odd, which is a physically reasonable demand, forces the momentum-space wave function to be complex.

As you likely expect, a similar argument works going from momentum-space to position-space. If we have a real-valued momentum-space wave function that is neither even nor odd, then the position-space wave function is complex valued.

Though I haven't gone through all the cases possible (i.e., wave functions that are imaginary valued & even, imaginary valued & odd, etc.), I think I can say I have essentially given you why position- and momentum-space wave functions are complex valued. The fourth and final foundational idea is the fact that to get all the shapes we want in one space, the conjugate space will often be complex valued.

Some of what I have described missed some details and didn't show equations. For the curious, google *Fourier series* for a good starting point.

Converting from one space to the other

Fourier transforms (FTs) are two integrals that let you go from a position-space function to a momentum-space function and vice versa. Importantly, they handle the phase shifts I mentioned. As well, they make the momentum-space wave function exist on the positive and negative axes of momentum space. I won't show you the integrals, but the idea is:

$$\Psi(x) \underset{\text{inverse FT}}{\overset{\text{FT}}{\rightleftharpoons}} \Psi(p_x) \tag{13.12}$$

The *FT* and *inverse FT* represent the process of using the integrals that transform functions. I am also considering these two functions to exist at a particular point in time, so I don't specify a time variable.

Normally, when we think of decompositions of the wave functions that make the transformations work, although we could, we don't think of separate series of sine and cosine functions. Instead, we think of one series of complex exponentials, which is mathematically equivalent to the two series. It doesn't matter if it's a position- or a momentum space wave function. Each exponential function in the superposition will have the form:

$$A_n \cdot e^{i \cdot \blacklozenge} = A_n \cdot (\cos \blacklozenge + i \cdot \sin \blacklozenge) \tag{13.13}$$

The n is an index that tells us which wave in the superposition, out of the unlimited number, we are looking at. Each function has a multiplying coefficient A_n that is complex valued and sets the maximum amplitude of the oscillations.

The left-hand side is what we write down. The right-hand side is an equivalent but more intuitive way to understand it. Do you remember, in the *Gallery* chapter, when I said when physicists see the *i* in the exponential function that they think of cycles in time or space? You can see that this idea comes from the cosine and sine functions, which are cyclic. For a comparison to what you've seen before, see Eqs. 10.3, 10.5, and 10.6 on pages 194, 195, and 197.

If one considers a position-space wave function, the diamond will have the form:

$$\begin{aligned} \blacklozenge &= \frac{1}{\hbar} \cdot (p_n \cdot x - KE_n \cdot t) \\ &= \frac{1}{\hbar} \cdot \left(p_n \cdot x - \frac{p_n^2}{2 \cdot m} \cdot t \right) \end{aligned} \tag{13.14}$$

This is the same expression as that in Eq. 13.10. Note though, that there won't be any phase shifts added to the present equation. Because A_n is complex valued, phase shifts aren't needed.

If we were dealing with a three-dimensional space Eq. 13.13 and Eq. 13.14 would be collectively called a *complex plane wave*, but the diamond would have a more complicated form with y and z variables along with the momentum versions of those. You can find more information about the three-dimensional form online.

I have explained how the two spaces are related mathematically. But no one can mentally decompose a wave they see into an infinitude of exponential functions and extract from that the shape of the other space's wave function. Fortunately for us, people have made tables of characteristic shapes and their Fourier transforms. They are called *Fourier-transform pairs.* We don't need to see many in this book. You can see a slew of them by searching for *Fourier-transform pairs graphed.*

I show two pairs in Fig. 13.4. One pair is the favorite example of every physicist, chemist, engineer, and mathematician: the Fourier transform of a Gaussian function is another Gaussian function with a different height and width. The other pair I show relates to single-slit interference. It is a box function (a single slit) and a sinc function.

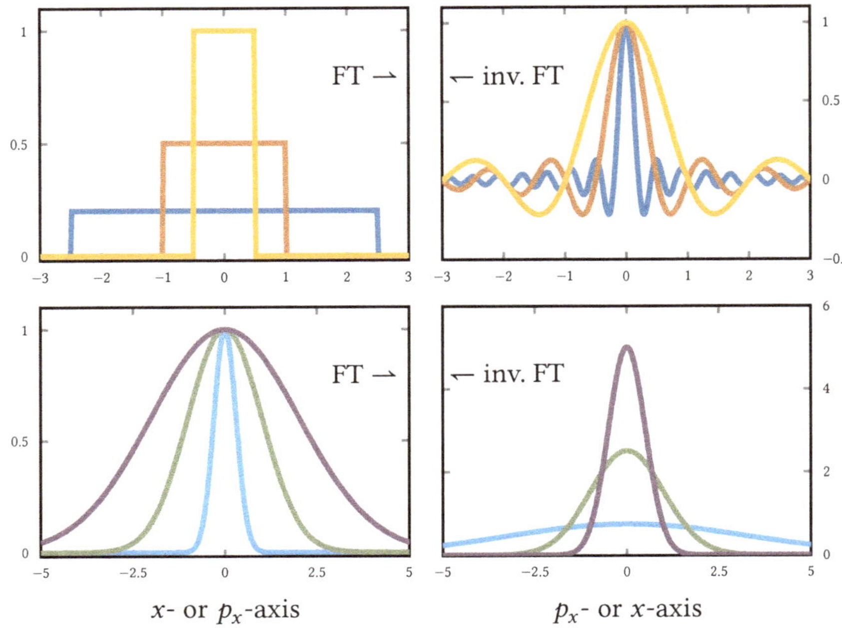

Figure 13.4: Two examples of real-valued Fourier-transform pairs. One column represents a particular space (your choice of momentum or position), and the other represents the conjugate space. The top row represents the box-sinc pair. The bottom row represents the Gaussian-Gaussian pair. I gave various heights and widths of one function to demonstrate the difference in its conjugate function. Note that what I show here is straight math. If these were wave functions, each function should be normalized, which means that the area under each curve would be the same. Additionally, for the rectangular wave function to be physical, we would have to say 1) the corners are smooth, 2) the vertical lines aren't exactly vertical, and 3) the regions of zero amplitude have tiny finite amplitudes that decay away. And all of those changes are imperceptible because we haven't zoomed in enough. The sinc function would then also have imperceptible deviations from a perfect sinc function.

In fact, all types of interference patterns that are far from the region of diffraction of the waves can be found via Fourier transforms (plus squaring of the wave function).

The flow function

The uncertainty principle doesn't concern itself directly with position- or momentum-space wave functions. Instead, we must consider the distribution of the matter density $|\Psi(x)|^2$ and the distribution of values in the flow function $|\Psi(p_x)|^2$. I show how these are related in the following equation:

$$\begin{array}{ccc} \Psi(x) & \underset{\text{inverse FT}}{\overset{\text{FT}}{\rightleftharpoons}} & \Psi(p_x) \\ \downarrow & & \downarrow \\ |\Psi(x)|^2 & & |\Psi(p_x)|^2 \end{array} \tag{13.15}$$

The uncertainty principle is about these two functions

I am compelled to remind you that for now I have left out the time variable. We are considering things at some particular point in time. It's important to note that the downward-pointing arrows don't have another arrow pointing upward. By squaring the magnitude of a wave function, we lose information on the phase of the wave function. Therefore, there is no unambiguous way to return to the original functions from $|\Psi(x)|^2$ and $|\Psi(p_x)|^2$.

To understand the flow function $|\Psi(p_x)|^2$, you must consider the steps we take to compose it. Let's consider a single point in momentum space. The amplitude A_n of a momentum-space wave function at a particular value of momentum is equal to the peak amplitude A_n of a single complex exponential function from position space that has that particular value of momentum. This is the first foundational idea I illustrated with cosines above.

When we square the absolute value of an amplitude in position space, we get a matter density; said differently, we get the fractional amount of the quantum object. We must realize that the amplitude used to get this fractional amount of the object is the same amplitude from the point in momentum space. In that space, its magnitude squared is the value of the flow function at a particular value of momentum. Therefore, at a single point in momentum space, the flow function $|\Psi(p_x)|^2$ tells us the fraction of the quantum object with that particular momentum. Different fractions of the quantum object will have different momenta values.

Because momentum is proportional to speed ($p = m \cdot v$), the flow function $|\Psi(p_x)|^2$ characterizes the speed with which fractions of the matter density distribution $|\Psi(x)|^2$ could potentially flow away from its current distribution in space.

Given that I said "could potentially", you should understand that the matter density will not flow away in every situation. Stationary states are states with unchanging matter densities and unchanging flow functions. In these states, the flow function, the matter density, and the potential energy function are matched in a way that prevents the matter distribution from changing. Though, once the potential energy is taken away, the matter is free to flow away.

Momentum standard deviation

Now that you understand momentum space and the flow function, the momentum standard deviation is easy to understand. It is the momentum analog of σ_x. A calculation of σ_{p_x} returns a number that is the standard deviation of the flow function distribution $|\Psi(p_x)|^2$. It is a "length" in momentum space (with units of momentum) measured from the mean value of the flow function. As an equation, it is:

$$\sigma_{p_x} = \sqrt{\langle \hat{p}_x^2 \rangle - \langle \hat{p}_x \rangle^2} \tag{13.16}$$

By the way, when a physicists calculates this, they will find the same value of σ_{p_x} whether they use $\Psi(x)$ and $\hat{p}_x = -i \cdot \hbar \cdot \partial_x$ or $\Psi(p_x)$ and $\hat{p}_x = p_x$ in the integrals of the expectation values of this equation. (These two representations of the momentum operator can be found in the table in the next chapter [p. 317].) It's a general rule with no exceptions that we will get the same expectation value for a physical quantity in different representations. However, you won't necessarily get the same range of numerical values returned after application of an operator. For example, in momentum space, the kinetic energy values can't be negative; this is the preferred-basis problem showing itself again.

With this, you now have an overview of the individual parts of the position-momentum uncertainty relation. In the context of a misconception, I will bring these parts together in the next section. Later, I will show you images of matter density and flow functions together.

Applying the uncertainty principle to atoms

In this short section, I show you the x-p_x uncertainty relation applied to the real-world example of atomic hydrogen. I do this in the context of an incorrect claim made by Richard Feynman.

Have a look at what Feynman said during an introductory lecture on electromagnetism:

> You may ask: "If this electrical force is so terrific, why don't the protons and electrons just get on top of each other? If they want to be in an intimate mixture, why isn't it still more intimate?" The answer has to do with the quantum effects. If we try to confine our electrons in a region that is very close to the protons, then according to the uncertainty principle they must have some mean square momentum which is larger the more we try to confine them. It is this motion, required by the laws of quantum mechanics, that keeps the electrical attraction from bringing the charges any closer together.

The problem with this claim is that the two types of standard deviations can be any values, *any values at all*, as long as their product is greater to or equal to the lower bound of $\hbar/2$.

To see what's going on, let's evaluate the standard deviations of the ground state of hydrogen. After some math involving expectation values, which I don't show you, the standard deviations are $\sigma_x = a$ and $\sigma_{p_x} = \hbar/(\sqrt{3}\,a)$, where a is the Bohr radius and is approximately

0.05 nm. Plugging the results into the uncertainty relation, we find that:

$$\sigma_x \cdot \sigma_{p_x} = a \cdot \frac{\hbar}{\sqrt{3}\,a} = \frac{\hbar}{\sqrt{3}} \geq \frac{\hbar}{2} \tag{13.17}$$

Because $\hbar/\sqrt{3}$ is indeed greater than $\hbar/2$, the ground state wave function of the hydrogen atom satisfies the uncertainty relation as it is required to.

But we must ask: if the uncertainty principle is a law changing the nature of wave functions, as many claim, why this value? Why slightly above the lower bound? Why doesn't the electric force take it to the lower bound as Feynman suggests it is wont to do? It's because appealing to the uncertainty principle is the wrong thing to do. Reading into what he says, it is apparent that he is implying that there is a balance between confinement and motion. Balance in mathematics is achieved with equal signs. He is trying to have a strict equality in the uncertainty inequality, and that is his mistake here. Near the beginning of the chapter, I warned you to watch out for exactly this.

The familiar claim is an enormous mistake

What do they say?

Have a read through the following quotes. They should sound familiar to claims you have heard many times before.

Auletta et al. (Quantum Mechanics):

> "The [x-p_x] relation ... states that when one tries to reduce the uncertainty of one of the two conjugate observables, then necessarily the uncertainty of the other increases."

Greene (The Elegant Universe):

> "Moreover, the more precisely you know one, the less precisely you know the other."

Cox & Forshaw (The Quantum Universe):

> "In other words, the more accurately we pin down the particle, the faster it moves away from its initial position." And: "And the larger we make [the position-space wave packet], the smaller the uncertainty in its momentum will be."

Griffiths & Schroeter (Introduction to Quantum Mechanics):

> "...but there is an inescapable trade-off here: the more precise a wave's position is, the less precise is its wavelength, and vice versa." And: "... the more precisely determined a particle's position is, the less precisely is its momentum."

Holzner (Quantum Physics for Dummies):

> "That is to say, the more accurately you know the position of a particle, the less accurately you know the momentum, and vice versa."

Looking past the exact words used, we can filter out the ideas of knowledge and measurement from those statements because they are things added on top of the uncertainty principle. In the language of this book, they are all basically saying, "The x-p_x uncertainty relation tells us that the more localized the matter density is (smaller standard deviation), the more spread out the flow function is (larger standard deviation), and vice versa." Again, these are all attempts to replace an inequality with an equality. This still is not the proper thing to do.

We must ask: how can physicists keep repeating those rubbish claims? It surely can't be that they don't understand the elementary school math of a greater-than-or-equal-to sign. Something must be short-circuiting their thinking processes. Without a doubt, I know the confusion surrounding Heisenberg's measurement-disturbance relation contributes to it. If you go to the sources of those quotes, you will find most of them ignorantly convolute the uncertainty principle and Heisenberg's measurement-disturbance relation. However, there is another source of confusion.

A theorem

They seem to be unconsciously invoking a theorem associated with Fourier transforms. It is called alternatively the **scaling theorem**, **shape theorem**, **dilation theorem** or **similarity theorem** (not a good name, in my opinion, because *similar* has a precise meaning in geometry). Often, it goes unnamed and is part of a list of properties of Fourier transforms. It is a standard part of the education of almost every mathematician, engineer, physicist, and spectroscopy expert. You have already seen it at work in the figure with the Fourier-transform pairs (Fig. 13.4). This theorem is a rule of mathematics that is already baked into Fourier-transform-paired functions. And when we use it in quantum mechanics, it is a rule of the universe that is already baked into the wave functions.

The theorem says if you pick a point in time and stretch out a function along its variable (for us, that's x or p_x), the Fourier transform of that function will be squeezed thinner along the other variable (for us, that's p_x or x) than it was before. The alternative operation also happens: if you squeeze in one space, it stretches out in the other. If you squeeze or stretch in one space, there is an exactly predictable response in the other space.

The general shapes remain the same in the two spaces. If you start with a rectangular shape in a particular space, you have to end up with a stretched or compressed rectangular shape in that same space. In other words, you can't change the form of the function's math expression involved. This is because expressions determine the general shapes of the numbers that functions output. If you have a starting function $\Psi(x)$ and want to use this theorem, the only thing you can do is multiply the variable. For example, it could be changed to $\Psi(5x)$, which compresses the function along x. If that is the case, then $\Psi(p_x)$ is stretched to $\Psi(p_x/5)$.

The function in the original space won't have its height change, but the conjugate space will have its height change due to the theorem (see Fig. 13.4). However, in quantum mechanics, the heights (amplitudes) of both functions will be adjusted anyway, via normalization, so that charge, mass, spin, and probability are conserved in $|\Psi(x)|^2$, so the mathematical theorem's height change doesn't concern us.

This description of the theorem may seem to correspond well to the quotations above. This behavior is exactly what they describe when they write those words. It is also the balance between confinement and motion that Feynman desired. Unfortunately, for anyone who has tried to link the scaling theorem to an uncertainty principle, there is no inequality, or lower bound, in the scaling theorem. Thus, this can't relate to the vagueness of a greater-than-or-equal sign in an uncertainty relation. Moreover, the scaling theorem and the uncertainty principle do not apply to the same mathematical objects. I make this explicitly clear in the following equation:

The scaling theorem relates to FT-paired functions

$$\begin{array}{ccc} \Psi(x) & \underset{\text{inverse FT}}{\overset{\text{FT}}{\rightleftharpoons}} & \Psi(p_x) \\ \downarrow & & \downarrow \\ |\Psi(x)|^2 & & |\Psi(p_x)|^2 \end{array} \tag{13.18}$$

The uncertainty principle is about these two functions

The ambiguity-inducing down arrow

In some cases, the quotations from the physicists above apply well to the standard deviations of $|\Psi(x)|^2$ and $|\Psi(p_x)|^2$. By gently tuning a potential energy function to induce a stationary state's matter density to be wider or thinner, one can stretch the flow function in the opposite sense. However, this does not justify their words. Each of them claims a universal principle from a misapplied math theorem. Something universal must have no exceptions.

In the next two figures, I show you one exception out of an infinite set of exceptions. It is a single scenario at three different points in

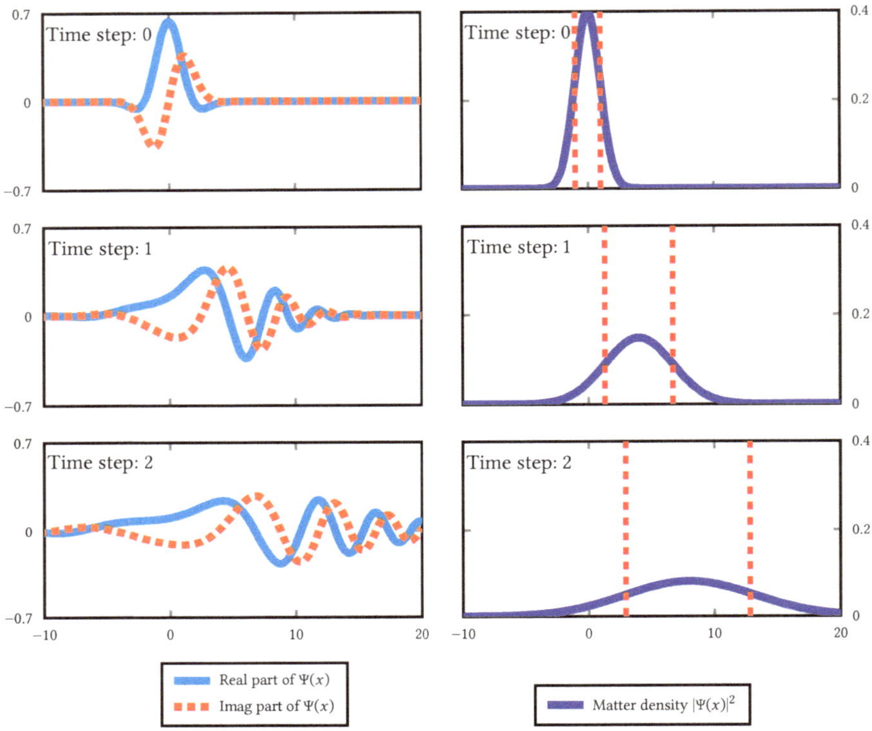

Figure 13.5: Position-space wave function and matter density at three points in time. It wasn't necessary to have a traveling particle, but it makes for more interesting wave functions. A red, vertical, dashed line is one standard deviation σ_x from the center of mass of the matter density. Notice σ_x increases with time.

time. First, $\Psi(x)$ and $|\Psi(x)|^2$ are shown after the quantum object was released from a potential with some momentum. Because there is no longer any potential energy function, what occurs is dispersion and as well as a translation of the center of mass at a constant speed. You can see that the standard deviation of the matter density is getting larger. (See Fig. 13.5.)

We should pause and consider the implication of a lack of a potential energy function on the flow function. Because potential energy functions change the momentum of objects, the lack of a potential

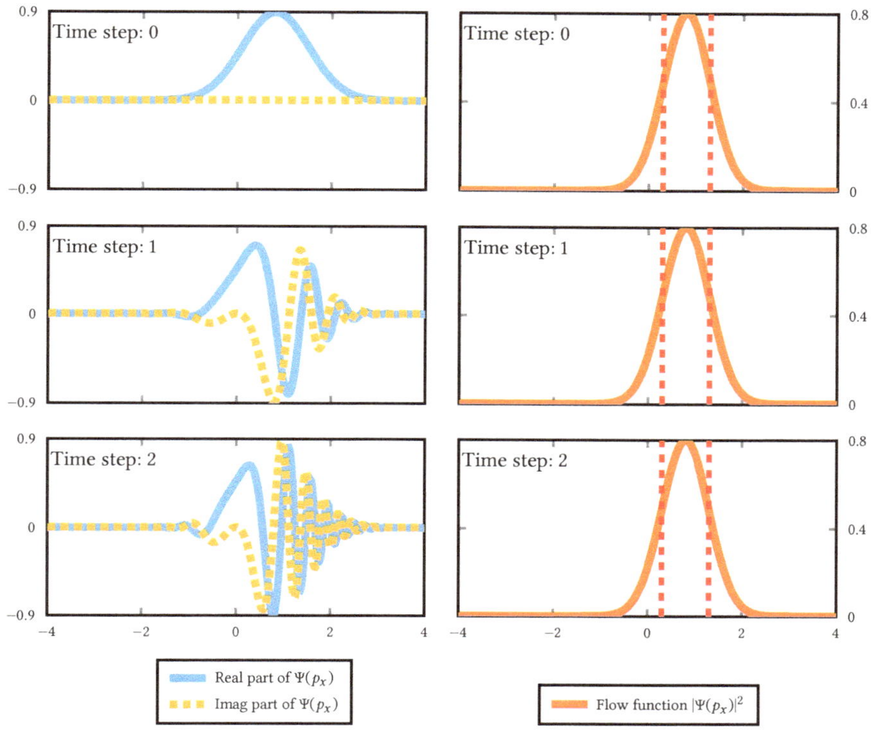

Figure 13.6: Momentum-space wave function and flow function at three points in time. The momentum-space wave function will stay centered on the same spot and accumulate oscillations as time goes on. The off-center functions cause the particle's center of mass to travel. With no external force on the particle, the flow function will not change and will always have the same σ_{p_x}.

energy function means there will be no momentum change. In the next figure, I show you the three points of time in momentum space. You can see the momentum-space wave function evolves with time, but the flow function never changes. Its standard deviation is constant (Fig. 13.6). Therefore, there is no "inescapable trade-off" as some of the above-quoted physicists characterize the uncertainty principle. At these three points in time, we have three values of σ_x but only one σ_{p_x}. An "inescapable trade-off", or any other description the authors use, describes a one-to-one mapping between two things. In reality, we have many-to-one situations. There's an infinitude of matter densities that correspond to the same flow function. There are also many-to-one mappings in the other direction. In the previous chapter, you saw two starting matter densities with the same σ_x (Fig. 12.3 on p. 260). Underlying and unseen were two flow functions of wildly different σ_{p_x}.

You may wonder why this can occur if it's one small mathematical step from two functions related by a scaling theorem to two functions related by an inequality. The reason for the disconnect is that the phase information is lost going from $\Psi(x)$ and $\Psi(p_x)$ to $|\Psi(x)|^2$ and $|\Psi(p_x)|^2$. If we were to change the spread of the matter density and that is all that we knew, we wouldn't know what would exactly happen to the flow function because we can't follow in the direction of arrows to get to the flow function in Eq. 13.18.

To actually get from the shape of the matter density to the shape of the flow function, we would start at the matter density in Eq. 13.18. To go against the down arrow, we would need to supply the phase information to get the position-space wave function. We can then continue algorithmically onward towards the flow function. But no one ever tells their audience these necessary steps. It seems they are unaware of them.

As for my gentle tuning of the width of matter density in a stationary state that I mentioned earlier, the phase information is implicitly known. For example, you know that if I gave you the magnitude of an atomic orbital of hydrogen, you could pick some phase for one location of the wave function and then the remainder of wave function is forced to have certain phases. In this carefully selected scenario, I know that when I tune the potential energy function and the matter density, the phase part of the position-space wave function will also change appropriately to narrow the flow function. But, to be clear, this is a particular scenario and is not general.

Even without the phase explanation, the pictures should leave you with no doubt that the *greater than* part of the symbol $\geq$ in the uncertainty relation is there for a reason. General attempts to use the relation as a strict equality are fools' errands. Given the value of one of the two standard deviations, the only thing one could say via the uncertainty principle is the smallest the other standard deviation could be. This shows that the Schrödinger equation remains king. There is no way for the uncertainty principle to usurp control over the behavior of wave functions.

Less than final words

I will repeat my warning from the beginning of the chapter. If a physicist invokes the uncertainty principle, that is a red flag warning you that what follows should be viewed with suspicion.

You can now judge them by what they say and don't say. If they talk of measurements, there's a good chance they aren't aware of the differences between the uncertainty relations, the observer effect, and Heisenberg's measurement-disturbance relations. If they start talking about knowledge or accuracy of position or momentum, they may be unaware that the scaling theorem applies only to the relationship between position- and momentum-space wave functions.

This is regrettably not the end of the problems with things called the *uncertainty principle*. Another one called the *energy-time uncertainty principle* is often used to justify supposed chaotic behavior in the quantum realm. I have relegated an analysis of that inequality to the final appendix due to the moderate amount of math it requires. There, I walk you step by step through some math to show you where the folly lies.

14 The postulates of quantum mechanics

> *And thirdly, the Code is more what you'd call guidelines than actual rules.*
>
> — Captain Barbossa on conformity to the Pirate's Code in *The Pirates of the Caribbean*

A working axiomatic system

Find x.

The purpose of an axiomatic system is to unerringly get to new truths, or theorems. To appreciate the state of the axiomatic system we have in quantum mechanics, it's informative to contrast it with a working axiomatic system. Euclid's axiomatization of geometry is the exemplar above all others. Unlike the state of affairs in quantum mechanics, his system is a model of logic, simplicity, and cohesion.

Euclid's terminology and grouping for the components of his axiomatic system are different from what we use today. Nowadays, we consider the components to be 1) unstated notions, 2) definitions, 3) postulates, and 4) theorems.

Anyone who has taken high school geometry has plenty of practice going from postulates to theorems or from several theorems to a more refined theorem. *Theorem* is a synonym for *truth* within the confines of the axiomatic system. A new theorem doesn't necessarily need to be profound. In a geometry class, this process for finding truths, commonly called a *proof*, would have started with lines on the page with limited information about the relations between them. Somewhere on the page, your teacher wrote, “Find *x*.” In a particular proof, you may have used a parallel-line theorem and said “side-angle-side” a few times. Your new theorem could then be summarized along the lines of “Given this oddly specific situation of lines and angles, the length of this line segment here is 2.”

Unstated notions and definitions

Proofs are a method of communication between two people. (For the most recluse of mathematicians, it could be between themselves and future selves.) We therefore need a common understanding.

To let the reader know how to understand what was being said, Euclid first defined the words that would appear in his postulates and proofs. Euclid wasn’t the only one to have a discourse on math, though. It is informative to look at the variety of definitions possible. From Prof. Timothy Peil’s webpage (MSU Moorhead), I quote:

point	Pythagoreans, “a monad having position”
	Plato, “the beginning of a line”
	Euclid, “that which has no part”
line	Proclus, “magnitude in one dimension”, “flux of a point”
	Euclid, “breadthless length”

One hopes that Plato had previously defined a line with no reference to a point. To understand the Pythagorean’s “monad”, perhaps you need to read some ancient Greek treatise that includes the subject.

For a self-contained work, one might consider it necessary to define the words we use in our first set of definitions. Obviously, that would open up the need to define the new words we use in the new set of definitions. And we would go on almost forever. That is moving backward away from the postulates when one wants to move forward and generate more complex knowledge. Therefore, we must accept certain concepts or notions as understood but unstated simply because we must start somewhere to move forward.

A good axiomatic system will have definitions and postulates that highly constrain the understanding of the unstated notions. For example, let's say, because we have a different understanding of the unstated notions, we misunderstood "that which has no part" to be the absence of a geometric entity. We think it is the blank parts of the page. Additionally, since there is an unstated notion of how to apply "breadthless" and "length" together, we interpret a line to be breadthless in two dimensions and think it refers to a point. With these two improper understandings, Euclid's next few definitions, which rely on an understanding of points and lines, are intractable messes. We are forced to go back and rethink the definitions. Euclid's system has constrained well the understanding of what he didn't explicitly say.

An axiomatic system doesn't have falsehoods

Physicists aren't complete dummies. The postulates they state must work some of the time to have survived this long. However, working part of the time is insufficient to classify it as an axiomatic system. Therefore, the postulates of quantum mechanics are more what you'd call guidelines.

In mathematics and philosophy, it is acceptable to posit different axiomatic systems and to see which truths exist in those universes of thought. In contrast, physics is a science. The acceptability of any so-called truths of a physical axiomatic system must agree with the universe we are in.

In the next section, I will state the ten postulates again. Some will be worded differently than before. To ensure they are understandable to you, they won't be as rigorous or as mathematical as you will see in a textbook. The general intent is to give you a broad overview with interesting forays into deeper details, so in no way is what I say complete. For each postulate, I will give commentary that is either clarifying, educational, or, at times, a bit digressive. I assume you have read all the previous chapters, so some information isn't repeated here.

Besides having incorrect postulates, axiomatic quantum mechanics has some ambiguity in the definitions of some terms and has a mass of unstated notions. Following the postulates, I will briefly discuss some problems with stated definitions. As for the unstated notions that I will discuss, the most obvious absences are regarding measurements (this is bad), but there is much physics that goes unstated too (this is not optimal, but is acceptable). I end the chapter with a minimal commentary on the suitability of applying the textbook axiomatic system to our universe.

The postulates of quantum mechanics

Some of the postulates in use today's textbooks are essentially unchanged from when Dirac and von Neumann suggested them almost a century ago. Other postulates were more like unstated notions until they were put into a list at some point. I make no attempt to give a historical account of them.

Not all postulates are required, and physicists are well aware of this. For example, the expectation-value postulate follows obviously from some unstated notions of averages and other postulates. They call it a postulate because it is used often. Perhaps, they should call it a central theorem in their discussion and justify it, but mislabeling it as a foundational postulate is inconsequential.

Some physicists concerned with the leanest axiomatic system possible have published papers showing that postulates can be grouped together or that one follows from others. I have even seen a paper claiming that only three postulates in total are needed. We, however, want to go in the opposite direction. We want to slice them thin so that they are understandable to you, so that is what I did.

The state postulate

> The state of a quantum object is represented by a mathematical wave function $|\Psi\rangle$.

By *wave function*, I mean the total wave function, which encompasses the position-space (or momentum-space) and spin parts. Technically, this is the math on the page. With the word *represented*, the math we write down is connected with the physical universe. All fields of physics use math to represent the universe, so this is uncontroversial. The language we use often doesn't differentiate between the universe and the math that describes the universe.

More mathematically-minded physicists mention that the mathematical wave function exists in an abstract Hilbert space. The vast majority of physicists, myself included, if caught off guard, could not tell you the exact mathematical requirements of a Hilbert space. It is not standard for physicists to receive an education in rigorous pure mathematics. We are posers in this regard. We use the math, yet we don't fully grasp what we are doing, unlike the ones who founded the mathematical tools.

That being said, if one wants to use some of the mathematical techniques invented by Dirac on wave mechanics, then the abstract space that must be used is outside of Hilbert space, and another space needs to be used. Some may say it's not too important for you or the average physicist to worry about the exact details of Hilbert or other spaces until one of us starts doing research-level, mathematical quantum mechanics. On the other hand, I think that if we all knew about these vector spaces approaching the level of a mathematician, we would be better able to see the limitations of the many declarations in the standard quantum mechanics curriculum.

The time-evolution postulate

When uninterrupted by a measurement, the time evolution of a state is governed by the law of Schrödinger's equation. If your wave function is at some time t, to find the wave function at a future time $t + \Delta t$, apply the *time-evolution operator* $\hat{U}$:

$$\hat{U} = e^{-i \cdot \hat{H} \cdot \Delta t} \tag{14.1}$$

Well, that's new. When applied to a ket, this looks like:

$$\hat{U} \cdot |\Psi(x,t)\rangle \;=\; e^{-i \cdot \hat{H} \cdot \Delta t} \cdot |\Psi(x,t)\rangle \;=\; |\Psi(x,t+\Delta t)\rangle \tag{14.2}$$

The U stands for *unitary process* and signals to experts that physical quantities are conserved and probabilities still make sense after applying the operator. I show this to you as a reminder that there is so much math that I've been dancing around in this book. In quantum textbooks, it takes a few pages of boring lines of math to derive this operator from Schrödinger's equation.

Recall the $\hat{H} = \hat{KE} + \hat{PE}$ is the Hamiltonian which defines the energy of the system. Exponentials can be a pain to work with. Exponentials with operators that have derivatives and functions are nearly impossible to work with. To apply that operator, we have to approximate it as a calculus series and cut off the series at our desired accuracy. So, quite a few steps are missing between the middle expression and the final expression in the equation above (Eq. 14.2).

When we looked at waves evolving in computer-like time steps, I came up with an equation for $\Delta\Psi$ (Eq. 9.15, p. 159). If I did just a little more to it, we would have the time-evolution operator cut-off at the second term and applied to $|\Psi(t)\rangle$. We would call the cut-off version the *infinitesimal time-evolution operator* if the time step were infinitesimally small.

The observable postulate

> For every physical quantity an object possesses in classical mechanics, there is a corresponding quantity in quantum mechanics. These physical quantities are called *observables*. In quantum mechanics, each observable has an associated mathematical object called an *operator*.

A table of operators corresponding to observables

Below, I show you a table of operators that correspond to observables and the expressions of their counterparts from classical physics (Table 14.1). This time, I show you the operators in the three-dimensional Cartesian-direction forms. Other three-dimensional forms for other coordinate systems (e.g., spherical and cylindrical coordinates) are not included in the table.

Note that the only place in this book where I show you the algebraic form of some wave functions is in the section on definitions below. Unless you know calculus and have a wave function on the page in front of you, the details in the table of operators are not so useful to you.

You should be aware of the distinction between similar-looking symbols, such as $\hat{x}$ and x; the first is the abstract operator, and the second is a variable from middle-school algebra. We could have used any other symbol underneath the hat of $\hat{x}$ as long as everyone knew we used it to label the position operator along the x-axis. When in the position basis, there is no choice but to use the variable x. (Well, I guess you could also relabel the variable everyone knows as x to some other symbol but nobody would like you.)

The wording I used for the nine other postulates stays fairly true to what other authors write. With the observer postulate, I've deviated from that pattern slightly. Using my definition of the observable postulate, it is only physical quantities that a classical object possesses that count as observables in quantum mechanics. Yet, I and everyone else include the position operator in tables like the one below. Classical objects don't *possess* position; their center of mass is *at* a position.

When written on the page, the position variable as well as the time variable correspond to the background in which objects exist, namely space and time. So, neither variable should have a corresponding operator in my table according to my wording of this postulate. So, I am inconsistent by including the position operator in the table below.

In other books, the observable postulate is worded so that the position operator is accepted in their tables more naturally. In fact, including a position operator makes *observable* a more appropriate word because a position can be observed. However, as you will see in the final appendix, time is excluded from being an observable because there can be no time operator when using the proper definition of an operator that uses the idea of a mapping from an abstract space back to the same space. That is a mathematical reason for not having time as an observable, but what is the physical reason? If we can observe an object at a point in space in quantum mechanics, why don't we have an operator for a point in time? I don't see a good physical reason for including one and not the other. So, it seems other books are also inconsistent about what gets included in their tables.

I say all this to admit that we physicists are often fuzzy about our reasons for doing things. The inclusion of the position operator and exclusion of time in the table is not because of nuanced differences in definitions (i.e., being generally observable vs. being a physical quantity of an object). Rather, I think it is because the position operator is useful, whereas a time operator is not needed.

You may wonder how we find the forms of all the operators in the table. It's standard in quantum textbooks to prove how to get the particular forms of the position and momentum operators. Unfortunately, that's too complex for this book. I however can tell you how to find some of the other operators.

We will look at the potential energy operator first. You will recall from the chapters on the Schrödinger equation that the situation under consideration dictates what the potential energy function will be in position space. Unfortunately, you will likely need a physicist to write down that one too. On the other hand, to find the potential energy function in momentum space, one needs to calculate the Fourier transform of the position-space version, which is doable by those with some calculus knowledge.

In the table, what remains are the kinetic-energy and orbital-angular-momentum operators. They can be found from the classical definitions plus the position and momentum operators. Let's see how that's done with the kinetic-energy operator in one dimension. Start with the classical expression for KE: $\frac{1}{2}mv_x^2$. Then, use the equation for

Table 14.1: Operators corresponding to physical quantities (observables) in the position and momentum bases. Technically, variables alone, like x, aren't operators. The operator is "multiplication by x".

Physical quantity	Classical variable	Position-space operator	Momentum-space operator
Position	x	$\hat{x} = x$	$\hat{x} = i \cdot \hbar \cdot \partial_x$
	y	$\hat{y} = y$	$\hat{y} = i \cdot \hbar \cdot \partial_y$
	z	$\hat{z} = z$	$\hat{z} = i \cdot \hbar \cdot \partial_z$
Momentum	p_x	$\hat{p}_x = -i \cdot \hbar \cdot \partial_x$	$\hat{p}_x = p_x$
	p_y	$\hat{p}_y = -i \cdot \hbar \cdot \partial_y$	$\hat{p}_y = p_y$
	p_z	$\hat{p}_z = -i \cdot \hbar \cdot \partial_z$	$\hat{p}_z = p_z$
Kinetic energy	$\text{KE} = \frac{1}{2} m \left(v_x^2 + v_y^2 + v_z^2\right)$	$\hat{\text{KE}} = -\frac{\hbar^2}{2m} \left(\partial_x^2 + \partial_y^2 + \partial_z^2\right)$	$\hat{\text{KE}} = \frac{1}{2m} \left(p_x^2 + p_y^2 + p_z^2\right)$
Potential energy	$\text{PE} = V(x, y, z, t)$	$\hat{\text{PE}} = V(x, y, z, t)$	$\hat{\text{PE}} = V\left(p_x, p_y, p_z, t\right)$
Total energy	$E = \text{KE} + \text{PE}$	$\hat{H} = \hat{\text{KE}} + \hat{\text{PE}}$	$\hat{H} = \hat{\text{KE}} + \hat{\text{PE}}$
Orbital angular	$L_x = y \cdot p_z - z \cdot p_y$	$\hat{L}_x = -i \cdot \hbar \left(y \cdot \partial_z - z \cdot \partial_y\right)$	$\hat{L}_x = i \cdot \hbar \left(\partial_y \cdot p_z - \partial_z \cdot p_y\right)$
momentum	$L_y = z \cdot p_x - x \cdot p_z$	$\hat{L}_y = -i \cdot \hbar \left(z \cdot \partial_x - x \cdot \partial_z\right)$	$\hat{L}_y = i \cdot \hbar \left(\partial_z \cdot p_x - \partial_x \cdot p_z\right)$
$\left(\vec{L} = \vec{r} \times \vec{p}\right)$	$L_z = x \cdot p_y - y \cdot p_x$	$\hat{L}_z = -i \cdot \hbar \left(x \cdot \partial_y - y \cdot \partial_x\right)$	$\hat{L}_z = i \cdot \hbar \left(\partial_x \cdot p_y - \partial_y \cdot p_x\right)$

momentum $p_x = m \cdot v_x$ to convert that into an expression for kinetic energy that uses momentum: $\frac{1}{2m} p_x^2$. To find an abstract form of the operator, put hats on any position or momentum variables you see. Next, to find the explicit form of the operator that you will apply to an actual wave function, you must know if you are working with position or momentum space; exchange the hatted operators with the explicit position and momentum operators from the table. You can do that procedure to $\frac{1}{2m} \hat{p}_x^2$ to find the explicit forms of the operator in position or momentum space. The answer will either be $\hat{\text{KE}} = \frac{\hbar^2}{2m} \partial_x^2$ or $\hat{\text{KE}} = \frac{1}{2m} p_x^2$.

On adjectives

When you get into the math of quantum mechanics approaching the level of a textbook, you will encounter either the word **Hermitian** or the word **self-adjoint** to describe a requirement of operators that correspond to observables. They have slightly different definitions, and I will delay discussing those differences until the section on definitions. For now, I will say *if* a physicist gives you a proper operator that corresponds to an observable, *then* you know that it will be either Hermitian or self-adjoint (I say "proper" so that the contentious $\hat{E}$ is excluded). The logic of that statement doesn't work in the other direction. Even if the observable postulate above included one of the words *Hermitian* or *self-adjoint*, it is most definitely not the case that if you have any ol' Hermitian or self-adjoint operator, you will be guaranteed to have an observable's operator.

I tell you this because I've seen the if-then conditional in the wrong direction. At one point in his book on quantum mechanics, Leonard Susskind lays out his first principle and ends with, "...the operators that represent observables are Hermitian." That is a fine statement. It is logically equivalent to my statement with the italicized *if* and *then* in the previous paragraph. Two chapters later, Susskind declares, "The unit operator I is officially an observable because it is Hermitian, but it's a very boring one." It's easy to see that the second quote doesn't follow from the first because we understand logic. For example, we know that all giraffes breathe, but not all things that breathe are giraffes.

The spin postulate

> Spin angular momentum is an intrinsic form of angular momentum and is analogous to the extrinsic form, namely orbital angular momentum. It can take on integer and half-integer values of quantum numbers. It is an observable (physical quantity) with corresponding operators.

This postulate is needed in non-relativistic quantum mechanics. Whereas, in relativistic quantum mechanics, spin angular momentum can be derived. Keeping with the spirit of the previous postulate, here are the spin-$\frac{1}{2}$ operators along the Cartesian directions:

$$\hat{S}_z = \frac{\hbar}{2}\begin{bmatrix} 1 & 0 \\ 0 & -1 \end{bmatrix}_z \quad \hat{S}_x = \frac{\hbar}{2}\begin{bmatrix} 0 & 1 \\ 1 & 0 \end{bmatrix}_z \quad \hat{S}_y = \frac{\hbar}{2}\begin{bmatrix} 0 & i \\ -i & 0 \end{bmatrix}_z \tag{14.3}$$

I put the label z on the right side of the matrix to tell us that we can only apply these matrices to column vectors in the z-basis. I don't recall any physicist ever labeling them like this, so don't expect to see them in the wild.

The word *intrinsic* means that spin is part of the fundamental character of the quantum object. A spin-s quantum object will always be a spin-s object unless it is destroyed or becomes part of another object. Conversely, the analogous quantum number l for orbital angular momentum can change for a particular quantum object via a change of state. Finally, note that the word *intrinsic* does not mean that the spin angular momentum is, in general, unchanging in magnitude. This is because there is also the spin-magnetic quantum number m_s. For example, a spin-1 particle can have angular momentum with a magnitude of $\hbar$ or 0 (corresponding to $m_s = \pm 1$ and 0).

The single-measurement postulate

> After applying an observable's operator to a ket, numerical values are returned. Should you do a measurement of that observable quantity, one of the numerical values will be the result of a measurement. The numerical values will be real numbers.

To extract the numerical values, you must work in the appropriate representation of the operator. For example, the operators in the third and fourth columns in the table above only apply to a wave function written as position- and momentum-space wave functions, respectively. That is easy to understand, but no one labels the spin operators' matrices as I do, so it's possible to be confused and apply them inappropriately. I will show you how to use them properly in the appendix on matrix mechanics.

You saw how the kinetic energy operator found absurd negative values and high positive values with no regard for energy conservation (p. 251 - 254). This is sufficient to know the single-measurement postulate is not true in all cases. Without generality, there is no longer a true axiomatic system for quantum mechanics.

Within the context of the Hamiltonian operator acting on hydrogen's ground state, I brought up the preferred-basis problem (p. 145 - 146): how does a measurement know to pick between an energy ket and a position ket in that example? The issue in that case was related to the projection postulate. However, the preferred-basis problem also touches the single-measurement problem because operators have different representations. For example, when we use the momentum-space version of the KE operator, it only returns non-negative numerical values (but the absurdly high positive values remain). Because the axiomatic system is silent on how to conduct measurements, it doesn't tell us how to do the momentum or position versions of these results. This silence goes beyond the kinetic energy operator. It is a general issue for all the operators.

Because these postulates are idealized and don't account for how actual experiments occur, this postulate also doesn't account for the pixelation of detectors, nor any other limitation to the resolution that can be achieved by an instrument. Experimental limitations are pretty odd things to not specify explicitly in the axiomatic framework since six out of ten of the postulates relate directly to measurements.

The probability postulate

> To find the relative probability of the result of a measurement, use the square of the magnitude of the quantum amplitude.

The math of matrix mechanics and wave mechanics do not inherently contain randomness. The experimental evidence suggests that there is randomness, so the probability postulate is imposed on top of the math.

I use *relative* because each quantum object has one wave function, but its strength of interaction is not the same with all types of particles. No one thinks using a photon and a neutrino, assuming both can be used in principle, to probe the location of an electron will return anything near the same probabilities. By far, the vast majority of neutrinos will neither interact significantly with an electron nor register in a detector, so the probability of measuring a result is unimaginably low (essentially zero). The probabilities of both types of probes are proportional to, but not equal to, the electron's squared magnitude of the wave function.

Finally, notice that the probability postulate doesn't say anything about the nature of the universe. It doesn't say nature is truly and inherently random. It's agnostic on the issue. It allows freedom for interpretation. Some, like Einstein, think that what we see is only apparently random because we do not yet have a complete theory. Others, like the Copenhagenists and those of the many-worlds interpretation, think that true randomness is what we have.

The projection postulate

The measurement process forces, or projects, the state of a quantum object into a new state or keeps it in the same state. Recall from the single-measurement postulate that the value measured in an experiment will correspond to a particular value returned by the application of some appropriate operator to an appropriate ket. Each numerical value has a ket that it came from. The particular value's ket is the new state.

Back in the *Math vs Measurement* chapter, this postulate failed in a measurement of the ionization energy of hydrogen and a measurement of spin in a Stern-Gerlach experiment. The reason why it failed in both cases is the observer effect. You saw in the *As a wave equation* chapter that it also fails with position measurements because it produces a non-function, which the Schrödinger equation cannot handle.

In *Quantum Mechanics Demystified*, during his explanation of the projection postulate, David McMahon let's his readers know that

> The reader who studies quantum theory more extensively will find that not all measurements can be described as projective measurements as we have done here.

This forthrightness, conciseness, and clarity of the limitations of the projection postulate is rare.

The expectation-value postulate

After numerous measurements on equivalently prepared states, the weighted average of the measured quantities will equal $\langle\Psi|\hat{O}|\Psi\rangle$, where $\hat{O}$ is the operator corresponding to the observable (physical quantity) that is measured. The bra-ket in the previous sentence is known as the *expectation value* of the operator $\hat{O}$.

In textbook quantum mechanics, this statement is unnecessarily classified as a postulate because it follows from the observable postulate, the single-measurement postulate, the probability postulate (if they were all true), and some unstated notions of weighted averages. As such, it should instead be called a theorem. It is included in the lists of postulates because of its importance in calculating results; a theorist will have a hard time deriving general theorems using individual and random results from equivalently prepared states. Expectation values, on the other hand, are predictable values.

In my reality interpretation, you could say there is an expectation-value postulate. An expectation value calculates a physical quantity that the wave function possesses. You saw this for the $\hat{KE}$, $\hat{PE}$, and spin operators in the *Spring cleaning* chapter (p. 254- p. 263).

I find it puzzling that anti-realists maintain their stance, considering that there is something called the *generalized Ehrenfest theorem* which allows us to calculate the time-evolution of an operator's expectation value. In my reality interpretation, this theorem demonstrates that energy, linear momentum, and angular momentum are conserved. At best, the anti-realists can claim these quantities are only conserved on average. I have seen experts (e.g., Griffiths and Schroeter) use expectation values to get back this weak version of conservation laws. But, in general, those who write the postulates do not emphasize the problems that textbook quantum mechanics has with conservation laws.

A problem with the weak version of conservation is what happens in actual scattering experiments. Atomic, nuclear, and particle physicists need energy and momentum to be conserved in all scattering events. Without conservation laws, they wouldn't have been able to do their jobs and calculate quantities as they have done. Moreover, physicists in the early years of quantum mechanics already considered the possibility that energy conservation could be violated. Bohr, Kramers, and Slater hypothesized that energy wasn't conserved in interactions of electromagnetic radiation with electrons. The Bothe-Geiger experiments of 1924/25 found that X-rays scattered from electrons conserved energy, and the physics community of the time was satisfied. These experiments won Bothe one-half of the 1954 Nobel Prize in Physics. In his Nobel lecture, even Heisenberg clearly states energy is conserved in measurements. So, some modern Copenhagenists are at odds with the co-creator of the Copenhagen interpretation. More famous than this example is Wolfgang Pauli's correct prediction, based on conservation of energy, that an unseen particle, the neutrino, is emitted in the random beta decay of radioactive nuclei.

The composite-system postulate

When a composite system of multiple quantum objects *of the same type* is described, tensor-product symbols must be used to connect all the input kets.

When a quantum object is described as being composed of spin and spatial parts, those parts are also joined by tensor-product symbols.

Operators then must also be tensor products to pick out the correct part of the wave function. For example, the following expression is a spin operator applied to the total wave function of a single particle:

$$(\hat{I} \otimes \hat{S}_z) \cdot |\Psi_{\text{spatial}}\rangle \otimes |\uparrow\rangle \tag{14.4}$$

We call $\hat{I}$ the *identity operator* or *unit operator*. It is the do-nothing operator. Like Susskind said above, it is a very boring operator. Since $\hat{I}$ is in the first position, it acts only on the spatial part and leaves the ket unchanged. The spin operator only acts on the spin part and asks for the magnitude of the spin. After applying the operators, the result looks like this:

$$(1 \otimes \frac{\hbar}{2}) \cdot |\Psi_{\text{spatial}}\rangle \otimes |\uparrow\rangle \tag{14.5}$$

To see the math like this is rare. Normally, spatial states and spin states are considered separately.

The symmetrization postulate

The total wave function $|\Psi_{AB}\rangle$ must follow a constraint when you exchange the notional labels of the particles of the same type:

$$|\Psi_{BA}\rangle = \pm|\Psi_{AB}\rangle \tag{14.6}$$

Recall (p. 271) that the notional labels are not the letters themselves, but are assigned to the letters by the order of the letters on the page. For bosons (integer-spin particles), the plus-sign constraint applies. For fermions (half-integer-spin particles), the minus-sign constraint applies.

There is an allowed approximation I subtly alluded to in a previous chapter (look for "overlap" on p. 270). Say we have a system of two particles of the same type (e.g., two electrons). When the wave functions have little overlap (colloquially: "the particles are far apart"), we can approximate them as independent and represent them as a superposition of the input kets $|\Psi_A\rangle$ and $|\Psi_B\rangle$, where the different letters indicate the wave functions are different. The more the wave functions overlap, the more the approximation breaks down. The current postulate is then needed.

In the following figure (Fig. 14.1), I illustrate how separated wave packets don't necessarily need the symmetrization postulate. This allows us to concentrate on a few particles and ignore the rest of the particles in the universe in our calculations. I warn you, though, not to be misled by this simple scenario. If well-separated wave packets approach each other, scatter significantly, and are subsequently well-separated again, the symmetrization postulate is needed if we measure the properties of the two particles. Some significant change is expected to occur because of the momentary overlap and interaction during the scattering.

Before ending this commentary, I must warn you about two words most other physicists use when describing the symmetrization postulate. The words are *indistinguishable* and *identical*. You will notice

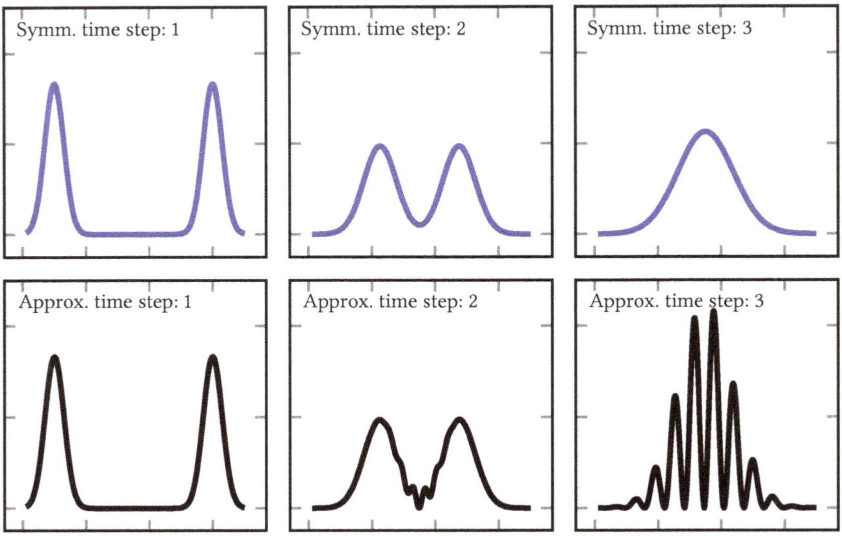

Figure 14.1: Illustrating that separated particles can be approximated without the symmetrization postulate This is the matter density of two particles with no potential-energy interaction traveling towards each other while also dispersing. I don't show the wave functions of the input kets nor the configuration space. Instead, I show the final results. The top row uses the symmetrization postulate. The bottom row uses the superposition principle only. In the first time step, the matter density bumps are well-separated. The two methods give density distributions that are essentially the same. In the next time step, the wave functions overlap to a small extent and the calculation of the matter densities shows a difference between the two methods. With significant overlap in the final time step, we can see a large difference between the two methods. The symmetrization postulate is needed because the approximation breaks down significantly when the wave functions of input kets overlap.

that I did not use them during this commentary nor when I first introduced the postulate to you in a previous chapter.

If you've got a good textbook in your hands and its author wants to use *indistinguishable*, the author will tell you that it refers to the particles with the same invariant properties of mass, charge, spin quantum number *s*, etc. For any electron, all those properties are the same. This is what I mean when I say "particles of the same type". A poor explanation, which is too common, would make its usage seem broader than intended. From these poor explanations, learners may get confused and think *indistinguishable* also applies to the wave function too.

You may see the word *identical* used by physicists in place of indistinguishable. These physicists aren't using the mathematical definition of the word, which means "the same in *every* respect." This can create confusion: you already know that Pauli's exclusion principle says that mathematically identical fermions don't exist (see Eq. 12.21 on p. 275).

Some of the language in textbooks and pop-sci books will make you think that, in all cases, we can't track particles, so they lose their identity. This is more likely to occur in books where *indistinguishable* and *identical* are used indiscriminately. These books often hold measurements up as the arbiter of what can be said. If, in place of that view, you accept that the mathematics can tell us about reality, then when the wave packets are separated, they can be treated, to a good approximation, as being distinct. In your mind's eye, you can keep track of wave packets until they overlap significantly (see again the figure) above. As one example of a chapter that I think will confuse many students, there is *Ch. 13 Identical Particles* in *Quantum Mechanics: A Paradigms Approach* by D.H. McIntyre which has as its opening sentences the following (bold as in original text):

> To study systems like multielectron atoms, we need to properly account for the fact that all fundamental particles like electrons and protons are **identical**. In classical physics, particles are not identical—we can always find a way to uniquely identify a particular particle. Even if we make two classical particles "the same" to the utmost level of precision, we can still find a way to identify the two particles without affecting their classical motion. For example, billiard balls behave identically, but can be identified by their numbers. In quantum mechanics, there is no way to identify two different electrons—they are **indistinguishable.** Two hydrogen atoms are identical no matter where they are in the universe.

Definitions

If you expect to open up a book on quantum mechanics and get a thoughtfully ordered list of definitions similar to Euclid's axiomatic system, you will be disappointed. In books written for students, definitions are brought up as needed for the reader as they learn.

You'll have more luck in the introductory sections of a research paper that aims to expand on or disprove some foundational aspect of quantum mechanics. Yet, in those papers, the authors expect you to be sufficiently informed that they can be brief and use technical, jargon-laden definitions.

With such a state of affairs and considering that I am not about to collate all the definitions in a textbook for analysis, I can't give a thoroughly researched comment on the completeness or soundness of the definitions presented to students. However, two closely related topics stick out in my mind: A) requiring the operators of observables to be either *Hermitian* or *self-adjoint* and B) what counts as an observable.

Over the next three-ish pages, there is no expectation that you, as you read along, will deeply comprehend the definitions and the math I only describe in outline. Many of you likely won't be doing calculations, so, instead, you should see what follows as more of a story of how professors, once again, make quantum mechanics inconsistent and confusing for students.

Hermitian vs. self-adjoint operators

In what follows, I will walk through two different requirements (Hermitian vs. self-adjoint) of the operators corresponding to observables. The differences are so severe that it creates two different axiomatic systems because it creates two different observable postulates. The system a student learns depends on which professor the student learns from. Some professors contribute to student confusion by using the words *Hermitian* and *self-adjoint* interchangeably.

A Hermitian operator is guaranteed to return a real-valued expectation value. This version is most often taught to physics students, and in the table several pages back, all the operators of observables are Hermitian. Since textbooks also say measurements are a fundamental part of quantum mechanics, they must dance around some limitations of this definition of an operator, such as imaginary numerical values that you will see below.

A self-adjoint operator returns real-valued numerical values. It is commonly said that, since measurement instruments show real values in their displays, this is the proper type of operator to use.

The two terms confuse many because they are closely related. A self-adjoint operator is also a Hermitian operator because the expectation value (i.e., weighted average) of real values will also be real. However, not every Hermitian operator is self-adjoint. As an example, consider a Gaussian wave function at a point in time when it is purely real: $\exp\left(-x^2\right)$. In this wave function, I've removed all the complicating constants. Applying the momentum operator $\hat{p}_x = -i \cdot \hbar \cdot \partial_x$, which is in the position-space representation, to the wave function will return a continuous function in front of the wave function that is purely imaginary (i.e., $-i \cdot 2\hbar \cdot x$). Therefore, in this case, it can't be a self-adjoint operator nor represent the results of measurements. On the other hand, the expectation value of this example is zero, a real-valued number, which is what is expected of a Hermitian operator.

Because the axiomatic system is supposed to predict real values as the results of measurements, you are probably wondering what the self-adjoint version of the momentum operator is. Well, we could try to use the momentum-space operator $\hat{p}_x = p_x$ on the Fourier transform of the original Gaussian wave function, which is the momentum-space wave function $\exp\left(-p_x^2\right)$ (this also has many physical constants thrown away for simplicity). The result is $p_x \cdot \exp\left(-p_x^2\right)$. The set of numerical values is then the real number line in momentum space p_x. Thus, $\hat{p}_x = p_x$ is self-adjoint when acting on this momentum space wave function.

I find the reasoning surrounding self-adjoint operators rather dubious because there is no obvious way to do things properly and it seems arbitrary to go to momentum space to discover the appropriate numerical values. It appears that I am in good company. In his book

The Road to Reality, Roger Penrose opines that self-adjointness is an "unreasonably strong requirement" because he is comfortable using complex numbers and they are used in classical situations too. (Note that what I call *self-adjoint*, he calls *Hermitian*.)

What is physical and what is simply math?

The next definitional ambiguity is about what counts as an observable and how many operators of observables there are. The first answer is the most physical and is the one I use in this book: an observable's operator is an operator that corresponds to a physical quantity in classical physics or is a spin operator. The second answer is from a subset of mathematical physicists who are disconnected from and who may not be so concerned with actual experiments. Susskind likely momentarily thought of this definition when he made the mistake I mentioned earlier in the chapter (p. 318). To this subset of mathematical physicists, every possible self-adjoint operator corresponds to an observable. This means they are using a different observable postulate and are working with a third axiomatic system that is very different from the other two I mentioned recently. If their use of the word *observable* still refers to a physical quantity, then they must agree that an experiment will be able to find the "fourness" of an object because the number four is a self-adjoint operator. No matter what situation you choose, applying the number four to a ket returns the number four as an eigenvalue. They must say the fourness of every object is four, and the fiveness of every object is five. And they better be down with the sixness, too.

Unstated notions

Some reasonably unstated notions

We now turn to some unstated notions in axiomatic quantum mechanics and start with those that are reasonably unstated. These notions concern the subset of physics we import from classical physics.

All of classical electromagnetism is used in non-relativistic quantum mechanics and goes unstated. For example, physicists assume that everyone knows about electric charges and the electric potentials they create. Assumed, too, is knowledge of Maxwell's equations and everything that goes along with them.

Additionally, many experimental facts, such as the magnitude of the fundamental unit of electric charge on the electron and proton, go unstated. In the lists of postulates, you won't find it stated that nuclear wave functions are so compact that they can be treated as infinitesimal volumes when creating the potential-energy functions for input into the Schrodinger equation when we solve for the electron wave functions of atoms. The list of unstated physics notions is, figuratively, endless.

The most problematic unstated notion

From the discussion of Euclid's axiomatic system, you can understand that I can't possibly give a comprehensive list of all the remaining unstated notions. Instead, I will emphasize what I have said a few times already and that researchers are well aware of: the axiomatic system taught to physics students, which has measurements as the arbiter of truth, does not tell us what a measurement is or how to properly conduct one.

Some may think that defining what a measurement is isn't needed because we all know what we mean by measurement. But this claim is problematic. As one example, let's say someone interprets the word *measurement* to mean "the determination of a physical quantity by an apparatus operated by a being with a consciousness." Mixing in some extreme philosophical views that I covered in the first chapter will lead to the ridiculous interpretation of quantum mechanics by Eugene Wigner and John von Neumann that consciousness causes reality to exist. (Wigner lived long enough to give up on this interpretation.) Today, it's common to hear that the majority of the physics community wholly rejects the idea that consciousness has a role in quantum mechanics. The possibility for such conclusions and the pushback from the physics community shows that the axiomatic system is, unlike Euclid's system, not well constrained.

We can look at the definitional gap in a more practical manner. Through my years in academia, I gained a fair amount of knowledge and experience in various types of experiments. Yet, as I've previously said to you, I don't know how to measure the absurd set of values that are predicted for an atom's kinetic energy. We physicists regularly fire particles at detectors and measure the deposition of their overall kinetic energy or portions of it (i.e., the $\langle \hat{KE} \rangle$ that the particle had in flight). Still, we never measure a negative KE nor an enormous and positive KE. We are one hundred years on from the founding of the field of quantum mechanics, and apparently we have not done either of those. Certainly, assuming the measurement is possible, a precise definition of a measurement would allow us to conduct the experiment and confirm this part of the theory.

The status of textbook axiomatic quantum mechanics

I have shown you that the observer effect and conservation of energy are not part of textbook axiomatic quantum mechanics. We could perhaps say these are rejected notions or anti-postulates because if they were included in the axiomatic system, they would lead to contradictions. If one accepts the observer effect and conservation laws as part of the physics of our universe, then logically, one must say that this axiomatic system doesn't exactly describe our universe. It also tells us that we can measure quantities that we have no idea how to measure.

Because of the vast number of accurate predictions it makes for our universe, the fictional universe it is modeling seems to approximate our own. Still, it is not exact. However, I insist the inexactness is not something that can't be admitted and explicated to students and the public. It would not be scandalous to detail the exact extent of the limitations of what is inside our textbooks and that some researchers are attempting to create a truly axiomatic system.

Afterword

> *If I have seen further, it is by standing on the shoulders of giants.*
>
> — Isaac Newton

Recap and some direction

In this book, I haven't taught you all that much. In the first chapter, I gave you permission to reject anti-reality and to picture the quantum realm. The rest of the book was devoted to disposing of incorrect views of the quantum realm and replacing them with the physics of wave functions and spins. During that you encountered ten postulates of axiomatic quantum mechanics and found that two of them are problematic and one of the two had three different versions. Throughout it all, you were introduced to a minimal amount of math.

Now that you have finished the book, you can understand the various sources of conflict between physicists' descriptions of quantum mechanics. The philosophical position of a physicist will greatly influence how they frame the theoretical and experimental evidence. The two big philosophical camps are those of the realists and those I have described as anti-realists, radical empiricists, or Copenhagenists. Among the members within a camp, there will be disagreements on how and which parts of the mathematical formalism touch on reality. This is how we chiefly get the various interpretations of quantum mechanics.

Beyond those more innocent differences, there are more reproachable things done that lead to conflicting views. Some physicists may ignore some phenomena, such as the observer effect. Or they may elect to use a principle, such as a universal projection postulate, that doesn't reflect how our universe functions. They may also be using seemingly innocuous assumptions that shade everything else about their description of the world. For instance, if one thinks that an object may only exist at a single point and no other, which is a world view of point particles, then that thinking is going to have a cascade of effects on how the math is viewed. Often, physicists parrot what they heard before without giving thought to what they claim. The parroting occurs too often with the descriptions of the various uncertainty inequalities. Nevertheless, you are now sufficiently informed to intelligently question what you are told by physicists when what they say is suspect.

Though, you must be warned of the limitations of your new education. There are many topics that fit the scope of this book that were not included because I needed to be discerning about what was included. I briefly mentioned entanglement but didn't go into an interesting discussion of it. To restrict myself to the simplest math, I didn't show any potential energy functions from magnetic fields or electric fields that varied with time. The so-called quantum Zeno effect, quantum Chesire cat experiments, and quantum eraser experiments could have each had a chapter deconstructing the mythicism around them. In the early days of writing this book, I wanted to include a chapter giving simple, logical reasons why the virtual particles in Feynman diagrams could not correspond to real particles (e.g., if virtual-particle diagrams represent real objects, how does a photon know to travel in a perfectly straight line to be in a location that an electron will be in the future? Why not have more diagrams for when it misses?). By excluding such topics, the book is better and more focused.

The amount of material from the wider fields that use quantum mechanics at their core could not have been included because their complexity is immense. There was nary a hint of the mathematical formalism of quantum field theory. Nor did you see what would be taught in standard courses of the quantum mechanical topics of atomic, chemical, nuclear, particle, or solid-state physics. Each of these topics could have a book of its own.

For now, I have only reoriented you, showed you that the Earth is a sphere, and pushed you in the right direction. My intention is to write more books covering the more specialized topics, and as the books are written I can build up much of the necessary vocabulary and mathematics so that I can write ever more precisely for you. Hopefully those comes sooner rather than later.

Appendix A: Atomic orbitals, their labels and transitions

Energy levels and the first quantum number

Hydrogen is the physicist's prototypical example of an atomic system because it is simple (only one electron and a proton) and is the only atom where we can find the nearly exact mathematical forms for the wave functions and **energy levels** (see Fig. A.1). I must say the forms are nearly exact because there are corrections due to several tiny effects that are better left for a book on atomic physics. In this appendix, I discuss labels for electron wave functions and barely mention the proton. Since the proton wave function is so compact, we ignore its details and consider it to be a positive charge in an infinitesimal volume. The electron wave functions are then the center of attention.

In this appendix, the electron wave functions we are primarily interested are classified as **stationary states**. If you've learned about *standing waves* before, these wave functions are analogous to those. A stationary state has an electron density distribution (i.e., $|\Psi|^2$) that doesn't change with time. These states have energies that are said to be discrete, or quantized. These energies only occur at the levels of the black lines in the figure, which are labeled by the various values of n.

The quantum number n is called the **principal quantum number**. The term **quantum number** is a fancy way of making a term for the label of a quantum state. These labels are closely associated with physical quantities. For example, the stationary state energies in the simple

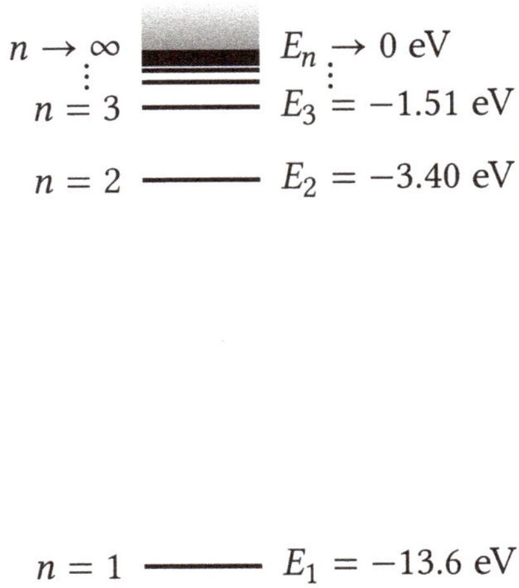

Figure A.1: Hydrogen energy levels. Vertical direction represents energy. Horizontal direction has no meaning. Transitions between n = 2 and $n \geq 3$ produce the Balmer lines (Fig. 1.2, p. 15).

model of hydrogen are found via $-13.6 \text{ eV}/n^2$. The **ground state** (labeled with $n = 1$) is the state with the lowest energy, to which all other bound states eventually decay. States higher than the ground state are called **excited states**.

It's not common for the following to be said so clearly: it's possible to have a state with an energy at *any* level in Fig. A.1 as long as it is above the ground state energy; it's not just stationary states that are allowed. When a state is not a stationary state, it is classified as a **non-stationary state**. In these states, the electron density will slosh around, spread (disperse), or travel away.

There is a further classification of states into **bound** and **unbound** states. Unbound states have energies above 0 eV in the figure. Bound states have energies from 0 eV down to the ground-state energy.

Non-stationary states that are also bound states are the ones that slosh around. By sloshing around, I mean, when a long period of time is considered, the electron density always "comes back" to the vicinity of the proton. The electron density is still bound to the proton. These states are made from superpositions of stationary states with *differ-*

ent energies. Non-stationary, bound states are generally shorter lived than excited, stationary states. In case there is any doubt, I will point out that bound, non-stationary states can have the same energies as stationary states if one chooses the superposition just right.

All unbound states are non-stationary states. These are the ones that have matter spread out or travel away from the proton. These states have more energy than any bound state. Unbound states are commonly called **scattered states**. You could say the electron wave function is scattered away by whatever mechanism that got it into that state. Then the electron is said to be free or is a *free electron*. The continuum of unbound states is represented as the fading gray box in the figure. It continues upward without limit.

Finally, it's time to emphasize an important distinction. We can't use the terms *energy levels* and *states* interchangeably. The only energy level to have one state associated with it is the $n = 1$ energy level (if we ignore spin). Multiple times in the book, I refer to the ground state or the 1*s* orbital; those are the $n = 1$ state. Every other energy level has multiple states associated with it. The multiplicity of energy levels requires that I teach you about the hierarchy of further quantum numbers and some terminology associated with the numbers.

An overview of the hierarchy

I use the word *hierarchy* because we say that shells have subshells, subshells have orbitals, and for each orbital, it can be in one of two spin states. The explanation that follows will go smoother if you first take several minutes to examine and ponder the following table.

We use the word **orbital** because we used to think that electrons were in circular or elliptical orbits around the nucleus. That was **Old Quantum Theory**. Now, we can't say they are in orbits, so we added the suffix *-al* to indicate that it was something vaguely of that idea. We now imagine an electron wave function filling all of space, at least with a minuscule amount of the wave function. More accurately, *orbital* is a synonym for a possible configuration of a *spatial wave function of a stationary state of one electron in an atom or molecule.* You

Table A.1: Hierarchy of stationary states. Specifying numbers for n, l, and m_l specifies a particular orbital of hydrogen. Size and shape of orbitals are discussed further in the *Gallery* chapter.

Name of quantum number (QN)	Symbol	Values	Comment
Principal QN	n	$1, 2, 3, \dots$	Labels energy levels in the basic model of hydrogen. Label for the **shell**. Tells us the relative size of the orbital.
Secondary QN (or orbital-angular-momentum QN or azimuthal QN)	l	$0, 1, \dots, n-1$ or $s, p, d, f, \dots$	Tells us the high/low values the orbital-magnetic QN can have. Label for the **subshell**. Tells us the distinctive set of “shapes” of the orbitals.
Orbital-magnetic QN	m_l	$-l, \dots, 0, \dots, l$	Tells us the actual amount of angular momentum a particular orbital has. Label for the **orbital**. Fine-tunes which “shape” of the orbital to picture.
Spin-magnetic QN	m_s	$\pm 1/2$	Label for the **spin state**.

saw the matter density of orbitals in the first image of the first chapter (p. 6).

As for *shell* and *subshell*, they are older remnants from Old Quantum Theory. It’s not important to know the history of them. After all, you don’t need to know the Latin origins of the words *family*, *genus*, and *species* to understand the taxonomic classification of organisms. You should think of *shell* and *subshell* as words that allow us to group orbitals and their quantum numbers in a hierarchy.

You may wonder, because a quantum number l leads to the m_l numbers, does a quantum number s lead to the m_s numbers? Yes. When we say an electron is a spin-$\frac{1}{2}$ particle, we are saying it is a spin-s particle with $s = 1/2$. The $s = 1/2$ tells us the highest and lowest values of m_s. But everyone and their cat knows that there are electrons in atoms and that they are spin-$\frac{1}{2}$, so we don’t mention it.

The numerology of quantum numbers

In the table, you likely weren't able to completely understand the *Values* column. This section goes over the numerology of that column. You should refer to Fig. A.2 as you work through this section.

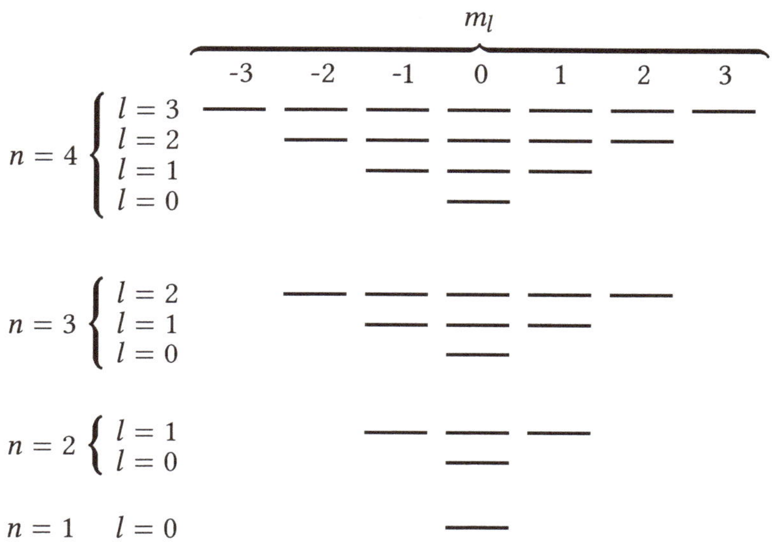

Figure A.2: Multiplicity of orbitals. In this diagram, the vertical axis is not energy. Instead, it is used for grouping according to quantum numbers. A horizontal line means "an orbital is allowed to exist at the intersection of these quantum numbers". In a chemistry class, an instructor would draw zero, one, or two arrows pointing up and down on those lines to represent when electrons are "occupying the orbitals" and the direction of their spins. Note that for each l there are $2 \cdot l + 1$ horizontal lines; this is a semi-important fact that everyone who does quantum mechanics knows. It will appear again in the spin chapter.

After the principal quantum number specifies the energy level or, equivalently, the shell, three other quantum numbers further specify the exact state we have in mind. The first two of the three are associated with orbital angular momentum, and the final one is due to spin angular momentum. The first three quantum numbers have integer values. Whereas, for the spin of an electron, they are $\pm 1/2$.

Quantum number l

The next quantum number l is the **secondary quantum number**. The letter l was chosen because uppercase L is the variable that is usually used with angular momentum in classical physics. In fact, it is often called the *orbital-angular-momentum quantum number*. I don't think it's a good name because it's actually telling you the maximum amount of orbital angular momentum that an orbital could have, not the amount it does have. For example, an $l = 1$ orbital might have no angular momentum.

The number of sub-shells in a shell equals the shell's principal quantum number n (remember to refer to the figure above for sentences like these). So, the ground state, with $n = 1$, has one subshell. We start counting at zero for the subshell's quantum numbers. The ground state's shell and subshell are then specified with the numbers $n = 1$ and $l = 0$. It is called the 1*s* subshell. Going up to −3.4 eV, we have the $n = 2$ shell, and it has the $l = 0$ and 1 subshells (aka the 2*s* and 2*p* subshells).

Due to a historical accident, the subshells are commonly specified using letters as above. The first four subshells are called *s*, *p*, *d*, and *f*, which correspond to $l = 1$, 2, 3, and 4. Originally, the letters were labels for spectral lines and came from the descriptions for sets of lines: *sharp*, *principal*, *diffuse*, and *fundamental*. When physicists learned of atomic states, the letters were ported over, and now those full names have no meaning in the description of states. Reusing the letters was a bad decision that we are now stuck with. After the *f* orbitals, the letters continue alphabetically with no repeats of the *s* and *p*. It goes *s*, *p*, *d*, *f*, *g*, *h*, *i*, *k*, *l*, *m*, *n*, *o*, *q*, *r*, *t* and so on. There is also no *j* for some reason.

In this book, I won't mention again any orbitals that come after the *s*, *p*, *d*, and *f* orbitals. We use the hydrogen orbitals as proxies to understand the orbitals in elements with more electrons. In other elements, the orbitals stack on top of each other in predictable patterns that are usually taught in an introductory chemistry class. Everyone who writes an introductory work stops at the *f* orbitals because no naturally occurring elements have a ground state with orbitals higher than f. However, if you fire a photon of the right energy at an atom, you can get to those higher orbital letters.

Finally, beware there are two ways s is used in this book and others. One way is as an orbital label s (which really means $l = 0$) and the other ways is as a spin quantum number s.

Quantum number m_l

The next quantum number m_l is also associated with orbital angular momentum. It is called the **orbital-magnetic quantum number**. The m stands for **magnetic** because a particle with electric charge and angular momentum will produce a magnetic field. The value of that angular momentum will be $m_l \cdot \hbar$, where $\hbar$ is Planck's reduced constant. The centered dot implies multiplication.

The number of orbitals in a subshell of a particular l is $2 \cdot l + 1$. The values of m_l run from $-l$ through 0 and on to l. You can see that the 1 in $2 \cdot l + 1$ is due to the $m_l = 0$, and the $2 \cdot l$ remaining states are due to the positive and negative quantum numbers of m_l.

The ground state of hydrogen has $l = 0$, so it will have $2 \cdot 0 + 1 = 1$ orbital. This single orbital is specified by the numbers $n = 1, l = 0, m_l = 0$.

Going up in energy, the next shell is $n = 2$. From the n, we know that there are two subshells: $l = 0$ and $l = 1$. The logic of this shell's $l = 0$ sub-shell (aka the $2s$-subshell) will be the same as the previous paragraph: there is one orbital in the $l = 0$ sub-shell. You can call it the $2s$ orbital. Looking at the $n = 2, l = 1$ subshell (aka the $2p$ subshell), we can see there will be $2 \times 1 + 1 = 3$ orbitals. The three will have quantum numbers $m_l = -1, 0, +1$. Collectively, they can be called the $3p$ orbitals.

A digression on chemists' atomic orbitals

If you've had an introductory chemistry class, notice that I never used p_x, p_y, or p_z. Your chemistry teacher didn't tell you the full story of where those come from. A p_z orbital could be said to be the same as $p_{m_l=0}$ orbital. But the p_x and p_y orbitals from chemistry class are superpositions of the orbitals I will draw for you in the *Gallery* chapter. The superpositions are

$$|p_x\rangle \text{ “=” } |p_{-1}\rangle - |p_{+1}\rangle \tag{A.1}$$

$$|p_y\rangle \text{ “=” } |p_{+1}\rangle + |p_{-1}\rangle \tag{A.2}$$

I switched the notation a bit here from the beginning of the paragraph to make the equations cleaner: e.g., p_{+1} means $p_{m_l=+1}$.

If you haven't read about Dirac notation in the *Mathematics vs. Measurement* chapter yet, don't worry. Just get a feel here for what is being added in a superposition. These two equations aren't an essential part of the book.

With regard to p_z, why did I say "could be said to be" and not "is"? It's because, by my count, there are three different ways chemists think of p orbitals (i.e., as wave functions, as the real part of wave functions, or as electron densities), and "is" only works when we're sure we're equating the same thing.

Note that, practically speaking, these superpositions can be considered stationary states because they have the same energy in the absence of a magnetic field. And if we are considering a hydrogen atom, the magnetic field is small. Technically, however, they are nonstationary states because of the proton's magnetic field.

As you might guess, the chemists' d_{xy}, d_{yz}, d_{xz}, d_{z^2}, and $d_{x^2-y^2}$ orbitals are also superpositions of the atomic orbitals I use. I'll say just a bit more on this general superposition topic in the *Gallery* chapter. Other than that, you won't see me using x, y, and z in the subscripts to describe orbitals in the rest of the book. They are primarily what chemists use to tell their folktales of chemical bonding to novices.

Quantum number m_s

The **spin-magnetic quantum number** m_s has two values: $1/2$ or $-1/2$. Those numbers come from dividing out Plank's reduced constant $\hbar$

from the possible values of its spin angular momentum: $\hbar/2$ and $-\hbar/2$. When you deal with other elementary particles or particles composed of multiple other particles, you can get spin numbers besides the two I mentioned. I touch on that in the chapter on spin and superposition. Until then, just remember $1/2$ and $-1/2$.

We often say that this final quantum number completes the four quantum numbers, and a stationary state of hydrogen is thus fully specified. To recap, the terms and the quantum numbers needed to specify that level of the hierarchy are: shell (n), subshell (n, l), orbital (n, l, m_l), and hydrogen's state (n, l, m_l, m_s).

It's not exactly true to say that the set of four quantum numbers is a complete specification of a stationary state. For it to be accurate, we must assume things such as the center of mass of the atom is at rest and there are no nuclear quantum numbers from the nucleus. In this book, I will use these two assumptions without stating them.

For simplicity, I always discuss spin angular momentum independent of the orbital angular momentum. Therefore, in the chapter where I discuss the hydrogen orbitals, the only quantum numbers needed are n, l, and m_l, and sometimes I say these label hydrogen's state.

Atomic transitions

In the second chapter, I introduce the sodium D doublet (see Fig. 2.2, p. 27) and assume the reader understands the **atomic transition** diagram that follows it (reproduced in Fig. A.3 below). That is what is typically drawn in a physics class and is a simplified *Grotrian diagram*. In an atomic physics research paper, the diagrams are more technical.

In this section, I will now explain atomic transitions. First, sodium isn't hydrogen, but, with a few exceptions due to the multiple electrons, most of the ideas transfer over to other atoms. The 1*s* orbital is the ground state orbital of hydrogen. It is the natural state of hydrogen. The ground state of sodium has one electron in the 3*s* orbital and all orbitals below 3*s* filled (two in each: spin up and spin down).

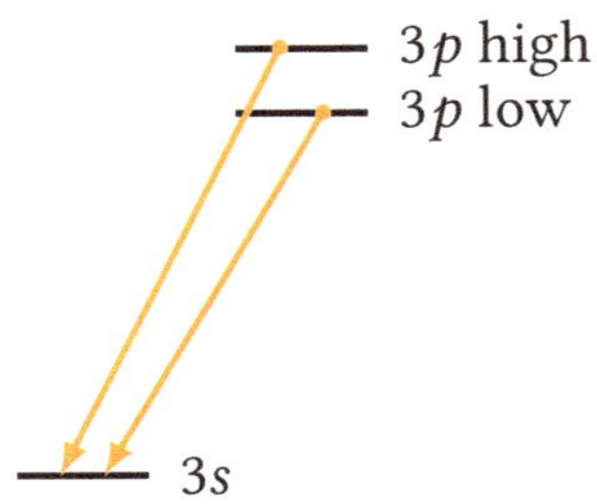

Figure A.3: Atomic transitions of the sodium D doublet. The short horizontal lines are no longer individual orbitals. They are, again, energy levels with multiple states represented by a single line. This means the vertical direction is energy. However, it is not to scale: the difference between the $3p$ high and the $3p$ low energy levels is 1/1000th that of between them and the $3s$ level.

An atomic transition is an atom going from one state to another. It can be to a higher or lower energy. A transition could be caused by an absorption (low to high) or emission (high to low) of a photon. If an electron is in a state above the atom's ground state, it is unstable and will randomly emit a photon. Transitions can also be caused by an atom colliding with another particle and exchanging energy and angular momentum. In the figure, I am showing transitions to lower energies, which involve the emission of yellow-orange photons. Transitions come with specific rules for the changing of quantum numbers. Some transitions are allowed, and others are forbidden. Examples of those and their detailed rules are more appropriate for a book on atomic physics.

The splitting of the $3p$ level is due to spin-orbit coupling and relativistic effects. The energy splitting is small because the magnetic field that arises from angular momentum is so weak. Due to the elaborate rules of quantum angular momentum, there are twice as many states in the higher level than in the lower level. To find the D doublet spectral line brightness, we take into account the number of states in each level, the transition rules, and the equal strengths of transitions. The result is that one peak is almost exactly double the other, as you will see or did see in Fig. 2.2 in the second chapter.

Appendix B: The postulates of quantum mechanics, a non-mathematical overview

Often it's better to learn something in an approximate way before learning it in a rigorous way. In this appendix, I give you a high-speed, low-drag version of the postulates so you can get an overview before we dive into the details. Two of the postulates (i.e., the single-measurement and projection postulates) are problematic because they aren't always true. I explain why this is so in particular chapters, so I don't go into the reasons here. I also resisted the temptation to give an extensive commentary on any of the postulates here. At best, I can provide a few clarifying words after a few of them.

If you don't see your favored quantum topic in the list below, then it is not a postulate. Instead, it should fall under one of three other parts of an axiomatic system (if we were working with a proper axiomatic system). It would either be 1) a definition, 2) an unstated notion that we all implicitly agree on (e.g., the electron has a specific amount of electric charge), or 3) be derivable from the postulates and the unstated notions (e.g., the uncertainty principle).

It is informative to compare the list here with the cursory statements of the postulates on the *Hyperphysics* website and the prolific descriptions on the Wikipedia page titled *Mathematical formulation of quantum mechanics.*

The state postulate: The state of a quantum object is completely specified by the mathematical wave function.

On one level, the state of a quantum object (e.g., an electron or an atom) can be thought of as what it's doing and what physical quantities it possesses. The purpose of this postulate is to make it possible for the other postulates to extract information from the mathematical wave function.

I used "mathematical" to make it clear that this refers to the math on the page. I do this because I tell you to think of the wave function as the quantum object, so the two uses should be distinguished here. After all, it wouldn't make much sense to say "the state of a wave function is completely specified by the wave function" (unless you knew one refers to math and the other to an object).

Often, it is also said that the mathematical wave function exists in a Hilbert space that is infinite-dimensional and has complex-valued amplitudes. This is a way to signal which math rules are to be followed when using the postulates. This math is where the principle of superposition is hidden among the postulates.

The time-evolution postulate: When uninterrupted by a measurement, the law of the Schrödinger equation governs how a state changes as time goes on.

The observable postulate: For every physical quantity an object possesses in classical mechanics, there is a corresponding quantity in quantum mechanics. These physical quantities are called observables. In quantum mechanics, each observable has an associated mathematical object called an *operator*.

The converse about operators is not true. Not every operator corresponds to an observable. For example, the number four is an operator, but it's not physically meaningful because we never talk about the fourness of an object.

When the postulate is written out rigorously, this is where it starts to get real mathy. Various mathematical terms, such as *linear map* and *self-adjoint*, are used to ensure that later postulates do proper things, like predict real values for measurements.

> **The spin postulate:** Spin exists and is an observable which has associated operators.

Spin angular momentum wasn't known to classical mechanics, so this postulate is needed to complete the set of observables.

> **The single-measurement postulate:** After applying an observable's operator to the mathematical wave function, a spectrum of possible real values for the physical quantity will be revealed. Should you do a measurement corresponding to that observable, one of those values will be the result of the measurement.

For example, if you have a quantum object in a superposition of spin-up and spin-down, the spectrum of physical quantities is $+\hbar/2$ and $-\hbar/2$. Only one will be seen in a single measurement.

> **The probability postulate:** The relative probability of measuring one of those values from the single-measurement postulate is determined by the squared magnitude of the quantum amplitude of the state.

Einstein was very vocal about his belief that nature was not truly random. Take note that the postulate is written so that it doesn't comment on whether nature is truly or only apparently random.

> **The projection postulate:** A measurement projects a quantum object into a new or the same state. This state has a correspondence with the result measured.

Continuing from the spin-up/down superposition above, this postulate says that the measurement process has forced the quantum object to be in the spin state corresponding to what was measured with your apparatus. It will no longer be in the previous superposition. Instead, it is now definitely either spin-up or spin-down.

> **The expectation-value postulate** Say we prepare a large number of quantum objects to be in equivalent states. We will perform the same type of measurement on each of them. The average of those values is predictable, whereas the individual measurements are random and unpredictable (but occur with a frequency determined by the probability postulate). A straightforward math expression finds the predictable average.

This postulate is useful because newly derived truths come from predictable quantities. By the way, the term *expectation value* is a statistician's way of saying *weighted average.*

> **The composite-system postulate:** When there is a system of two or more quantum objects, one does not simply write the total wave function as a superposition of the single-particle wave functions. Instead, we must utilize tensor multiplication of the wave functions.
>
> A single-particle wave function has a spin part and a spatial wave function part. Those two parts are also joined by tensor multiplication.

> **The symmetrization postulate:** All quantum objects can be classified as bosons or fermions. There is a law of the universe that constrains the mathematical form of the wave function when the state is made of bosons or fermions of the same type (e.g., a wave function of all electrons). Bosons and fermions have slightly different constraints.

A consequence of this postulate is the *Pauli exclusion principle*: No two electrons can exist in the *exact same* state simultaneously.

Appendix C: Matrix mechanics and spin

Introduction

The language of matrix mechanics is linear algebra. In contrast, the language of wave mechanics is calculus. Because of the ease with which I can show you how to use it, I will give you some of the algorithms for using matrix mechanics on spin states. Beware, though, this appendix is neither mathematically nor physically complete. I give you just enough so that you will have a small head start if you study quantum mechanics somewhere else. A university textbook, on the other hand, would give you more rules and tools and would also derive much of what I simply declare to be true.

Kets as column vectors

As a reminder, here is how we wrote a spin-up particle in the three "natural" representations (from Eq. 11.12 on p. 213):

$$|+z\rangle = \underbrace{|+x\rangle}_{\sqrt{50\%}} + \underbrace{|-x\rangle}_{-\sqrt{50\%}} = \underbrace{|+y\rangle}_{\sqrt{50\%}} + \underbrace{|-y\rangle}_{i\cdot\sqrt{50\%}} \tag{C.1}$$

As column vectors, those kets are represented as follows:

$$\begin{bmatrix} 1 \\ 0 \end{bmatrix}_z = \begin{bmatrix} \sqrt{50\%} \\ -\sqrt{50\%} \end{bmatrix}_x = \begin{bmatrix} \sqrt{50\%} \\ i \cdot \sqrt{50\%} \end{bmatrix}_y \tag{C.2}$$

In the column vectors, the quantum amplitudes are all that remain explicitly. Which ket each amplitude came from is implied by its position in the column vector. The convention is that the most "plus" basis ket is at the top and we work down to the most "minus" basis ket on the bottom. Beware that, out in the wild, physicists don't put the basis letter as a subscript because they are usually only working in one basis. When they are working with multiple bases, you must keep in mind the descriptions they write to keep things straight. They have the curse of knowledge and don't remember how confusing it was for them before they were experienced.

To make reading the vectors easier, we often pull common factors outside the brackets as multipliers:

$$|+z\rangle = \begin{bmatrix} 1 \\ 0 \end{bmatrix}_z = \sqrt{50\%}\begin{bmatrix} 1 \\ -1 \end{bmatrix}_x = \sqrt{50\%}\begin{bmatrix} 1 \\ i \end{bmatrix}_y \tag{C.3}$$

Take note that no one in the world except me uses this percentage convention. The vectors are normally written as:

$$|+z\rangle = \begin{bmatrix} 1 \\ 0 \end{bmatrix}_z = \frac{1}{\sqrt{2}}\begin{bmatrix} 1 \\ -1 \end{bmatrix}_x = \frac{1}{\sqrt{2}}\begin{bmatrix} 1 \\ i \end{bmatrix}_y \tag{C.4}$$

Next, have a look at the following and try to deduce which physical spin state (i.e., which single ket) is represented:

$$\begin{bmatrix} 0 \\ 0 \\ 1 \end{bmatrix}_z = \begin{bmatrix} \sqrt{25\%} \\ -\sqrt{50\%} \\ \sqrt{25\%} \end{bmatrix}_x = \begin{bmatrix} -\sqrt{25\%} \\ i \cdot \sqrt{50\%} \\ \sqrt{25\%} \end{bmatrix}_y \tag{C.5}$$

I haven't memorized quantum amplitudes, so I can't use the second or third column vector to figure out the AM vector. However, the left-most column vector pretty easily gives up the answer. That column vector is in the z-basis. The fact that there are three entries tells us that there are three basis kets. From that number, we know it is a spin-1 particle via $2s + 1 = 3$ (recall p. 236). The bottom entry in the column vector of '1' tells us that the quantum amplitude is 100% in front of the most "minus" ket. Combining all the information, we deduce that it represents the $|1, -1\rangle_z$ state, a spin-1 particle pointing in the negative z-direction (ket representations: Eq. 11.36 on p. 243).

Spin operators

Introducing the operators

Recall that when applied to a ket, operators make a ket return a value, change its state, or both. The simplicity of spin operators makes them suitable to show you how to do matrix multiplication.

Operators in matrix mechanics are square matrices with the numbers of rows and columns equal to the number of kets in the basis. The spin-$\frac{1}{2}$ operator for the z-direction in the z-basis is:

$$\hat{S}_z = \frac{\hbar}{2}\begin{bmatrix} 1 & 0 \\ 0 & -1 \end{bmatrix}_z \tag{C.6}$$

The subscript on $\hat{S}_z$ does not tell us the basis. Instead, it tells us which axis the operator is asking about (in a Cartesian coordinate system we have agreed on). In contrast, the z on the matrix tells us the basis. If the two z's seem confusing, it will make more sense later in the section on the numerical values returned by operators. It is also not common for physicists to write basis subscripts on the matrices; this causes a lot of confusion for students when working in multiple bases.

The matrix entries would change if I were to change the basis, but I won't do that often. Except to explain how to find an observable's numerical values, everything we need in this appendix is in the z-basis.

The spin-1 operator for the z-direction is:

$$\hat{S}_z = \hbar \begin{bmatrix} 1 & 0 & 0 \\ 0 & 0 & 0 \\ 0 & 0 & -1 \end{bmatrix}_z \tag{C.7}$$

Now, I have just used the same symbol (i.e., $\hat{S}_z$) for two different things. Since it's obvious which matrix corresponds to spin-$\frac{1}{2}$ and which corresponds to spin-1, we will say I am allowed to do that.

Applying the operators

Spin-$\frac{1}{2}$

Back in the *Math vs Measurement* chapter (p. 102), I showed you the following eigenvalue equation:

$$\hat{S}_z \cdot |+z\rangle = \frac{\hbar}{2} \cdot |+z\rangle \tag{C.8}$$

Now, I can show you the intervening math used to get from the left to the right side. As a first step, we can retain the left side and substitute in the matrix and column vector we now know. You know the final answer will be the eigenvalue multiplying a ket, so I will also give that with a column vector of unknown entries:

$$\hat{S}_z \cdot |+z\rangle = \frac{\hbar}{2} \begin{bmatrix} 1 & 0 \\ 0 & -1 \end{bmatrix}_z \cdot \begin{bmatrix} 1 \\ 0 \end{bmatrix}_z = \frac{\hbar}{2} \cdot \begin{bmatrix} \textcolor{pink}{\blacksquare} \\ \textcolor{cyan}{\blacksquare} \end{bmatrix}_z \tag{C.9}$$

In English, we read from left to right, top to bottom. Figuratively, that is what matrix multiplication is about. To find the top entry of the final column vector ($\textcolor{pink}{\blacksquare}$), we apply the two values in the top row of the matrix to the two values in the column vector. The first multiplies with the first, and the last multiplies with the last. You then add those. The math you do is $1 \cdot 1 + 0 \cdot 0 = 1$. Inside the equation, that looks like this:

$$\hat{S}_z \cdot |+z\rangle = \frac{\hbar}{2}\begin{bmatrix} 1 & 0 \\ 0 & -1 \end{bmatrix}_z \cdot \begin{bmatrix} 1 \\ 0 \end{bmatrix}_z = \frac{\hbar}{2} \cdot \begin{bmatrix} 1 \cdot 1 + 0 \cdot 0 = 1 \\ \blacksquare \end{bmatrix}_z$$

The next entry (■) is found by reading the next row in the matrix:

$$\hat{S}_z \cdot |+z\rangle = \frac{\hbar}{2}\begin{bmatrix} 1 & 0 \\ 0 & -1 \end{bmatrix}_z \cdot \begin{bmatrix} 1 \\ 0 \end{bmatrix}_z = \frac{\hbar}{2} \cdot \begin{bmatrix} 1 \\ 0 \cdot 1 - 1 \cdot 0 = 0 \end{bmatrix}_z$$

And we have the result we were expecting, an eigenvalue equation:

$$\hat{S}_z \cdot |+z\rangle = \frac{\hbar}{2} \cdot \begin{bmatrix} 1 \\ 0 \end{bmatrix}_z = \frac{\hbar}{2} \cdot |+z\rangle \tag{C.10}$$

You may have noticed that the -1 in the matrix didn't do anything in this case. You should try applying $\hat{S}_z$ to $|-z\rangle$ to see the purpose of the -1 in the matrix.

Spin-1

To ensure you understand how the matrix multiplication algorithm scales, we will look at an abbreviated example with spin-1 and $m_s = -1$. We start by applying the spin operator for the z-axis and put everything in matrix form. There are now three rows in the matrix and three entries in the column vector, but our algorithm doesn't change:

$$\hat{S}_z \cdot |1, -1\rangle = \hbar \begin{bmatrix} 1 & 0 & 0 \\ 0 & 0 & 0 \\ 0 & 0 & -1 \end{bmatrix}_z \cdot \begin{bmatrix} 0 \\ 0 \\ 1 \end{bmatrix}_z = \hbar \cdot \begin{bmatrix} 1 \cdot 0 + 0 \cdot 0 + 0 \cdot 1 \\ 0 \cdot 0 + 0 \cdot 0 + 0 \cdot 0 \\ 1 \cdot 0 + 0 \cdot 0 - 1 \cdot 1 \end{bmatrix}_z \tag{C.11}$$

Once practiced, you can quickly see the first two entries will be zero and the final entry is -1 in your head. So we get:

$$\hat{S}_z \cdot |1,-1\rangle \;=\; \hbar \cdot \begin{bmatrix} 0 \\ 0 \\ -1 \end{bmatrix}_z \;=\; -\hbar \cdot \begin{bmatrix} 0 \\ 0 \\ 1 \end{bmatrix}_z \;=\; -\hbar \cdot |1,-1\rangle \qquad \text{(C.12)}$$

As expected, we applied the spin operator corresponding to the z-axis to a spin-1 particle in a state with its spin pointed down and we found that its angular momentum is pointed downward via the minus sign.

But you may wonder how I knew to take the negative sign out of the column vector and assign it to the eigenvalue rather than keep it with the quantum amplitude in the column vector. Well, the unsatisfying answer is that I knew to do that because I know the final answer. I know the physics of the situation and how it should look. Without knowledge of the physics, you can't know whether the negative sign belongs to the amplitude or the eigenvalue.

Let's do the matrix multiplication again with a spin-1 ket that has no angular momentum so that I can show you a must-know trick:

$$\hat{S}_z \cdot |1,0\rangle \;=\; \hbar \begin{bmatrix} 1 & 0 & 0 \\ 0 & 0 & 0 \\ 0 & 0 & -1 \end{bmatrix}_z \cdot \begin{bmatrix} 0 \\ 1 \\ 0 \end{bmatrix}_z \;=\; \hbar \cdot \begin{bmatrix} 0 \\ 0 \\ 0 \end{bmatrix}_z \qquad \text{(C.13)}$$

What we have done is mathematically legitimate. We have found what is called a *zero vector*, or *null vector*. There seems to be another ambiguity here. Is the numerical value $\hbar$? Perhaps $\hbar$ gets applied to the zero vector, and we have a zero vector multiplied by a numerical value of one or maybe zero?

With the way the postulates are normally presented (and the way I have presented them to you), after applying an operator corresponding to an observable quantity, the numerical values and the kets tell you about the possible measurement results and the possible kets that the quantum object will be projected into.

To be consistent with that, we need a little trick. At this point, quantum mechanics and math diverge slightly: we redefine the zero-vector results. We say the zero is the numerical value and its ket is the original ket. Therefore, the calculation goes like this:

$$\hat{S}_z \cdot |1,0\rangle \;=\; \cdots \;=\; \hbar \cdot \begin{bmatrix} 0 \\ 0 \\ 0 \end{bmatrix}_z \;\rightarrow\; \hbar \cdot 0 \cdot \begin{bmatrix} 0 \\ 1 \\ 0 \end{bmatrix}_z \;=\; 0 \cdot |1,0\rangle \qquad \text{(C.14)}$$

This shows us that the $\hat{S}_z$ operator gives us the anticipated eigenvalue of zero spin angular momentum from the original ket.

Rotations operators and change of bases

Introducing the operators

For every axis in three-dimensional space, there is a spin rotation operator that transforms the state about that axis. When dealing with classical objects, we can apply the same mathematical rotation to a baseball and a bear. The same is not true with matrix mechanics. Spin is represented by column vectors of varying "dimensions" depending on the spin quantum number s. The number of "dimensions" is the number of basis kets. Therefore, there are separate sets of rotation operators for our spin-$\frac{1}{2}$ and spin-1 examples.

For each axis of a spin-s particle, there is a rotation matrix. In the examples that follow, we will be rotating vectors that start pointing somewhere in the xz-plane and end in the same plane. That means the rotation axis is the y-axis. The spin-$\frac{1}{2}$ rotation operator about the y-axis is:

$$\hat{R}_y(\theta) \;=\; \begin{bmatrix} \cos\frac{\theta}{2} & -\sin\frac{\theta}{2} \\ \sin\frac{\theta}{2} & \cos\frac{\theta}{2} \end{bmatrix}_z \qquad \text{(C.15)}$$

The subscript on the R tells you the axis of rotation, and the subscript on the matrix tells you the basis. The trig functions inside the matrix will look simpler once we choose an angle θ for the rotation. We will say a positive θ is a counterclockwise rotation and a negative θ is a clockwise rotation when, in our mind's eye, we are looking back at the positive y-axis vector pointing at us. This follows the right-hand rule.

As we increase the spin quantum number s, entries in the matrices become more complicated. The spin-1 rotation operator about the y-axis is:

$$\hat{R}_y(\theta) = \begin{bmatrix} (1+\cos\theta)/\sqrt{2} & \sin\theta & (1-\cos\theta)/\sqrt{2} \\ -\sin(\theta) & \sqrt{2}\,(\cos\theta - 1) & \sin\theta \\ (1-\cos\theta)/\sqrt{2} & -\sin(\theta) & (1+\cos\theta)/\sqrt{2} \end{bmatrix}_z \tag{C.16}$$

I won't show you this operator in action, but after reading this appendix, you can try it yourself.

In the chapter on spin, I introduced you to the idea of a phase ambiguity and the acquiring of a -1 in phase when any half-integer spin particle is rotated by 360 degrees. I mentioned that it is due to the argument in the trig functions. You can see here in the spin-$\frac{1}{2}$ operator that it is due to $\theta/2$. The integer spin particles don't have this feature because the arguments of their trig functions have the full θ.

Active and passive rotations

Active rotations

Before we apply these spin rotation matrices, we need to see what it means to rotate an object or our point of view in physical space. First, I want you to imagine you are looking at a photo of a skyscraper on a wall. It has a pointed spire and is like a vector pointing to the ceiling of the room you are in. We will say that direction is the positive z-direction. To your right will be the positive x-direction. We won't include the third space dimension in the vector.

The skyscraper's current direction is represented as:

$$|\text{skyscraper } up\rangle \;=\; \begin{bmatrix} 1 \\ 0 \end{bmatrix}_{\text{original basis}} \tag{C.17}$$

The top entry of the column vector is the z-direction component of the vector, and the bottom entry is the x-direction component.

Now, physically rotate the picture 90° clockwise. This is an **active rotation** because an object changes. The skyscraper's spire is now pointing in the positive x-direction, so the bottom entry of the column vector will be filled and the top entry will have a zero:

$$|\text{skyscraper } right\rangle \;=\; \begin{bmatrix} 0 \\ 1 \end{bmatrix}_{\text{original basis}} \tag{C.18}$$

This explanation shows that we can rotate an object and put it into a new state of being. If this were a spin-$\frac{1}{2}$ particle in a spin-up state, it would correspond to a physical rotation by 90° clockwise into a new state:

$$\hat{R}_y(+90^\circ) \cdot |+z\rangle \;=\; |+x\rangle \tag{C.19}$$

Of course, you would have to use a quantum rotation operator and the spin's column vector. A look back at Fig. 11.12 on p. 228 and Fig. 11.14 on p. 231 should help you interpret the spin equation above.

Passive rotations

Next, we look at **passive rotations**. They are passive because the objects of interest do not change. We will start over with the skyscraper's spire pointing at the ceiling again.

This time, you will imagine holding a transparent plastic sheet in your hands with your arms outstretched in front of you. On this sheet, you have drawn a coordinate system with the x and z Cartesian directions on it. This transparent sheet will be rotated. To start, you will hold it so that the z-axis is aligned with the skyscraper's spire.

At this moment, you would say the skyscraper is pointing in that direction. Given all that, we are back to the starting point of our first skyscraper equation (Eq. C.17).

Next, imagine you want to lie on a bench on your left side to look at the skyscraper photo. You will hold your arms out rigidly to look through the transparent plastic the whole time. To accomplish this, your body rotates 90 degrees counterclockwise (CCW) around the y-axis. As you look through the sheet, it appears that the skyscraper rotates clockwise with respect to your coordinate system.

At one point, the coordinate system is 45° from where it was originally. As a column vector, at this point in time, it is:

$$|\text{skyscraper } up\rangle = \begin{bmatrix} \frac{1}{\sqrt{2}} \\ \frac{1}{\sqrt{2}} \end{bmatrix}_{45^\circ \text{ CCW basis}} \tag{C.20}$$

Notice that we didn't change the state of the skyscraper, so the left side didn't change. The right side values can be found using the Pythagorean theorem, noting that the hypotenuse is a length of 1 because the skyscraper photo didn't change its length.

Your view continues to change until you come to rest. In your view, you now see a skyscraper pointing along the x-axis on your sheet. This is represented as:

$$|\text{skyscraper } up\rangle = \begin{bmatrix} 0 \\ 1 \end{bmatrix}_{90^\circ \text{ CCW basis}} \tag{C.21}$$

This shows that you must pay attention to the notation. The column vector looks similar to the column vector for the photo that was actually rotated (Eq. C.18). However, the ket label tells you the photo's state didn't change. The basis label reminds you we have changed bases.

When comparing the similar-looking equations of Eq. C.18 and Eq. C.21, make a mental note that to get each of these, in both cases, we started with an object in the same state. We then either rotated the object clockwise or rotated a coordinate system counterclockwise.

Applying the spin rotation operators

All of this converts over to spins. We can imagine starting with a spin-up particle represented in the z-basis. If we want to put it in a *different state*, we tell ourselves our column vector will stay in that basis and a rotation operator corresponds to a forcible and physical change. If we want a *different representation* of the current state, we tell ourselves the column vector's basis will change via a rotation.

Active rotation of spin-$\frac{1}{2}$

If we want to change the $|+z\rangle$ state to the $|+x\rangle$ state, we need to apply a clockwise rotation operator of positive 90 degrees. Via Eq. C.15, that rotation operator looks like:

$$\hat{R}_y(+90^\circ) = \begin{bmatrix} \frac{1}{\sqrt{2}} & -\frac{1}{\sqrt{2}} \\ \frac{1}{\sqrt{2}} & \frac{1}{\sqrt{2}} \end{bmatrix}_z = \frac{1}{\sqrt{2}}\begin{bmatrix} 1 & -1 \\ 1 & 1 \end{bmatrix}_z \tag{C.22}$$

With that, we can apply the operator to the state to get:

$$\hat{R}_y(+90^\circ) \cdot |+z\rangle = \frac{1}{\sqrt{2}}\begin{bmatrix} 1 & -1 \\ 1 & 1 \end{bmatrix}_z \cdot \begin{bmatrix} 1 \\ 0 \end{bmatrix}_z = \frac{1}{\sqrt{2}}\begin{bmatrix} 1 \\ 1 \end{bmatrix}_z \tag{C.23}$$

We can convert the final answer back to ket notation to see what happened.

$$\frac{1}{\sqrt{2}}\begin{bmatrix} 1 \\ 1 \end{bmatrix}_z = \frac{1}{\sqrt{2}}\Big(|+z\rangle + |-z\rangle \Big) = |+x\rangle \tag{C.24}$$

I use round brackets for grouping things because square brackets are used for matrix algebra in this appendix.

Passive rotation of spin-$\frac{1}{2}$

If we want to make a passive rotation of 90 degrees counterclockwise to change from the z-basis to the x-basis, then we set $\theta = -90$ degrees. I will also add a p to indicate this is a passive rotation. The rotation operator then becomes:

$$ {}_p\hat{R}_y(-90^\circ) = {}_p\begin{bmatrix} \frac{1}{\sqrt{2}} & \frac{1}{\sqrt{2}} \\ -\frac{1}{\sqrt{2}} & \frac{1}{\sqrt{2}} \end{bmatrix}_z = \frac{1}{\sqrt{2}}\,{}_p\begin{bmatrix} 1 & 1 \\ -1 & 1 \end{bmatrix}_z \tag{C.25} $$

With that, we can apply the operator to the state to get:

$$ {}_p\hat{R}_y(-90^\circ)\cdot|+z\rangle = \frac{1}{\sqrt{2}}\,{}_p\begin{bmatrix} 1 & 1 \\ -1 & 1 \end{bmatrix}_z \cdot \begin{bmatrix} 1 \\ 0 \end{bmatrix}_z = \frac{1}{\sqrt{2}}\begin{bmatrix} 1 \\ -1 \end{bmatrix}_x \tag{C.26} $$

Since this is a passive rotation, the state is still spin-up. What's new is that we have revealed what it is mathematically equivalent to:

$$ \begin{aligned} |+z\rangle &= {}_p\hat{R}_y(-90^\circ)\cdot|+z\rangle \\ &= \frac{1}{\sqrt{2}}\begin{bmatrix} 1 \\ -1 \end{bmatrix}_x = \frac{1}{\sqrt{2}}\Big(|+x\rangle + (-1)|-x\rangle\Big) \end{aligned} \tag{C.27} $$

This agrees with what you know from the first equation of this appendix and from the chapter on spin.

Operators and multiple discrete values

Say you have a spin-$\frac{1}{2}$ particle with spin-up (i.e., the spin is along the positive z-direction) and you want to find the possible results of measurements of spin angular momentum along the y-axis. So, you apply the $\hat{S}_y$ operator to the $|+z\rangle$ to find the operator's numerical values for this state. Impulsively, you choose the spin matrix and column vector in the z-basis. Here is the $\hat{S}_y$ operator in the z-basis is:

$$\hat{S}_y = \frac{\hbar}{2}\begin{bmatrix} 0 & i \\ -i & 0 \end{bmatrix}_z \tag{C.28}$$

Using this as we said, you will get:

$$\hat{S}_y \cdot |+z\rangle = \frac{\hbar}{2}\begin{bmatrix} 0 & i \\ -i & 0 \end{bmatrix}_z \cdot \begin{bmatrix} 1 \\ 0 \end{bmatrix}_z = \frac{\hbar}{2} \cdot \begin{bmatrix} 0 \\ -i \end{bmatrix}_z = -i \cdot \frac{\hbar}{2} \cdot |-z\rangle \tag{C.29}$$

This is mathematically legitimate, but it makes no physical sense to us. The operation has changed the state. This is the physics mistake I mentioned on p. 92. Coming to your senses, you remember you want to find numerical values in front of two kets in the y-basis.

You are probably asking what the matrix for the y-basis is, then. The 'trick' is that it is the $\hat{S}_z$ matrix in the z-basis (Eq. C.6) but relabeled for the y-basis:

$$\hat{S}_y = \frac{\hbar}{2}\begin{bmatrix} 1 & 0 \\ 0 & -1 \end{bmatrix}_y \tag{C.30}$$

We also need to use a spin-up column vector in the y-basis (Eq. C.4). Putting the two together, we get:

$$\begin{aligned} \hat{S}_y \cdot |+z\rangle &= \frac{\hbar}{2}\begin{bmatrix} 1 & 0 \\ 0 & -1 \end{bmatrix}_y \cdot \frac{1}{\sqrt{2}}\begin{bmatrix} 1 \\ i \end{bmatrix}_y = \frac{\hbar}{2} \cdot \frac{1}{\sqrt{2}}\begin{bmatrix} 1 \\ -i \end{bmatrix}_y \\ &= \frac{1}{\sqrt{2}}\left[\frac{\hbar}{2} \cdot |+y\rangle + \frac{-\hbar}{2} \cdot (i) \cdot |-y\rangle\right] \\ &= \frac{\hbar}{2} \cdot \underbrace{|+y\rangle}_{\sqrt{50\%}} + \frac{-\hbar}{2} \cdot \underbrace{|-y\rangle}_{i\cdot\sqrt{50\%}} \end{aligned} \tag{C.31}$$

As expected, you have equal chances to measure the spin along the positive y-direction (because $\hbar/2$ is positive) or along the negative y-direction (because $-\hbar/2$ is negative). In this case, it was obvious that the i belongs to the quantum amplitude because the numerical values are required to be real.

Expectation values

In the *Spring cleaning* chapter, I showed you the integral that allows you to calculate the expectation value of kinetic energy. The integral is the math of wave mechanics. I also said to find an expectation value of spin, we use matrix mechanics. Matrix multiplication is so straightforward that I can show you the full calculation. Let's find the expectation values of the operators $\hat{S}_z$ and $\hat{S}_y$ of a spin-$\frac{1}{2}$ particle pointing up.

The first case

The first step is to write down the expectation value of $\hat{S}_z$ applied to a spin-up state:

$$\langle \hat{S}_z \rangle = \langle +z | \hat{S}_z | +z \rangle \tag{C.32}$$

You know that the ket is written as a column vector:

$$|+z\rangle = \begin{bmatrix} 1 \\ 0 \end{bmatrix}_z \tag{C.33}$$

Transforming this into a row vector gives us its **bra**:

$$\langle +z| = \begin{bmatrix} 1 & 0 \end{bmatrix}_z \tag{C.34}$$

(When the values are complex, there is another step when creating a bra, which you'll see in a moment.) You can see that the row and column vectors are in the z-basis, so the matrix operator must be in that basis too. Putting it all together, we find:

$$\langle \hat{S}_z \rangle = \begin{bmatrix} 1 & 0 \end{bmatrix}_z \cdot \frac{\hbar}{2} \begin{bmatrix} 1 & 0 \\ 0 & -1 \end{bmatrix}_z \cdot \begin{bmatrix} 1 \\ 0 \end{bmatrix}_z \tag{C.35}$$

Next, apply the matrix to the column vector. We will find that:

$$\langle \hat{S}_z \rangle = \begin{bmatrix} 1 & 0 \end{bmatrix}_z \cdot \frac{\hbar}{2} \begin{bmatrix} 1 \\ 0 \end{bmatrix}_z \tag{C.36}$$

Finally, you need to know how a row vector multiplies with a column vector. It's always the same rule in matrix algebra: first multiplies first, and last multiplies last. Then add them up:

$$\langle \hat{S}_z \rangle = \frac{\hbar}{2} \cdot (1 \cdot 1 + 0 \cdot 0) = \frac{\hbar}{2} \tag{C.37}$$

The end result is a number. This is the expectation value of the S_z operator in a spin-up state. However, you might be wondering, "But weren't kets and bras implicitly or invisibly in there? How do you get rid of those?" Yes. You are right. But, we are using a quick and compact notation that hides some things.

I will rewrite Eq. C.36 to explicitly show the bras and kets:

$$\begin{aligned} \langle \hat{S}_z \rangle &= \frac{\hbar}{2} \cdot \Big(\langle +z| + 0 \cdot \langle -z| \Big) \cdot \Big(|+z\rangle + 0 \cdot |-z\rangle \Big) \\ &= \frac{\hbar}{2} \cdot \langle +z \,|\, +z \rangle \end{aligned} \tag{C.38}$$

I remind you that I am using round brackets to group things; I don't want you to think that these are row vectors. We do normal middle school multiplication across the round brackets.

The $\langle +z \,|\, +z \rangle$ equals one because we are using an *orthonormal basis*: when the labels in the bra and kets are the same, the result equals one. When the bras and kets are different (e.g., $\langle +z \,|\, -z \rangle$ and $\langle -z \,|\, +z \rangle$) the results are zero.

The second case

Now, we calculate the other expectation value of the spin-up state:

$$\begin{aligned}
\langle \hat{S}_y \rangle = \langle +z | \hat{S}_y | +z \rangle &= \begin{bmatrix} 1 & 0 \end{bmatrix}_z \cdot \frac{\hbar}{2} \begin{bmatrix} 0 & i \\ -i & 0 \end{bmatrix}_z \cdot \begin{bmatrix} 1 \\ 0 \end{bmatrix}_z \\
&= \begin{bmatrix} 1 & 0 \end{bmatrix}_z \cdot \frac{\hbar}{2} \begin{bmatrix} 0 \\ -i \end{bmatrix}_z \\
&= \frac{\hbar}{2} \cdot (1 \cdot 0 + 0 \cdot -i) \\
&= 0
\end{aligned} \tag{C.39}$$

If you calculate the expectation value of $\hat{S}_x$, you will also find zero. (I give you that operator on p. 319.) In my reality interpretation, this means that the spin angular momentum of a spin-up state has no components in the x- and y-directions, and the spin is pointing purely up if it is left undisturbed by fields, other particles, or a measurement.

Before I end this appendix, I must show you how to make a proper bra from a ket with complex values. In the y-basis, the spin-up column vector is:

$$|+z\rangle = \frac{1}{\sqrt{2}} \begin{bmatrix} 1 \\ -i \end{bmatrix}_y \tag{C.40}$$

With the previous examples, you simply laid the column vector on its side, so to speak, to make it a row vector. With complex entries, the additional step is to change the imaginary unit vector i to $-i$ wherever you see it. This step is called making the *complex conjugate*. The bra for this state is then:

$$\langle +z| = \frac{1}{\sqrt{2}} \begin{bmatrix} 1 & i \end{bmatrix}_y \tag{C.41}$$

It would be good practice for you to find the expectation value of $\hat{S}_y$ in the y-basis and find again that it is zero. You now have all the pieces to do that.

A gift for some

Some motivated readers may want to play with rotation matrices, so below, I give you the general form of the spin-$\frac{1}{2}$ rotation matrix:

$$R_{\vec{n}}(\theta) = \begin{bmatrix} \cos\frac{\theta}{2} - i \cdot n_z \cdot \sin\frac{\theta}{2} & (-i \cdot n_x - n_y) \cdot \sin\frac{\theta}{2} \\ (-i \cdot n_x + n_y) \cdot \sin\frac{\theta}{2} & \cos\frac{\theta}{2} + i \cdot n_z \cdot \cos\frac{\theta}{2} \end{bmatrix}_z \tag{C.42}$$

Note that it is in the z-basis. You already know you must choose a rotation angle. In addition, you must choose an axis for that rotation, which is defined by a vector $\vec{n} = (n_x, n_y, n_z)$. This vector follows the right-hand rule, which tells us which is the positive and negative rotation direction. This vector must be normalized, which means it has a length of 1. This is easy to do. Once you've chosen your vector components, divide each one by:

$$\sqrt{n_x^2 + n_y^2 + n_z^2} \tag{C.43}$$

Before you try any other vector, I suggest using $\vec{n} = (0, 1, 0)$ to confirm you can find the rotation matrix of Eq. C.15.

Appendix D: The energy-time uncertainty relation doesn't exist

A variety of claims

Have a look at the putative energy-time inequality that some physicists claim as meaningful:

$$\sigma_E \cdot \sigma_t \geq \frac{\hbar}{2} \tag{D.1}$$

In this appendix, I will show that it is a meaningless, garbage set of symbols. Beyond the refutation of the inequality, the content in this appendix is useful because it gives you some initial exposure to some important mathematics. But precisely because it is more mathematical (and without any associated figures), I've relegated the material here to be optionally read.

From other physicists, what were you told the energy-time uncertainty relation means? Does it mean that virtual matter-antimatter pairs of particles are allowed to come into existence out of nothing because they can borrow energy for short periods of time? Or does it mean that the longer we measure a quantum system, the better we know the energy? Perhaps it tells us the half-life of a metastable energy state. These are all explanations given by people with Ph.D.'s in physics. The variety of answers should give you pause. Why don't they agree? That lack of consensus is certainly a warning sign.

If the inequality refers to creating energy and particles out of nothing, suppose then we have a pair of particles of some energy (rest mass energy + kinetic energy) that pop into existence. In this case, I guess the σ_E is interpreted as how much the energy of the universe changes. But because this is an inequality, we could have a σ_t of any enormous duration. It could be as long as the age of the universe. The virtual particles would never need to disappear. To get around this, the authors of this bold claim must be mentally replacing the $\geq$ sign with an $=$ sign that we saw in a previous chapter. This is again obviously the wrong thing to do.

The half-life claim is as easily dismissed since it has the same equality vs. inequality problem. If a metastable state emits some particle with some range of energies σ_E, you can't move variables around in the expression to get an exact half-life of the particle without changing the inequality to an equality.

If, instead, the inequality refers to the time spent measuring a system, how does that work in general? Let's say I measure in my detector a photon emitted from an excited atom. If I wait two more minutes after the detection, do I get a more precise result? How do we control the amount of time for detection if we don't even understand the collapse of the wave function?

I have thus shown you that what physicists say of the inequality is nonsensical. However, I haven't shown the inequality itself is untrustworthy. To truly illustrate to you that the symbols above are nonsense, I will show you three ways to get something like them. (It is coincidental that there are three explanations above and that I will have three attempts to make the inequality make sense.) I imagine these are what many physicists attempted in private and discarded in the early days of quantum mechanics.

Before we get to my goal, an intermediate section educates you on a few topics. You will see the **generalized uncertainty principle**, the **commutator operator**, and some quick examples of how they work when applied to position and momentum, which are examples of how to do things right.

The generalized uncertainty principle

The things we will use over and over

Let's get the pain over with and rip off the band-aid right away. The generalized uncertainty principle is:

$$\sigma_A \cdot \sigma_B \;\geq\; \left| \frac{1}{2 \cdot i} \cdot \left\langle \left[\hat{A}, \hat{B} \right] \right\rangle \right| \tag{D.2}$$

The $\hat{A}$ and $\hat{B}$ are placeholders for the operators that one substitutes in. They don't need to correspond to observables (which also tells us that these inequalities aren't fundamentally about measurements). The operators that are substituted in do, however, need to be Hermitian because that property is used to derive this inequality.

To start, you only need to be concerned with:

$$\left[\hat{A}, \hat{B} \right] \tag{D.3}$$

This is the commutator operator composed of the operators $\hat{A}$ and $\hat{B}$. It's definition is:

$$\left[\hat{A}, \hat{B} \right] \;=\; \hat{A} \cdot \hat{B} \;-\; \hat{B} \cdot \hat{A} \tag{D.4}$$

You might look at the definition of the commutator operator and think that it's clearly zero, but this is operator algebra, not standard algebra. Operators only have meaning when they get applied to states. Sometimes the result of the application of a commutator is zero, and sometimes it is not.

The first important example

To show you how the commutator and generalized uncertainty principle work, I will derive a new position-momentum uncertainty principle. It's also a very easy case because we will find a zero. This first example will allow you to understand the first attempt at deriving an energy-time uncertainty inequality later.

Looking at the generalized uncertainty principle, you can see the angled brackets. They tell us we are to calculate an expectation value. You know that a state is implied by the notation (p. 257). Since an operator's sole purpose is to act on a state, the commutator operator would have also tipped you off to the implied state.

In the following explanation, I will give you the tiniest insight into the calculus of derivatives. I need to show that the wave function depends on the variable x to do this, so the state notation will include the variable: $|\Psi(x)\rangle$. Making the expectation value notation more explicit, it is:

$$\langle\Psi(x)| \cdot \left[\hat{A}, \hat{B}\right] \cdot |\Psi(x)\rangle \tag{D.5}$$

I have decided we will find the uncertainty relation for y and p_x. In other words, we will see how the spread of the matter density in the y-direction is related to the spread of the flow function in the x-direction. We now apply the commutator to the ket. Because I know where this is going (we're going to find zero), I will tell you we can ignore the bra on the left for the moment:

$$[\hat{y}, \hat{p_x}] \cdot |\Psi(x)\rangle \;=\; \left(\hat{y} \cdot \hat{p_x} \;-\; \hat{p_x} \cdot \hat{y}\right) \cdot |\Psi(x)\rangle \tag{D.6}$$

But, we must stop being abstract and input the explicit math in place of the abstract operators. Since this is a position-space wave function, the operators are $\hat{y} = y$ and $\hat{p_x} = -i \cdot \hbar \cdot \partial_x$ (from p. 317). Putting those in gives us:

$$-i \cdot \hbar \cdot \left(y \cdot \partial_x \;-\; \partial_x \cdot y\right) \cdot |\Psi(x)\rangle \tag{D.7}$$

I took the opportunity to pull the constants outside the parentheses. Next, I will distribute the ket across the terms in parentheses. It's like normal multiplication:

$$-i \cdot \hbar \cdot \left(y \cdot \partial_x \cdot |\Psi(x)\rangle \;-\; \partial_x \cdot y \cdot |\Psi(x)\rangle\right) \tag{D.8}$$

Because I know how things end up, I can say that we can start by concentrating on the second term in parentheses. Recall that operators act on the thing to the right of them. The second term has the variable y acting on $|\Psi(x)\rangle$. But that operation is just multiplication, so the result is nothing changes.

Next, we must consider how ∂_x acts on $y \cdot |\Psi(x)\rangle$. Here's your calculus lesson. The thing about ∂_x is that it is not impartial. It is partial to things with x's. It has a severe bias for x's. In fact, the name of ∂_x is *the partial derivative with respect to x*. It is going to ignore y, move passed it, and act on $|\Psi(x)\rangle$:

$$-i \cdot \hbar \cdot \Big(y \cdot \partial_x \cdot |\Psi(x)\rangle \; - \; y \cdot \partial_x \cdot |\Psi(x)\rangle\Big) \qquad \text{(D.9)}$$

The only thing I did was move the ∂_x one position to the right.

It doesn't matter that we have a generic ket and don't know what answer the derivative would return. That's because the two terms in parentheses have a difference of zero, so the above expression is zero. Since the ket could be of any state, the result you will see in a moment will be completely general.

Recall that the expression immediately above (which is equal to zero) is a particular case of the expectation value in Eq. D.5 with the bra taken out. Returning the bra to the left of the expression immediately above, tells us that the expectation value in our uncertainty principle will be zero:

$$\Big\langle [\hat{y}, \hat{p_x}] \Big\rangle \; = \; 0 \qquad \text{(D.10)}$$

Looking back at the generalized uncertainty principle, you can see that the entire right side of our uncertainty principle will be zero, and the final answer is:

$$\sigma_y \cdot \sigma_{p_x} \; \geq \; 0 \qquad \text{(D.11)}$$

You can replace y and x in this inequality with any axes that are 90° to each other and the right-hand side will always come out to zero. Therefore, there is no connection between the spread of the matter density in one direction and the spread of the flow function 90° away.

I can't resist teaching you a tiny bit more that is not so important. Everything I did above was just straight math, except for when I substituted in the forms of the operators. I would call that part physics. We can make the above inequality a little better if we consider the physics of wave functions. Since the wave function can never be completely localized to a point (in position or momentum space), the standard deviations can never be zero. They will always be positive. Therefore, the most accurate form of the inequality is:

$$\sigma_y \cdot \sigma_{p_x} \;>\; 0 \tag{D.12}$$

The second important example

This second example will make the second energy-time uncertainty inequality easier to understand. Don't worry. This is the last example before we look at energy and time.

We will derive the x-p_x uncertainty inequality, which was the center of attention in the chapter on uncertainty. Let's start by doing the easy substitutions into the generalized uncertainty principle to give us our uncertainty principle:

$$\sigma_x \cdot \sigma_{p_x} \;\geq\; \left| \frac{1}{2 \cdot i} \cdot \left\langle [\hat{x}, \hat{p_x}] \right\rangle \right| \tag{D.13}$$

You know the right side will be $\hbar/2$. Therefore, examining the right side for a bit, you can see that the expectation value needs to give $i \cdot \hbar$. Everything I do below will be to find that the expectation value has this value.

Let's first concentrate only on the commutator applied to the ket as we did in the last example. Since you know how it works, I can jump to the point where you would get stuck:

$$\begin{aligned} [\hat{x}, \hat{p_x}] \cdot |\Psi(x)\rangle \;=& \\ & -i \cdot \hbar \cdot \Big(x \cdot \partial_x \cdot |\Psi(x)\rangle \;-\; \partial_x \cdot x \cdot |\Psi(x)\rangle \Big) \end{aligned} \tag{D.14}$$

The right-hand side is equivalent to Eq. D.8 with slightly different symbols. In the end, we will find that the math in parentheses equals

$-\,|\Psi(x)\rangle$. Because I know where this is going, I say once more we only need to focus on the second term in parentheses for the moment.

Like in the last example, x operating on the ket is just multiplication, so nothing changes, and we think of it as one thing that the derivative will act on. We, therefore, only need to find the result of this:

$$\partial_x \cdot x \cdot |\Psi(x)\rangle \tag{D.15}$$

You need to know two rules of calculus to solve this.

The first is called the *product rule*. This time, the partial derivative won't ignore the x in front of the ket. Roughly speaking, the product rule says to act on the first factor it sees and leave the second thing untouched. Then do it again, but leave the first untouched and act on the second factor. Finally, stick a plus sign between those two things you've just done:

$$\partial_x \cdot x \cdot |\Psi(x)\rangle \;=\; \Big(\partial_x \cdot x\Big) \cdot |\Psi(x)\rangle \;+\; x \cdot \Big(\partial_x \cdot |\Psi(x)\rangle\Big) \tag{D.16}$$

The parentheses let you know what the derivative is specifically acting on.

The second calculus rule is easy: $\partial_x \cdot x = 1$. That makes the first term on the right side of Eq. D.16 a lonely ket. This equation was once "the second term in parentheses" in Eq. D.14. Let's put this back where it belongs. It makes things simpler:

$$\begin{aligned}
&[\hat{x}, \hat{p}_x] \cdot |\Psi(x)\rangle \\
&\qquad = -i \cdot \hbar \cdot \Big(x \cdot \partial_x \cdot |\Psi(x)\rangle \;-\; |\Psi(x)\rangle \;-\; x \cdot \partial_x \cdot |\Psi(x)\rangle\Big) \\
&\qquad = i \cdot \hbar \cdot |\Psi(x)\rangle
\end{aligned} \tag{D.17}$$

We must remember that this was part of an expectation value, so we should apply the bra to the left side. We will do that in a moment. Since we are done with the derivatives, I'll drop the x when I write the wave function so that things look more streamlined.

One more thing you need to know is that $\langle\Psi|\Psi\rangle = 1$ as long as the bra and ket have identical labels, which means they represent the same state. The math this is hiding is an integral of the matter density over all space. Because the matter density is normalized, its integral over all of space equals one.

Applying the bra, we find that:

$$\begin{aligned} \left\langle [\hat{x}, \hat{p_x}] \right\rangle &= i \cdot \hbar \cdot \langle \Psi | \Psi \rangle \\ &= i \cdot \hbar \end{aligned} \tag{D.18}$$

The bra moved past the i and $\hbar$ because they are constants. This is exactly what we expected to find. Therefore, we have derived the x-p_x uncertainty relation:

$$\sigma_x \cdot \sigma_{p_x} \geq \frac{\hbar}{2} \tag{D.19}$$

Having seen all that math work on examples that we can trust now gives you the knowledge to understand my attempts at deriving an energy-time uncertainty relation.

The three energy-time inequalities

The inequality with a zero

For the first inequality, we will use the Hamiltonian operator $\hat{H} = \hat{KE} + \hat{PE}$ and the operator t. As a reminder, $\hat{H}$ is the operator that asks the ket for the total energy. To make this easier to look at, I will set PE equal to zero. We would get the same answer in the end had I included it, which you can check for yourself to see how that comes about.

Our first step is to put the operators into the big inequality:

$$\sigma_E \cdot \sigma_t \geq \left| \frac{1}{2 \cdot i} \cdot \left\langle \left[\hat{KE}, t \right] \right\rangle \right| \tag{D.20}$$

We need to use the explicit form and not the abstract form for the kinetic energy operator:

$$\hat{KE} \;=\; -\frac{\hbar^2}{2m}\partial_x^2 \tag{D.21}$$

However, I know the constants won't matter in the end, so I'm going to drop them and pretend that:

$$\hat{KE} \text{ “=” } \partial_x^2 \tag{D.22}$$

The next step is to focus on the commutator applied to the ket:

$$\begin{aligned}\left[\partial_x^2, t\right] \cdot |\Psi(x)\rangle \;=& \\ \partial_x^2 \cdot t \cdot |\Psi(x)\rangle \;-\; t \cdot \partial_x^2 \cdot |\Psi(x)\rangle&\end{aligned} \tag{D.23}$$

The ∂_x^2 will ignore the t and move past it. You can see that this expression will be zero. Therefore, we have our answer:

$$\sigma_E \cdot \sigma_t \;\geq\; 0 \tag{D.24}$$

With the zero on the right side, this isn't the form of the energy-time uncertainty principle that any author puts forward for their claims, so we could move on to the next attempt. However, I want to show you something interesting first. To understand the inequality physically, we need to know what σ_t represents. Is it the distribution of time in some function? Perhaps we should try to calculate it. In the uncertainty chapter, I showed the equation for this. We can simply substitute in the t:

$$\sigma_t \;=\; \sqrt{\langle t^2\rangle \;-\; \langle t\rangle^2} \tag{D.25}$$

For those familiar with calculus, using the integral definition of expectation values (recall p. 261 and 256), it's easy to show that this standard deviation is zero because the integration variable is x or p_x and because the t and t^2 come outside the integrals. Therefore, for any wave function whatsoever, this uncertainty relation says that:

$$0 \geq 0 \tag{D.26}$$

Therefore, this is not going to reveal any secrets of physics to us.

An inequality with $\hbar/2$

In an act of desperation, someone might think to use the "energy" operator $\hat{E} = i \cdot \hbar \cdot \partial_t$ along with time t. We know the drill now. The first steps are to substitute these into the generalized uncertainty principle and evaluate the commutator acting on a ket. This time, I'll explicitly show the time dependence of the ket because the partial derivative will act on it:

$$\begin{aligned}\left[\hat{E}, t\right] \cdot |\Psi(x,t)\rangle \\ &= i \cdot \hbar \cdot \Big(\partial_t \cdot t \cdot |\Psi(x,t)\rangle - t \cdot \partial_t \cdot |\Psi(x,t)\rangle\Big) \\ &= -i \cdot \hbar \cdot \Big(t \cdot \partial_t \cdot |\Psi(x,t)\rangle - \partial_t \cdot t \cdot |\Psi(x,t)\rangle\Big)\end{aligned} \tag{D.27}$$

This is similar to the form you saw in the second x-p_x example above (Eq. D.14). The difference is that t is now in place of x. The subsequent math operations are going to be similar. We can skip to the expectation value:

$$\begin{aligned}\left\langle\left[\hat{E}, \hat{t}\right]\right\rangle &= i \cdot \hbar \cdot \langle\Psi|\Psi\rangle \\ &= i \cdot \hbar\end{aligned} \tag{D.28}$$

Putting this into the uncertainty inequality, we end up with:

$$\sigma_{\text{“}E\text{”}} \cdot \sigma_t \geq \frac{\hbar}{2} \tag{D.29}$$

This looks like what physicists want. But hold up, we have a big problem. We already know from the previous example that σ_t makes the left side zero. That leads to a mathematical absurdity of:

$$0 \geq \frac{\hbar}{2}$$

We will find the same mathematical absurdity if we consider a stationary state because it is an energy eigenstate. As such, it has no spread in the numerical values of total energy that are returned by the application of $\hat{E}$, which means $\sigma_{\text{“}E\text{”}} = 0$.

The fact that this inequality is violated tells us that a mistake was made somewhere. The problem with this and the previous example is that neither $\hat{E} = i \cdot \hbar \cdot \partial_t$ nor t meet the definition of a quantum-mechanical operator. The fact that there is no mathematical absurdity in the final result of the previous example (namely $0 \geq 0$) is a lucky, or perhaps unlucky, accident.

Quantum operators must map a ket back to Hilbert space, which the two aforementioned operators can't do. Why can't they do this mapping? I haven't mentioned before that there is no time dimension t in the functions of Hilbert space in quantum mechanics. An instance of a Hilbert space and its functions/vectors must be considered at some fixed point in time. Applying the time t to a ket produces a new function/vector object that is in a vector space larger than Hilbert space. Therefore, the application of t can't be a quantum operator according to the definition.

Interestingly, it's not the case that the application of $i \cdot \hbar \cdot \partial_t$ to a ket makes an unwanted mapping. The application of these symbols to Hilbert space is not mathematically defined because there is no time dimension in Hilbert space. We would be attempting to take the derivative of a dimension that isn't there.

Returning to the energy-time inequalities, as far as I can tell, we finally encounter a problem because the derivation of the generalized uncertainty principle uses what is called the *Cauchy-Schwarz inequality* and Hermitian operators. Both of these require the use of operators that follow the proper definition of quantum operators. Mathematical theorems are only valid if you propagate forward the assumptions and definitions that were used to derive them.

If this new information seems terribly convoluted to you, it was for me too. The average undergraduate- or graduate-level quantum mechanics class doesn't cover in detail what I have just told you about these two non-operators (I had to do much research to educate myself enough to explain this to you). On the surface, it appears that we are sometimes flipping a switch and activating the use of Hilbert space and other times deactivating it when we want to use time (e.g., with Schrödinger's equation or the derivation of the time-evolution operator). As I see it, in this case, the actual problem is probably that professors are doing a poor job of educating the next generation on Hilbert spaces as well as the nuances between the technical ideas of dimensions, variables, and parameters in the math we use.

A common and not so bad *E*-*t* inequality

I will end this appendix with the most legitimate form of an *E*-*t* inequality that you will find in some textbooks. I won't derive it. Instead, I will summarize the results using the notation from Griffiths' textbook.

The final form of the inequality looks different than what I have been showing you:

$$\Delta E \cdot \Delta t \;\geq\; \frac{\hbar}{2} \tag{D.30}$$

You can see that Griffiths has used uppercase deltas, which are usually the physicist's way of indicating some difference or change. He uses it because he doesn't want you to think that this inequality has the same legitimacy as the position-momentum inequality. You will understand what I mean soon enough.

The first symbol is easy to understand:

$$\Delta E \;=\; \sigma_E$$

This uncertainty in energy E is found by using the Hamiltonian operator $\hat{H}$. The next symbol is rather convoluted:

$$\Delta t = \frac{\sigma_Q}{\left| \mathrm{d}\langle \hat{Q} \rangle / \mathrm{d}t \right|} \tag{D.31}$$

The $\hat{Q}$ is Griffiths' way of writing a generic Hermitian operator. In the numerator is the standard deviation of that operator. In the denominator is the rate of change of the expectation value of that operator. Note that the symbol Δt in this case has little mathematical content. Despite the fact that Δ is typically used for difference or change and t usually represents time, most any other symbol could be used in place of Δt.

The vexing thing for anyone who wants to use this *E*-*t* inequality is that it involves any generic Hermitian operator. It's your choice of operator, and your choice will affect the value of Δt. So, the inequality is hardly fundamental. It seems that whoever first defined Δt obviously did it out of desperation to get the units of time.

A second vexing thing is that this inequality can't be used in any scenario where:

$$\frac{\mathrm{d}\langle \hat{Q} \rangle}{\mathrm{d}t} = 0$$

This condition says the expectation value of the operator doesn't change with time. It can't be used because a zero would be put in the denominator and thus make the Δt undefined. When one chooses a conserved quantity for Q, this particular *E*-*t* inequality only applies to states where some external potential is changing the quantity $\langle \hat{Q} \rangle$.

Finally, because the state evolves with time, ΔE and Δt are changing and need to be evaluated at a particular point in time, which really seems to be at odds with what the misleading label of Δt is trying to evoke: uncertainty in time.

Glossary of technical terms

Bullet points (•) in a glossary entry indicate contrasting uses of a term. If there are no bullet entries, all sentences work together to explain the term. The use of sans serif text like this inside a glossary entry indicates there is an entry for that term in the glossary. Even though there are glossary entries for them, I won't use the font on every instance of certain, more common terms, such as *measurement, particle, quantum object, wave function, quantum mechanics*, etc.

Absolute value (of a complex number) See magnitude.

Angular momentum In the classical physics view, it is a conserved quantity associated with rotational motion. It is a vector quantity whose magnitude and direction is calculated from the distribution of matter in space about an axis and each mass element's rotational velocity about that axis.

angular momentum, Orbital (OAM) The quantum analog to classical angular momentum. A vector quantity whose magnitude and direction are contingent on how the values of the spatial wave function are distributed in space. With spin angular momentum, they are together a conserved quantity.

angular momentum, Quantum (QAM) Encompasses both types of angular momentum: OAM and spin.

angular momentum, Spin aka spin An intrinsic type of quantum angular momentum. With orbital angular momentum, they are together a conserved quantity. Although it is okay to say spin-$\frac{1}{2}$ particles have a permanent angular momentum because the magnitude is always $\hbar/2$, it's not true in general because higher spins can have the magnitude of their spins change. The intrinsic quantity is actually the spin quantum number s.

Axiomatic system A consistent, logical system that starts with definitions, unstated notions, and postulates that lead to further truths, or theorems, within the system. Despite what some students are taught, quantum mechanics does not have a true axiomatic system.

Basis functions/kets/states/vectors/ In quantum mechanics, a special set of mathematical objects that are multiplied by quantum amplitudes and are then superposed to give an overall state. The set is characterized by having all the states/vectors/kets orthogonal to each other. See also completeness. A set is not unique; one must pick a representation. An example of a complete set of basis vectors for our 3D space, which is just one particular representation is: a vector of some length that points right, a vector of some length that points up, and a vector of some length that points forward; these would define a Cartesian (i.e., an x, y, z) coordinate system.

Boson An elementary or composite particle with an integer spin quantum number s. See also Fermion.

Bra $\langle\square|$ (The blank square is replaced by some label.) First, see ket. Has the same information as its ket and also represents a state. Used for various mathematical purposes in quantum mechanics, but only appears briefly in the appendix on matrix mechanics.

Classical physics The physics before quantum mechanics, which approximately describes the macroscopic world. It includes the various types of mechanics, fluid dynamics, electrodynamics, and (usually) relativity. The term and idea are abused by pop-sci physicists. As an example, consider this question "What does classical physics say happens as the electron goes through the slit?" That is an anachronistic question. Classical physics never knew the true nature of an electron. We must supply the information. If a physicist (erroneously) implies the electron is a tiny little ball, then the (erroneous) answer without the pageantry and misdirection is "Classical physics says this tiny little ball travels like a tiny little ball", which to me seems tautological and pointless.

Collapse of the wave function Hypothesized solution to the measurement problem. Many imagine the process to be discontinuous (i.e., the change is instantaneous); others imagine it to be rapid but not instantaneous. • Visualized as a spread-out wave function becoming localized into a randomly determined and smaller volume. • Other authors may mean collapse to a zero-dimensional point when they use the term.

Commutator or **commutator operator** It is simply a mathematical object made of two operators: $[\hat{A}, \hat{B}] = \hat{A} \cdot \hat{B} - \hat{A} \cdot \hat{B}$. In a proper textbook, a lot can be done with a commutator. However, don't let anyone tell you that a commutator has to do with measurements. The application of operators in succession does not correspond to a measurement nor to two measurements.

Completeness • When you have a set of all the basis functions that could make every possible superposition that would correspond to any physical wave function that the present quantum object could find itself in, the basis functions are said to be complete or to span the vector space. • Incompleteness/completeness of 3D/2D space in Euclidean geometry: a vector pointing to the right and a vector pointing up is not a complete basis set for 3D space because you can't scale them and make a superposition in order to specify every point in the third dimension. On the other hand, the two vectors are a complete basis for a particular 2D plane.

Configuration space A term imported from classical physics. Heisenberg's disparaging term for the multi-dimensional space a spatial wave function is described in at a particular point in time. He uses it because wave functions can have more variables than there are spatial dimensions. The spatial part of Hilbert space is made by assigning a position ket to every point in configuration space.

Copenhagen interpretation An opinion on reality that is partially characterized 1) by belief in a real macroscopic world that is discerned by our human senses and 2) by agnosticism or denial of the existence of quantum objects when measurements (by people) are not taking place.

Correspond I use the word *correspond* numerous times in this book. I do this because the mathematical objects written on the page are not the objects or quantities that exist in the real world. For example, I can't truly say the ket *is* the state because a ket is some symbols on a page. However, kets correspond to the states of the physical objects under consideration, so I often write as if the kets are the states. This should not be surprising; every field of science assumes their math on the page corresponds to the real world.

Decomposition Breaking down a ket/function/vector into a mathematically equivalent superposition of kets/functions/vectors. The decomposition does not need to form a basis set.

Density See the types of distributions.

Diffraction • Angular spreading of waves after encountering an obstacle. • In the most general and historic sense, it is any deviation of a wave from what it was doing before. Therefore, mirror reflection is a special case of diffraction.

dipole moment, Magnetic A quantity that tells us the strength, orientation, and shape of a magnetic field. The shape is that of a tiny, idealized bar magnet's field because *dipole* refers to two magnetic poles: north and south. Only occurs when a quantum object has both electric charge and non-zero angular momentum.

dipole moment, Electric (EDM) First, see magnetic dipole moment. An EDM tells us the strength and orientation of electric fields that are in the shape of a bar magnet's field. In quantum mechanics, the only known EDMs are produced by a wave function's charge density distribution (in molecules or solids), and that distribution must not have a particular type of symmetry (i.e., can't have inversion symmetry through center of mass).

Dirac notation Also known as braket or bra-ket notation. A notation in QM that writes math as abstractly as possible. Eventually, the abstraction is left behind, and either matrix mechanics or wave mechanics is chosen to make the final calculation. See also bra and ket.

Dispersion The spreading of a wave function or matter density in space. Matter wave functions that are in some non-stationary states will have their position-space wave functions naturally disperse. This is because each component of the infinite mathematical superposition of complex exponential waves that makes up the matter wave function has a different momentum and things with mass have different velocities when they have different momenta, so the components get out of phase with each other and the superposition must change as time goes on. Stationary states are special situations where the amplitudes of the components of the superposition are all just right to not have dispersion. Note that making an atom to emerge from a tiny slit causes it to spread out as it travels. Therefore, the electromagnetic interaction with the barrier containing the slit has caused the atom to be in a non-stationary state. *In vacuum*, a photon wave function's components all travel at the same speed. A physicist can create examples where there is (e.g., spherical waves) or is not (e.g., a wave packet traveling in a distinct direction) dispersion of photon wave functions.

Distribution In this book, it describes a physical quantity or probability spread out in position or momentum space. The word is rarely used in this book but is meant implicitly whenever I use *density*; *density* should be understood as a density distribution.

distribution, Charge-, Mass-, or Spin-density The distribution of those physical quantities in space. All are directly proportional to the matter-density distribution: $q \cdot |\Psi(x)|^2$, where q is the overall charge, mass, or spin of a quantum object with rest mass.

distribution, Electron-density I use this term in place of matter density when I talk about electrons specifically.

distribution, Matter-density The distribution throughout space of a quantum object that has rest mass. It is specified by $|\Psi(x)|^2$, which is the squared magnitude/absolute value of the wave function at every point in space.

distribution, Momentum-space probability-density In this book, I have named it the flow function.*
*Beware: no one else uses this term.

distribution, Photon-density The distribution throughout space of a photon. It is specified by $|\Psi(x)|^2$, which is the squared magnitude/absolute value of the photon wave function at every point in space. Rarely used in this book because photons are not the focus.

distribution, Probability-density Governs the random nature of measurement results. In some sources, it is also called a probability wave. In this book, it is mathematically equivalent to matter-density distribution or photon-density distribution.

Eigen- In German, *eigen* can mean *peculiar*, which should be read as a particular case of something.

Eigenvalue When an operator acts on a single ket and what is returned is a single value multiplying an unchanged ket, that single value is an eigenvalue.

Energy level Probably most often brings to mind the energies of stationary states. However, that is too restrictive; see Appendix A.

Entanglement Not a topic of focus in this book. Entanglement occurs when quantum objects can be treated as a single quantum object, and when measurements "break apart" the object, the measured properties of the separate objects are strongly correlated because of conservation laws.

Expectation value • Statisticians' word for *weighted average*. • For anti-realists, it is the weighted average of a large number of identical measurements. • In the reality interpretation of this book, it is way to calculate the physical quantity that a quantum object possesses.

Fermion An elementary or composite particle with a half-integer spin quantum number s. See also Boson.

Feynman's path-integral formulation Gives the same answers as Schrödinger's equation. Often harder to use.

Flow function My idiosyncratic name for probability density distribution in momentum space. It characterizes how much momentum each fraction of the matter density has.

Forbidden region or **classically forbidden region** or **energetically forbidden region** A misleading term that makes quantum mechanics seem more mysterious. It is said that when the KE of a classical object doesn't exceed some PE, the region where PE > KE is then inaccessible, or forbidden, to the particle. But then pop-sci physicists switch to the so-called quantum realm and imply that wave functions can, almost like magic, travel through forbidden regions. Wave functions can do this, but without the idea of the region being "forbidden", the physics would seem mundane.

Function • To a mathematician, it is a rule that uses one set of numbers to produce another set where there is only one output for every input (e.g., $y = x^2$ is a rule that takes as an input the real number line x and outputs the numbers y of a parabola.) • With regards to the wave function, I believe most physicists don't think of the wave function as a rule; instead, they think of it as the set of results that comes from the rule (i.e., the set of smoothly varying quantum amplitudes at every point in space). Beyond this, the meaning that each physicist sees in those numbers gives us their interpretations of quantum mechanics.

infinitesimal, Mathematician's An idea that every mathematician and many physicists agree is logically absurd and thus is not a mathematical or physical thing. It is defined as greater than zero and smaller than the smallest real value (this leads to a contradiction/absurdity).

Infinitesimal volume A region of space that is tiny compared to the situation/problem at hand. I also call this the physicist's infinitesimal. Often confused with the mathematician's infinitesimal and point particles.

infinity, Actual/Completed The infinity of set theory. Imagined instance of having more elements in a set than any finite set. It is a concept and not a number. It can't be used in place of numbers.

infinity, Potential The infinity of calculus or counting upward without end. It is a process.

Integral Informally, it is a sum of the values of a function taken at many points multiplied by the infinitesimal length/area/volume that each point represents. This description is technically wrong, but good enough for a conceptual understanding. The sum description is how computers numerically approximate integrals.

Interference The result of superposition. The result when complex numbers (aka quantum amplitudes) are added.

Ket $|\Box\rangle$ (The blank square is replaced by some label.) As a lone ket: an abstract mathematical representation of a state. Or: a part of a superposition. Also thought of as a vector in Hilbert space. See also bra and Dirac notation.

ket, Input My name for the kets that contribute to a product state. Often, each input represents a single particle. In the literature, input kets are often called *single-particle states/kets*, or whatever else an author chooses (e.g., *orbitals* is another common name).

Macroscopic world I primarily use this term because the anti-reality/Copenhagenist view splits our universe in two (the other half is the quantum realm). When I use it, I usually mean the apparently classical world that you see through your faulty eyes. (If you aren't aware of how faulty our perceptions are, consider that rainbows don't have cyan or magenta and that there are no wavelengths of light corresponding to those colors. Those colors are invented by your photon receptors and your brain.)

Magnitude of a complex number Imagine a grid, or plane, made by the real and imaginary number lines at 90 degrees to one another (google: *complex plane* or *Argand diagram*). A complex number $(a+i{\cdot}b)$ is a point (a, b) on this grid. The magnitude c of this complex number is the distance from the origin to that grid point and is found by the Pythagorean theorem: $c = \sqrt{a^2 + b^2}$.

mechanics, Matrix A mathematical method of predicting/calculating the states and quantities of quantum systems that uses vectors and matrix math. See also wave mechanics and Dirac notation.

mechanics, Wave A mathematical method of predicting/calculating the states and quantities of quantum systems which uses functions that are defined over all of position or momentum space. See also matrix mechanics and Dirac notation.

Measurement Undefined in axiomatic quantum mechanics. There is no universally accepted definition that covers all cases in the larger field of quantum mechanics.

Measurement-disturbance relation A formula for how much the state of a quantum object is disturbed by the measurement process (see Observer effect). In this book, I only consider Heisenberg's version, which is a very rough approximation to the situation he imagined. Heisenberg's measurement-disturbance relation is commonly and wrongly referred to as the uncertainty principle. Historically, the term *uncertainty principle* was first applied to Heisenberg's relation, but now refers more properly to a relation due to Kennard and Weyl.

Measurement problem The biggest mystery of quantum mechanics. We don't know exactly how state/wave functions get turned into values in our measurement apparati. See also collapse of the wave function.

Normalization The integral of the probability density over all space needs to be equal to 1 if a scientist wants to calculate probabilities. This is the mathematical process to do that. I don't show you how to do the calculation in this book. Often, my figures are not normalized to give more comprehensible and aesthetically pleasing numbers.

Numerical values* (that are returned after an operator is applied to a ket) When an operator acts on a ket or kets, there will be a set of numerical values returned in front of the ket or kets, whether there is one value, many discrete values, or many values in a continuum. If the operator corresponds to an observable, these values are supposed to be the possible values one will measure of the observable quantity.

*Beware: no one else uses this term. There is no standard term that reflected perfectly what I wanted to say. The closest you will see in other places is something like *sets of eigenvalues*. *Spectrum of an operator* would be a great term, but it means something else very closely related.

Observable An ill-considered name for a technical term. In normal speech, it is an adjective. In quantum mechanics, it is a noun and means *physical quantity* or an *observable quantity.*

Observer effect To observe the state of anything, classical or quantum, an interaction with that thing (e.g., an atom) or something emitted from that thing (e.g., a photon) must occur. The interaction or the emission will change the state of the thing of interest. In quantum mechanics, the effect is not always small and often needs to be considered.

Old Quantum Theory A title given to the attempts at modeling quantum phenomena before the mathematical formalisms of matrix mechanics and wave mechanics were discovered.

Operator • A mathematical object that acts on a ket to return a value beside the ket, change the ket, or both. In the most abstract form, they have hats (circumflexes) on top of them. For example, the abstract kinetic energy operator is $\hat{KE}$. • A more precise definition of an operator as used in quantum mechanics will talk about mapping state vectors from a Hilbert space back to the same Hilbert space. In this book, this definition can be ignored until the final appendix.

Orbital A position-space wave function of a stationary state of one electron of an atom or molecule. With hydrogen, it is specified by the quantum numbers n, l, and m_l.

Orthogonal A mathematical term that roughly means vectors are 90° from each other. Orthogonality means that there is no amount of east in a direction due north (directions would be confusing otherwise). It is an important condition that ensures each member of a basis set isn't a mixture, or superposition, of other members.

Particle It is a very ambiguous and abused term. • Many say, in classical physics, it is a geometric-point object. Others think of it as a rigid, corpuscular object (has a volume). • In quantum mechanics books, rarely defined outright. Good luck figuring out what each author means. The word may have one of the meanings in the previous bullet. • In this book, a particle in QM is synonymous with my term quantum object.

particle, Composite A quantum object (e.g., molecules, atoms, ions, protons, neutrons, nuclei, etc.) that is made up of other quantum objects. Thought of as a "microscopic" object, so metals aren't called particles even though they are composite quantum objects.

particle, Elementary A quantum object (e.g., electrons, photons, quarks, neutrinos, etc.) that is not made up of other quantum objects.

Pauli exclusion principle A rule of the universe that says two fermions cannot both be in the exact same state. In an atom, no two electrons may have the same four quantum numbers n, l, m_l, and m_s. This principle is a consequence of the more general symmetrization postulate.

Phase and **Phase angle** Not the same thing, but are directly related. Physicists will use them almost interchangeably. Phase is the most relevant quantity in calculations, but it's easier to think in terms of phase angles. Specifying a magnitude and a phase fully describes a complex number. (See Magnitude of a complex number.) Phase angle is the angle, in the complex plane, from the positive real number line to the hypotenuse measured counterclockwise. The phase of a complex number is the complex number divided by the magnitude of the number.

Photon Not a topic of focus in this book. Requires relativistic quantum mechanics to handle. In introductory quantum mechanics, treated semiclassically. Often called the *quantum of light.* I believe there are two uses of the word *photon* that lead to confusion: • In quantum field theory, a photon is said to be an elementary excitation of the field and thus is not localized. • The traditional usage sees it as a wave packet, which QFT says is made up of a continuum of elementary excitations.

Plank's constant / reduced Plank's constant ($h/\hbar$) Has units of angular momentum (alternatively units of action, but I don't get into that). When you see this, you are guaranteed to be working with quantum things. Many theorists drop it from their equations, so you won't always see it when it's really there.

Point in space When I use this term, it is often a shorthand for something like "an infinitesimal volume centered on a point in space" when an integral is involved.

Postulate A foundational statement of an axiomatic system that is taken to be unerringly true.

Quantum (Plural: quanta) Invokes the idea of a discrete thing or property. For example, electric charge and angular momentum each come in multiples of a fundamental amount; there isn't a continuum of quantities possible. The stationary states of hydrogen are discrete; transitions between the states are informally known as quantum jumps.

Quantum amplitude In many other books, called probability amplitude, but I think that confuses novices. In wave mechanics, it is the complex-valued amplitude of the wave function. It is the number I write in a different color underneath kets in the *Spin* chapter. The squared magnitude of the quantum amplitudes gives the probability or probability density.

Quantum jump See transition.

Quantum mechanics An ambiguous term. Some people make a distinction between the terms *quantum mechanics* and *quantum physics.* In practice, most physicists don't. In this book, the quantum mechanics is non-relativistic. • In the early literature, sometimes it was a synonym for matrix mechanics. Nowadays, it could mean: • the physics of quantum objects (which is independent of our imperfect understanding of said physics), • the current concepts and mathematical formalism used to describe our understanding of the physics of quantum objects, or • the aforementioned concepts and mathematical formalism plus someone's physical interpretation of the formalism.

Quantum number An index used to label the state of a quantum object so that, for example, you don't need to write out the full position-space wave function when communicating with a colleague. Some quantum numbers are conserved quantities or are closely related to conserved quantities (e.g., angular momentum, electric charge, etc.)

Quantum object A thing that carries mass, charge, spin, etc., and has a total wave function described by a position-space wave function (or equivalently a momentum-space wave function) and a spin state. (Mass and charge could be zero, which would indicate a photon [or a gluon].) The singular thing ("it") that is verbally pointed to when someone erroneously says, "It is sometimes a particle and sometimes a wave." Every object in the universe, macro or micro, matter or photon, is a quantum object.

Quantum realm/world • I use this term because the anti-reality/Copenhagenist view splits our universe in two via Heisenberg's *cut/schnitt.* (The "shifty split" that J.S. Bell [and other realists] detest). It refers to things on the micro/nano/attoscopic scale. It contrasts with the macroscopic world that we perceive with our faulty senses. • I don't think there is a *schnitt.* To most realists, everything is quantum.

Quantum system One or many quantum objects that are under consideration. Together, those objects are another more complex quantum object. Could include the objects plus an externally applied potential, so the term has broad meaning.

Quantum tunneling • A dynamic process where a position-space wave function flows from one region to another under the condition that the total KE of the wave function is less than the PE function of the region it flows through. This region it flows through is called a *potential barrier* or a (classically) forbidden region. • Many authors equivocate by using the above definition as well as one that invokes a collapse of the wave function so that a wave function appears mostly on the opposite side of the barrier.

Reality interpretation • Could refer to any one of many diverse interpretations that don't deny reality in the quantum realm. • In this book, it refers to my interpretation, where matter and photon density distributions exist in the quantum realm.

Representation • Most often, this word has its common meaning. • It is the specific amplitudes and kets used to describe a state. A particular quantum state can be described mathematically in more than one way. A position-space wave function of a quantum object contains the same information as the object's momentum-space wave function; each one is a mathematical representation of the spatial part of the state. A quantum object in a spin-up state is described with single ket in the z-axis representation (a "lone ket"). The same spin state can be described as a superposition of spin-left and spin-right kets, which are in the x-axis representation. An operator's representation must match the representation of the kets it acts on.

rotation, Active Corresponds to a physical rotation of a state.

rotation, Passive A rotation of a coordinate system so that the state written on the page appears in a different form.

Schrödinger's cat Every object you see is already a superposition of innumerable electron and nuclear wave functions. Schrodinger's cat is not in a superposition of alive and dead. Meow.

Schrödinger's equation An equation that describes how wave functions change as time goes on when uninterrupted by collapse of the wave function. Gives the same answers as Feynman's path integral formulation.

space, Abstract/Hilbert A mathematical construct. Not easy for novices to understand. It is defined by math rules and the basis kets/vectors that partially define position-space wave functions, momentum-space wave functions, or spin states. By assigning a complex-valued number to each ket allows us to say "the wave function (or spin state) exists in a Hilbert space." The Euclidean space of three dimensions of high school geometry happens to be an example of a Hilbert space but is not general enough for quantum mechanics; however, knowing that Euclidean space is an example of a Hilbert space will allow you some intuition into quantum-mechanical Hilbert spaces.

space, Momentum A mathematical space that uses momentum coordinates rather than position coordinates. See also momentum-space wave function and flow function.

space, Physical • The three spatial dimensions objects exist in. • While technically not true, I also say the space where we do the math for some physical quantities, such as angular-momentum and light-polarization vectors, are physical spaces.

space, Position Could refer to: • The three spatial dimensions you and quantum objects exist in. • Or, a higher- or lower-dimensional version of the previous bullet point.

space, Spin A mathematical space that allows us to see the spin state decomposition in a particular basis.

Spin-$\frac{1}{2}$ polarization principle A name I have adopted (and updated) from Susskind and Friedman. The orientation, or polarization, of a spin-$\frac{1}{2}$ particle's angular-momentum vector is not uncertain. It always points in some direction in three-dimensional space.

Standard deviation A particular measure of how spread out a set of numbers is. When applied to matter density or a flow function, it is a length, in their respective spaces, measured from their mean value.

state, Bound A state of a composite object where the component objects of the system have less KE than the PE needed to overcome the force binding the composite object together. As an alternative description, let's consider a hydrogen atom, which is composed of an electron and a proton. The electron is bound to the proton when the most extreme distance the center of mass of the electronic wave function has from the proton is a particular finite value; the electronic wave function could be in some pathological, non-stationary state but its center of mass does not "escape" and eventually "turns back" toward the proton. See also unbound state.

state, Excited A stationary or non-stationary state with greater energy than the ground state.

state, Ground The state of lowest energy a quantum object can have. It is the most natural condition because the object will eventually end up in this state from a higher *excited state* if left alone.

state, Non-stationary See also stationary state. A state with a matter density or spin expectation value that changes with time. Examples are unbound states and superpositions of stationary states of differing energies.

state, Product The state that is represented by any number of input kets connected by tensor product symbols.

state, Spin $|\chi\rangle$ (The χ is replaced with more specific symbols when not being general.) Colloquially, a complicated way to say which way the spin angular momentum points. It can be mathematically represented by a column vector with quantum amplitudes in its entries.

state, Stationary A state with a matter density and spin expectation values that do not change with time. In the hydrogen atom, these states are the ones that are labeled by a single set of quantum numbers. Contrast with non-stationary state.

state, Total A complete description of 1) which quantum object is under consideration (if the specificity is required) and 2) what it is doing. The "which" part is usually handled by stating the type of particle (e.g., an electron) or by giving the invariant properties of charge, mass, and spin quantum number. The "what" part is covered by the spin state plus either the position-space wave function or the momentum-space wave function.

state, Unbound/Scattered Consider a quantum object that is composed of two other objects (e.g., a hydrogen atom). An unbound state of such a composite object is one in which you can always find a later time t_2 where the distance between the centers of mass of the component objects is larger than at some other earlier time t_1. This is because the system of two objects has more KE than the PE needed to overcome the force binding the two together. Practically speaking, eventually the two component objects are so far apart that one of them runs into the rest of the universe, and the original idealization of a lone hydrogen atom breaks down. See also bound state.

Superposition A special case of addition. The addition of many states/vectors/functions to give an overall state/vector/function. When applied to wave mechanics, it is the addition of waves (results in constructive and destructive interference). When applied to the matrix mechanics, the addition of row or column vectors. See also decomposition.

Tensor product $\otimes$ In mathematics, there are many types of tensor products. When physicists say *tensor product*, they mean a Kronecker product. It joins kets more intimately than the plus sign of a superposition. A product state created by tensor products of spatial wave functions has a number of spatial variables equal to the sum of the spatial variables from all the input kets. This means that tensor products create a multi-dimensional wave function, or configuration space.

Transition (quantum jump) A change of state to a higher or lower energy. Commonly associated with going from one atomic energy level to a higher or lower energy level (an atomic transition). When *quantum jump* is used the idea is that the transition is discontinuous. My guess is that the colloquial *quantum jump* was made to be a play on the Latin words *natura non facit saltus* (nature does not make jumps) said by Leibniz in regard to calculus; the idea being that nature (as it was known by Leibniz) smoothly changes because everything is continuous and not abrupt.

Uncertainty principle A mathematical inequality that compares the spread (standard deviation) of quantities of a state to some limiting value. Commonly refers specifically to the x-p_x uncertainty relation. Often confused with the observer effect and Heisenberg's measurement-disturbance relation.

uncertainty principle, Generalized An inequality that quantum operators can be inserted into to make a new uncertainty principle/inequality.

Volume element, synonym for infinitesimal volume.

Wave function Represented with Ψ or $|\Psi\rangle$. At times, an ambiguous term. Could refer to either the total wave function or the position-space wave function. • To anti-realists, it is only a collection of complex numbers used to predict probabilities and outcomes of experiments. • To realists, the math on the page corresponds to an element of reality that is likely near the fundamental level of the universe.

wave function, Total The spin state plus the position-space wave function $|\Psi(x,t)\rangle \otimes |\chi\rangle$ (or: spin state plus the momentum-space wave function $|\Psi(p_x,t)\rangle \otimes |\chi\rangle$). Note that the time variable t doesn't "belong" to the spatial part. You are seeing one way of grouping things.

wave function, Macroscopic The word *macroscopic* denotes a size comparable to the human scale of centimeters. However, when physicists use the word, they also unconsciously mean wave functions with substantial mass, like cats and baseballs, because the double- and single-slit experiments already deal with macroscopic, but lightweight, wave functions.

wave function, Momentum-space $|\Psi(p_x,t)\rangle$ Exists in a mathematical momentum space. This first appears in the chapter on the uncertainty principle. Can be found via a mathematical transform of the position-space wave function and contains the same information but emphasizes the momentum information.

wave function, Position-space $|\Psi(x,t)\rangle$ A wave function with a time variable plus position variables that correspond to our three-dimensional space. There is a proliferation of position variables as the number of quantum objects in the system increases (see configuration space).

wave function, Spatial • I introduced it to you with terms and pictures that described a position-space wave function. However, if you read the book again, the term could also be said to encompass the momentum-space wave function. A quantum object's position- and momentum-space wave functions both contain the same information, which is the distribution of matter in physical space and the amount of momentum that various fractions of the matter distribution have.

Wave packet A function that is localized in some region of space. It does not repeat forever, like a sine or cosine function. But it is composed of an infinitude of complex-valued exponential functions that do repeat forever; interference causes the forever repetitions to cancel out.

Wave-particle duality An abused term. It's better to learn about wave functions, the measurement problem, and the collapse of the wave function than to look for profoundness in the works of people discussing this duality. If nothing else, replace this term in your mind with *wave-particle simultaneity*.

Bibliography

While writing this book, my fact-checking benefited to greater and lesser degrees from some general sources: a dozen quantum textbooks, the Stanford Encyclopedia of Philosophy, various Stack Exchanges (physics, chemistry, and math), and Physics Forums. I was glad to find during the editing stage that a fair bit of what I say is in agreement with a website of frequently asked question on theoretical physics by Prof. Arnold Neumaier (who is a prominent contributor to Physics Stack Exchange and Physics Forums). Below, I include references to quotes and some more specific things I mention.

General sources

G. Auletta, M. Fortunato, and G. Parisi, *Quantum Mechanics*, Cambridge University Press, (2009).

L.E. Balantine, *Quantum Mechanics: A Modern Development*, World Scientific, (2000).

D. Bohm, *Quantum Theory*, Dover Publications, (2015).

C.E. Burkhardt and J.J. Levanthal, *Foundations of Quantum Physics*, Springer, (2008).

P.A.M. Dirac, *The Principles of Quantum Mechanics*, Snowball Publishing, (2019).

R. Eisberg and R. Resnick, *Quantum Physics of Atoms, Molecules, Solids, Nuclei, and Particles*, 2nd ed., John Wiley & Sons, (1985).

R. Feynman, R. Leighton, and M. Sands, *The Feynman Lectures on Physics*, (2013), <https://www.feynmanlectures.caltech.edu/>.

D.J. Griffiths and D.F. Schroeter, *Introduction to Quantum Mechanics*, 3rd ed., Cambridge University Press, (2019).

• *Note that I will refer to this book as "Griffiths" because the first and second editions were written by him alone.*

D.H. McIntyre, *Quantum Mechanics: A Paradigms Approach*, Pearson, (2012).

A. Peres, *Quantum Theory: Concepts and Methods*, Kluwer Academic Publishers, (2002).

J.J. Sakurai and J. Napolitano, *Modern Quantum Mechanics*, Cambridge University Press, (2020).

R. Shankar, *Principles of Quantum Mechanics*, 2nd ed., Kluwer Academic/Plenum Publishers, (1994).

D.V. Schroeder, *Notes on Quantum Mechanics*, draft copy, (2022), <https://physics.weber.edu/schroeder/quantum/QuantumBook.pdf>.

J.S. Townsend, *A Modern Approach to Quantum Mechanics*, 2nd ed., University Science Books, (2012).

N. Zetelli, *Quantum Mechanics: Concepts and Applications*, 2nd ed., John Wiley & Sons, (2009).

Stanford Encyclopedia of Philosophy, <https://plato.stanford.edu/index.html>.

Physics Stack Exchange, <https://physics.stackexchange.com>.

Chemistry Stack Exchange, <https://chemistry.stackexchange.com>.

Math Stack Exchange, <https://math.stackexchange.com>.

Physics Forums, <https://www.physicsforums.com>.

A. Neumaier, *A theoretical physics FAQ*, <https://arnold-neumaier.at/physfaq/physics-faq.html>.

Preface

`"Science is"`
R. Feynman, *The Pleasure of Finding Things Out: The Best Short Works of Richard P. Feynman*, Basic Books, (2001), p. 187.

`"When people thought"`
I. Asimov, *The Relativity of Wrong*, The Skeptical Inquirer 14.1 (1989): 35-44.

Ch. 1 - On interpretations

"Judging by its"
A. Neumaier, *Coherent Quantum Physics: a Reinterpretation of the Tradition*, de Gruyter, (2019), §1.3.

"shut up and calculate"
N.D. Mermin, *Could Feynman Have Said This?*, Physics Today, (2004) 57 (5), <https://pubs.aip.org/physicstoday/article/57/5/10/412592/Could-Feynman-Have-Said-This>.

"The most embarrassing graph"
S. Carrol, (15 Nov 2024), <https://www.preposterousuniverse.com/blog/2013/01/17/the-most-embarrassing-graph-in-modern-physics/>.

The graph Carrol mentions and the percentages I mention:
M. Schlosshauer, J. Kofler, and A. Zeilinger, *A Nnapshot of Foundational Attitudes Toward Quantum Mechanics*, <https://arxiv.org/pdf/1301.1069>.

"Whole Sort of General Mish Mash"
D. Adams, *Mostly Harmless*, Harmony Books, (1992), p. 26.

Historians and philosophers:
J, Faye, *Copenhagen Interpretation of Quantum Mechanics*, The Stanford Encyclopedia of Philosophy (Summer 2024 Edition), Edward N. Zalta & Uri Nodelman (eds.), <https://plato.stanford.edu/archives/sum2024/entries/qm-copenhagen/>.

Solidified Copenhagen Interpretation:
W. Heisenberg in *Niels Bohr and the Development of Physics*, Ed. W. Pauli, Pergamon Press, (1962), 12-29.

Heisenberg's ``public-relations campaign" and ``invention":
A. Becker, *What is Real? The Unfinished Quest for the Meaning of Quantum Physics*, New York: Basic Books, (2018), p. 207.

• *I suggest reading this book. It is more thoroughly researched and honest than many books that try to reconstruct the history of quantum mechanics.*

Heisenberg as a high-schooler (Ch.1) and on space and time not existing (p. 80):
W. Heisenberg, *Physics and Beyond: Encounters and Conversations*, Harper & Row, (1972).

"The very fact"
W. Heisenberg, Nobel Lecture, Nobel Prize Outreach AB 2024, (26 Nov 2024), <https://www.nobelprize.org/prizes/physics/1932/heisenberg/lecture/>.

Heisenberg's Chicago lectures:
W. Heisenberg, *The Physical Principles of the Quantum Theory*, Courier Corporation, (2013), see mainly Ch. 1, §1 and §3 of Ch. 3, and §8 of the Appendix.

This book was also consulted:
W. Heisenberg, *Physics and Philosophy: The Revolution in Modern Science*, Harper & Row, (1958).

This book was also consulted:
W. Heisenberg, *The Physicist's Conception of Nature*, Hutchinson & Co., (1958).

Schrödinger's review article:
E. Schrödinger *An Undulatory Theory of the Mechanics of Atoms and Molecules*, Physical Review 28.6 (1926): 1049.

Ch. 2 - A first look at quantum angular momentum

"two-valuedness not describable classically"
W. Pauli, Nobel Lecture, Nobel Prize Outreach AB 2024, (26 Nov 2024), <https://www.nobelprize.org/prizes/physics/1945/pauli/lecture/>.

Neutron charge distribution:
G.A. Miller, *Charge Density of the Neutron*, <https://arxiv.org/abs/0705.2409>.

Data for the sodium spectrum came from:
A. Kramida, Y. Ralchenko, J. Reader and NIST ASD Team (2024), *NIST Atomic Spectra Database* (version 5.12), [Online]. Available: <https://physics.nist.gov/asd>. National Institute of Standards and Technology, Gaithersburg, MD.

Ch. 3 - Misconceptions about the electron in particular

"a particle is like a BB"
D. Lincoln, *Quantum mechanics and the double slit experiment*, Fermilab channel, (4 Aug 2021), <https://youtu.be/nmxwVU88Bd8>, 3:03 min:sec.

"as long as"
D.J. Griffiths and D.F. Schroeter, *Introduction to Quantum Mechanics*, 3rd ed., Cambridge University Press, (2019), p. 197.

"you can picture"
B. Greene, *The Fabric of the Cosmos: Quantum Leap*, (2011), <https://www.pbs.org/video/nova-the-fabric-of-the-cosmos-quantum-leap/>, 14:35 min:sec.

H. Dehmelt, *A Single Atomic Particle Forever Floating at Rest in Free Space: New Value for Electron Radius*, Physica Scripta, T22 (1988): 102.

H.G. Dehmelt, Nobel Lecture, Nobel Prize Outreach AB 2024, (3 Dec 2024), <https://www.nobelprize.org/prizes/physics/1989/dehmelt/lecture/>.

IUPAC Gold book, <https://goldbook.iupac.org/>.

"this is not true"
G. Scharf, *Finite Quantum Electrodynamics: The Casual Approach*, Dover Publications, (2014), p. 236.

"yet so erroneously"
O. Steinmann, *Perturbative Quantum Electrodynamics and Axiomatic Field Theory*, Springer, (2000), p. 2.

Various quotes on point particles:
B. Cox, and J. Forshaw, *The Quantum Universe: (And Why Anything That Can Happen, Does)*, Da Capo Press, Incorporated, (2013), p. 116.

"straightjacket of a particle picture"
A. Neumaier, *Coherent Quantum Physics: a Reinterpretation of the Tradition*, de Gruyter, (2019), §1.3.

"Conventionally, people think"
Ghosh, Pallab, *Electron particle's shape revealed*, BBC News, (25 May 2011), <https://www.bbc.com/news/science-environment-13545453>.

The science article that is the subject of the news on egg-shaped electrons:
J. Hudson, D. Kara, I. Smallman, et al., *Improved measurement of the shape of the electron*, Nature 473, 493–496 (2011). <https://doi.org/10.1038/nature10104>. Open-access version: <https://core.ac.uk/reader/76996880>.

Permission to fudge Dirac notation is given by Gieres and Dirac:
F. Gieres. *Mathematical surprises and Dirac's formalism in quantum mechanics*, <https://arxiv.org/pdf/quant-ph/9907069>, p. 34.

Dirac deltas:
D.J. Griffiths and D.F. Schroeter, *Introduction to Quantum Mechanics*, 3rd ed., Cambridge University Press, (2019), §2.5.2.

C.E. Burkhardt and J.J. Levanthal, *Foundations of Quantum Physics*, Springer, (2008), §4.3.2.

R. Shankar, *Principles of Quantum Mechanics*, 2nd ed., Kluwer Academic/Plenum Publishers, (1994), §1.10.

Ch. 4 - Waves and superposition

"easiest person to fool"
R.P. Feynman. *Surely You're Joking, Mr. Feynman! Adventures of a Curious Character*, W.W. Norton & Co., (1997), p. 343.

Ch. 5 - The cat, the keg, and the cut

A. Einstein, B. Podolsky, N. Rosen, *Can Quantum-Mechanical Description of Physical Reality be Considered Complete?*, Physical Review 47.10 (1935): 777.

Schrödinger's cat is on p. 328 of
J.D. Trimmer. *The Present Situation in Quantum Mechanics: A Translation of Schrödinger's 'Cat Paradox' Paper*, Proceedings of the American Philosophical Society, Vol. 124, No. 5 (Oct. 10, 1980), pp. 323-338.

The tunneling of Polonium-212:
C.R. Nave. Hyperphysics, (25 Nov 2024), *Modeling Polonium-212 Alpha Half-life*, <http://hyperphysics.phy-astr.gsu.edu/hbase/Nuclear/alptun2.html>.

"exploded and not-exploded"
A. Fine. *The Einstein-Podolsky-Rosen Argument in Quantum Theory*, The Stanford Encyclopedia of Philosophy (Summer 2020 Edition), Edward N. Zalta (ed.), <https://plato.stanford.edu/archives/sum2020/entries/qt-epr/>.

"shifty split"
J. Bell, *Against 'measurement'*, Physics World 3.8 (1990): 33.

- *This is a good article that should be read in full.*

E. Crull & G. Bacciagaluppi, *Translation of: W. Heisenberg, "Ist eine deterministische Ergänzung der Quantenmechanik möglich?"*, (2 May 2011), <https://philpapers.org/rec/CRUTOW>

Baseball in a superposition:
D.J. Griffiths and D.F. Schroeter, *Introduction to Quantum Mechanics*, 3rd ed., Cambridge University Press, (2019), p. 492.

Ch. 6 - Mathematics vs. Measurement

"non-working cat"
D. Adams, Speech given to Digital Biota 2 in Cambridge U.K., (September 1998), <https://digitalspace.com/biota.org/people/douglasadams/>.

Ch. 7 - Wave-particle duality and the double-slit experiment

"names don't constitute knowledge"
R. Feynman. Originally on Yorkshire Television, UK: *Take the World from Another Point of View*, (1973), Transcript at <https://calteches.library.caltech.edu/35/2/PointofView.htm.

Dismissing wave-particle duality:
D.J. Griffiths and D.F. Schroeter, *Introduction to Quantum Mechanics*, 3rd ed., Cambridge University Press, (2019), p. 463.

Historical remark about wave-particle duality:
J.J. Sakurai, *Advanced Quantum Mechanics*, Addison-Wesley Publishing, (1967), p. 37.

No mention of wave-particle duality
J.J. Sakurai and J. Napolitano, *Modern Quantum Mechanics*, Cambridge University Press, (2020).

Feynman's double-slit explanation:
R. Feynman, R. Leighton, and M. Sands, *The Feynman Lectures on Physics*, Vol 1. Lecture 37, (2013), <https://www.feynmanlectures.caltech.edu/I_37.html>.

I checked the preparation Feynman's students had by checking a particular lecture recording, the list of lectures, and the course handouts.
R. Feynman, R. Leighton, and M. Sands, *The Feynman Lectures on Physics*, (2013), <https://www.feynmanlectures.caltech.edu/flptapes.html> and <https://www.feynmanlectures.caltech.edu/flphandouts.html>.

J. Al-Khalili, *Double Slit Experiment explained! by Jim Al-Khalili*, The Royal Institution, (1 Feb 2013). <https://www.youtube.com/watch?v=A9tKncAdlHQ>.

Images in Royal Institution presentation are from Ch. 1 in:
J. Al-Khalili, *Quantum: A Guide for the Perplexed*, Weidenfeld & Nicolson, (2003).

The section title ``speakable and unspeakable" is from a book title:
J.S. Bell, *Speakable and Unspeakable in Quantum Mechanics: Collected Papers on Quantum Philosophy*, Cambridge University Press, (2004).

O. Carnal and J. Mlynek, *Young's double-slit experiment with atoms: A simple atom interferometer*, Physical Review Letters 66.21 (1991): 2689.

Ch. 8 - The mysterious hieroglyphics of Schrödinger's equation —As an energy balance

"great triumphs of physics"
R. Feynman, R. Leighton, and M. Sands, *The Feynman Lectures on Physics*, Vol 3. Lecture 19, (2013), <https://www.feynmanlectures.caltech.edu/III_19.html>.

"exact and rather tiresome proof"
E. Schrödinger, *An Undulatory Theory of the Mechanics of Atoms and Molecules*, Physical Review 28.6 (1926): 1049.

Preferred-basis problem:
B. Galvan, *On the preferred-basis problem and its possible solutions*, <https://arxiv.org/abs/1008.3708>.

Ch. 9 - The mysterious hieroglyphics of Schrödinger's equation —As a wave equation

"right track"
J.J. Sakurai and J. Napolitano, *Modern Quantum Mechanics*, Cambridge University Press, (2020), p. 457.

Feynman's textbook on path integrals:
R. Feynman and A.R. Hibbs, *Quantum Mechanics and Path Integrals*, emended ed., Dover Publications, (2005).

Ch. 10 - Gallery of atomic orbitals

E.W. Dijkstra, *A first exploration of effective reasoning*, E.W. Dijkstra Archive, Center for American History, University of Texas at Austin, <https://www.cs.utexas.edu/~EWD/transcriptions/EWD12xx/EWD1239.html>.

Ch. 11 - Spins, superpositions, and abstract spaces

"in a laboratory"
A. Peres, *Quantum Theory: Concepts and Methods*, Kluwer Academic Publishers, (2002), p. xi.

Confused quantum amplitudes and eigenvalues in the second edition of his text:
N. Zetelli, *Quantum Mechanics: Concepts and Applications*, 2nd ed., John Wiley & Sons, (2009), p. 166 - 177.

How to find rotation operators:
T.L, Curtright, D.B. Fairlie, and C.K. Zachos, *A compact formula for rotations as spin matrix*

polynomials, SIGMA: Symmetry, Integrability and Geometry: Methods and Applications 10 (2014): 084. Open-access version:<https://arxiv.org/abs/1402.3541>.

"fuzzy and uncertain"
B. Greene, *The Fabric of the Cosmos: Quantum Leap*, (2011), <https://www.pbs.org/video/nova-the-fabric-of-the-cosmos-quantum-leap/>, 27:25 min:sec.

Is Greene an anti-realist or realist?
B.R. Greene, *The Fabric of the Cosmos: Space, Time, and the Texture of Reality*, Alfred A. Knopf, (2004).

Spin-polarization principle:
L. Susskind and A. Friedman, *Quantum Mechanics: The Theoretical Minimum*, Basic Books, (2015), §3.8.

Neutrons and 720 degrees:
H. Rauch, A. Zeilinger, G. Badurek, A. Wilfing, W. Bauspiess, and U. Bonse, *Verification of coherent spinor rotation of fermions*, Physics Letters A 54.6 (1975): 425-427.

Ch. 12 - Spring cleaning

"more fundamental than measurement?" J.S. Bell,*Speakable and Unspeakable in Quantum Mechanics*, Cambridge University Press, (1993), p. 117.

"a set of function is **complete**"
D.J. Griffiths and D.F. Schroeter, *Introduction to Quantum Mechanics*, 3rd ed., Cambridge University Press, (2019), p. 93.

Ch. 13 - The uncertainty principle is not what you think it is

"You keep using that word"
"Inigo Montoya" in *The Princess Bride* (film), Act III Communications and Buttercup Films, (1987).

Kennard and Weyl contributions explained in:
R.F. Werner and T. Farrelly, *Uncertainty from Heisenberg to Today*, <https://arxiv.org/abs/1904.06139>.

"isn't about measurements"
S. Carrol, *Quanta and Fields: The Biggest Ideas in the Universe*, Penguin Random House, (2024), p. 55.

English translation of Schrödinger's uncertainty relation paper by A. Angelow and M.-C. Batoni:
E. Schrödinger, *About Heisenberg Uncertainty Relation*, <https://arxiv.org/abs/quant-ph/9903100>.

"average error"
W. Heisenberg, *The actual content of quantum theoretical kinematics and mechanics* found at <https://ntrs.nasa.gov/citations/19840008978> and translated from *Über den anschaulichen Inhalt der quantentheoretischen Kinematik und Mechanik.* Zeitschrift für Physik, 43 (3): 172–198.

"according to the uncertainty principle"
R. Feynman, R. Leighton, and M. Sands, *The Feynman Lectures on Physics*, Vol 2. Lecture 1, (2013), <https://www.feynmanlectures.caltech.edu/II_01.html>.

"then necessarily"
G. Auletta, M. Fortunato, and G. Parisi, *Quantum Mechanics*, Cambridge University Press, (2009), p. 84.

"Moreover, the more precisely"
B. Greene, *The Elegant Universe: Superstrings, Hidden Dimensions, and the Quest for the Ultimate Theory*, Random House, (2011), p. 114.

"pin down" / "smaller the uncertainty"
B. Cox and J. Forshaw, *The Quantum Universe: (And Why Anything That Can Happen, Does)*, Da Capo Press, Inc., (2013), p. 70 and p. 85.

"inescapable trade-off" / "more precisely"
D.J. Griffiths and D.F. Schroeter, *Introduction to Quantum Mechanics*, 3rd ed., Cambridge University Press, (2019), p. 19.

"more accurately you know"
S. Holzner, *Quantum Physics for Dummies*, Wiley, (2013), p. 21.

Ch. 14 - The postulates of quantum mechanics

"the Code is more what you'd call guidelines"
"Captain Barbossa" in *Pirates of the Caribbean: The Curse of the Black Pearl* (film), Walt Disney Pictures and Jerry Bruckheimer Films, (2003).

Euclid's axiomatic system can be found at:
Website by D.E. Joyce, copyright 1998, <http://aleph0.clarku.edu/~djoyce/java/elements/elements.html>.

G. Carcassi, L. Maccone, and C.A. Aidala, *Four Postulates of Quantum Mechanics Are Three*, Phys. Rev. Lett. 126, 110402, (2021). Open-access version: <https://arxiv.org/abs/2003.11007>

P.A.M. Dirac, *The Principles of Quantum Mechanics*, 4th ed., Snowball Publishing, (2019).

J. von Neumann, *Mathematical Foundations of Quantum Mechanics: New Edition*, Princeton University Press, (2018).

`An example of an author admitting the projection postulate is not universal:`
D. McMahon, *Quantum Mechanics DeMystified*, McGraw-Hill, (2006), p. 216.

`Griffith's weak conservation (see warning below):`
D.J. Griffiths and D.F. Schroeter, *Introduction to Quantum Mechanics*, 3rd ed., Cambridge University Press, (2019).

• *A warning: A quick scan of Griffiths (and Schroeter) will make it seem like he believes in the full version of the conservation laws. He has dozens and dozens of mentions of conservation that make it seem this way. However, at best, he is inconsistent. At worst, his is the weak version of conservation. On page 112, he says "Nowhere does quantum mechanics license violation of energy conservation, and certainly no such authorization entered into the derivation of Equation 3.76." But, on p. 30, in the context of the probabilities of getting particular results in experiments and expectations values, he tell us that "These are manifestations of energy conservation in quantum mechanics." Therefore, his version of energy conservation only works on an average over measurements. At the end of the book, he states something which reinforces my assessment: "In this book...particles do not in general possess specific dynamical properties (position, momentum, energy, angular momentum, etc.) until an act of measurement intervenes".*

`Bohr, Kramers, and Slater theory:`
N. Bohr, H.A. Kramers, and J.C. Slater, *The Quantum Theory of Radiation*, The London, Edinburgh, and Dublin Philosophical Magazine and Journal of Science, (1924), 47:281.

`Bothe and Geiger experiment:`
W. Bothe and H. Geiger, *Ein Weg zur experimentellen Nachprüfung der Theorie von Bohr, Kramers und Slater*, Zeitschrift für Physik, (1924), 26 (1): 44.

`Heisenberg and conservation laws`
W. Heisenberg, Nobel Lecture, Nobel Prize Outreach AB 2024, (26 Nov 2024), p. 299. <https://www.nobelprize.org/prizes/physics/1932/heisenberg/lecture/>.

`“all fundamental particles like electrons and protons are identical”`
D.H. McIntyre, *Quantum Mechanics: A Paradigms Approach*, Pearson, (2012), p. 410.

`“unreasonably strong requirement”`
R. Penrose, *The Road to Reality: A Complete Guide to the Laws of the Universe*, Vintage Digital, (2005).

Afterword

`“shoulders of Giants”`
I. Newton, found in H.W. Turnbull ed., *The Correspondence of Isaac Newton: 1661–1675*, Volume 1, London, UK: Published for the Royal Society at the University Press, (1959), p. 416.

• *The actual spelling of the quote is "If I have seen further it is by standing on ye sholders of Giants."*

Appendix A: Atomic orbital labels and atomic transitions

`Data for the sodium energy levels came from:`
A. Kramida, Y. Ralchenko, J. Reader and NIST ASD Team (2024), *NIST Atomic Spectra Database* (version 5.12), [Online]. Available: <https://physics.nist.gov/asd>. National Institute of Standards and Technology, Gaithersburg, MD.

Appendix D: The energy-time uncertainty relation doesn't exist

`Griffiths explaining a so-called energy-time uncertainty principle:`
D.J. Griffiths and D.F. Schroeter, *Introduction to Quantum Mechanics*, 3rd ed., Cambridge University Press, (2019), §3.5.3.

Index

www.ingramcontent.com/pod-product-compliance
Ingram Content Group UK Ltd.
Pitfield, Milton Keynes, MK11 3LW, UK
UKHW022032041026
15591UKWH00012B/650